THE MAKINGS OF A CLINICAL PROTOCOL

THE MAKINGS OF A CLINICAL PROTOCOL

A JOURNEY FROM PURE TO APPLIED BIOMEDICAL RESEARCH AND BEYOND

Edited by

BRUCE K. KOWIATEK

Allied Health Sciences,
Blue Ridge Community and Technical College,
Martinsburg, WV, United States

ACADEMIC PRESS

An imprint of Elsevier

Academic Press is an imprint of Elsevier
125 London Wall, London EC2Y 5AS, United Kingdom
525 B Street, Suite 1650, San Diego, CA 92101, United States
50 Hampshire Street, 5th Floor, Cambridge, MA 02139, United States
The Boulevard, Langford Lane, Kidlington, Oxford OX5 1GB, United Kingdom

ISBN: 978-0-323-95749-6

For information on all Academic Press publications visit our website at
https://www.elsevier.com/books-and-journals

Publisher: Stacy Masucci
Acquisitions Editor: Rafael E. Teixeira
Editorial Project Manager: Aleksandra Packowska
Production Project Manager: Sajana Devasi PK
Cover Designer: Christian J. Bilbow

Typeset by TNQ Technologies

Working together
to grow libraries in
developing countries

www.elsevier.com • www.bookaid.org

Contents

Contributors   *ix*

About the editor   *xi*

Preface   *xiii*

Acknowledgment   *xv*

1. Rooted tRNAomes and evolution of the genetic code   **1**

Daewoo Pak, Nan Du, Yunsoo Kim, Yanni Sun and Zachary F. Burton

Abbreviations   1

Introduction   1

Results   4

Discussion   15

Methods   21

Acknowledgments   23

References   23

2. Aminoacyl-tRNA synthetase evolution and sectoring of the genetic code   **27**

Daewoo Pak, Yunsoo Kim and Zachary F. Burton

Abbreviations   27

Introduction   27

Results   30

Discussion   45

Methods   53

Acknowledgments   54

References   54

3. Evolution of life on earth: tRNA, aminoacyl-tRNA synthetases and the genetic code   **59**

Lei Lei and Zachary F. Burton

Introduction   59

This report   60

A working model for evolution of the genetic code   62

Life on another planet   85

Predictions   86

The frozen accident   88
Author contributions   88
References   89

4. **Type II tRNAs and evolution of translation systems
   and the genetic code**   **95**
   Yunsoo Kim, Bruce K. Kowiatek, Kristopher Opron and Zachary F. Burton

   Abbreviations   95
   Introduction   95
   Results   98
   Discussion   106
   Materials and methods   112
   Conclusions   112
   Author contributions   113
   References   113

5. **An additional, complementary mechanism of action for
   folic acid in the treatment of megaloblastic anemia**   **117**
   Bruce K. Kowiatek

   Abbreviations   117
   Introduction   117
   Experimental   119
   Results and discussion   122
   Acknowledgments   123
   References   123

6. **Non-enzymatic methylation of cytosine in RNA by
   S-adenosylmethionine and implications for the
   evolution of translation**   **125**
   Bruce K. Kowiatek

   Abbreviations   125
   Introduction   125
   Experimental   129
   Results and discussion   131
   Prospectus   134
   Acknowledgments   134
   References   134

7. Early evolution of transcription systems and divergence of Archaea and Bacteria   **137**

Lei Lei and Zachary F. Burton

Abbreviations	137
Introduction	137
Evolution of 2-DPBB RNAPs and DNAPs	138
Evolution of archaeal and bacterial GTFs	150
Divergence of Archaea and Bacteria	158
References	159

8. The 3-minihelix tRNA evolution theorem   **163**

Zachary F. Burton

Models for tRNA evolution	163
Numbering of tRNAs	164
References	175

9. Methylating agents as adjunct therapy to chemotherapeutic alkylating medications for improved outcomes in chronic lymphocytic leukemia: a case study   **179**

Bruce K. Kowiatek

Introduction	179
Methylating agents	181
The Kowiatek protocol	181
Case study involving the Kowiatek protocol	181
CBC highlights immediately post-chemo (12/13/2019)	181
03/30/2020: change in methylating agent	182
CBC highlights of first follow-up allowed by easing of COVID-19 restrictions (04/29/2020)	182
Statistical analysis of 4/29/2020 CBC highlights	182
Discussion of results	182
CBC highlights 4/13/2021	183
4/13/2021 CBC analysis	183
Updates: 8/23/2021 and 3/14/2022	183
Initial protocol prospectus, July/August 2019	183
U.S. nationwide implementation of protocol first-line as per the U.S. FDA	183
U.S. nationwide implementation of protocol second-line as per the U.S. FDA	184
U.S. nationwide implementation of protocol third-line as per the U.S. FDA	184
Nationwide implementation of protocol first-line as per the U.S. FDA for breast cancer stage 1	186

Latest research August 2021   187
Micelle self-division   188
On the horizon   188
Acknowledgments   190
References   190

10. Chaos, order, and systematics in evolution of the genetic code   191

Lei Lei and Zachary F. Burton

Introduction   191
Artificial intelligence in evolution of life   194
The prelife to life transition: evolution of translation   195
Evolution of the genetic code (overview)   202
Evolution of ribosomes   204
Aminoacyl-tRNA synthetases (aaRS)   210
Pre-life to LUCA   218
Polyglycine world   220
Evolution of the genetic code (a working model)   222
The great divergence   233
Models to describe genetic code evolution   233
Polyglycine world (a working model)   235
Conclusions   239
Acknowledgments   240
References   241

11. Evolution of the genetic code   247

Lei Lei and Zachary F. Burton

Introduction   247
Methods   248
Prelife evolution of transcription, metabolism, and translation   249
Evolution of tRNA   251
Evolution of aaRS enzymes   259
EF-Tu and coding degeneracy   262
Evolution of the genetic code   264
Discussion   276
Author contributions   280
References   280

Index   289

Contributors

Zachary F. Burton
Department of Biochemistry and Molecular Biology, Michigan State University, East Lansing, MI, United States

Nan Du
Department of Computer Science and Engineering, Michigan State University, East Lansing, MI, United States

Yunsoo Kim
School of Information, University of Michigan, Ann Arbor, Michigan, MI, United States; IMC Trading, Chicago, IL, United States

Bruce K. Kowiatek
Allied Health Sciences, Blue Ridge Community and Technical College, Martinsburg, WV, United States

Lei Lei
School of Biological Sciences, University of New England, Biddeford, ME, United States

Kristopher Opron
Bioinformatics, Michigan State University, Ann Arbor, MI, United States

Daewoo Pak
Center for Statistical Training and Consulting, Michigan State University, East Lansing, MI, United States; Department of Computer Science and Engineering, Michigan State University, East Lansing, MI, United States

Yanni Sun
Department of Computer Science and Engineering, Michigan State University, East Lansing, MI, United States

About the editor

Dr. Bruce K. Kowiatek earned his bachelor's in English Writing from the University of Pittsburgh in 1998 and his doctorate in Pharmacy and master's in Business Administration from Shenandoah University in 2002. In addition to researching and investigating the origins of life for over 20 years, he has also published books and papers on a variety of topics, worked as a clinical pharmacist, and is currently a full-time Assistant Professor at Blue Ridge Community and Technical College in Martinsburg, WV.

Preface

An ancient Chinese proverb once declared that a journey of a thousand miles begins with a single step. The first, single step in this years-long journey started in the realm of pure, sometimes called fundamental science, research performed to further understanding of how the universe works. In this case, it was looking into the evolution of the genetic code and transfer RNA (tRNA), the intracellular molecule that, as its name implies, transfers genetic information from DNA via messenger RNA (mRNA) to the sequences of amino acids that define the proteins of all living organisms and viruses.

Not long into the journey, this pure research began pointing in the direction of not one, but several biomedical therapies applicable to the treatment of various cancers, thus now entering the realm of applied science, or research performed with the goal of developing more practical applications. While pure scientific research is still currently performed worldwide to a certain degree, the majority of present-day research appears to have a more applied bent to it. Although many invaluable biomedical treatments and therapies began as applied research only, with applied research definitely occupying its own needed position in science, the contention is made here that pure scientific research casts a much wider net, as it were, potentially leading to many more practical applications than beginning with more limited applied scientific research alone.

The journey outlined in the gradually building and branching research and applications contained in the chapters of this textbook therefore attempts to make that case. The biomedical treatments and therapies herein include those protocols for megaloblastic anemia, chronic lymphocytic leukemia (CLL), and solid cancerous tumors in general, with perhaps more forthcoming. This is why the pure research beyond the applied must also continue, to not only find out as much as possible about nature, but to apply that newfound knowledge to as many therapeutic modalities and protocols as possible, helping as many people as possible.

Bruce K. Kowiatek, PharmD, MBA, BA
06/12/2022

Acknowledgment

I especially thank my loving family for their generous gifts of time and understanding in the editing of this textbook, as well as my fellow faculty and staff at Blue Ridge Community and Technical College (BRCTC). Also, very special thanks to Dr. Zachary Burton, Professor Emeritus, at Michigan State University (MSU), on whose work much of this textbook is based.

CHAPTER 1

Rooted tRNAomes and evolution of the genetic code*

Daewoo Pak[1], Nan Du[1], Yunsoo Kim[2], Yanni Sun[1] and Zachary F. Burton[3]

[1]Department of Computer Science and Engineering, Michigan State University, East Lansing, MI, United States; [2]IMC Trading, Chicago, IL, United States; [3]Department of Biochemistry and Molecular Biology, Michigan State University, East Lansing, MI, United States

Abbreviations

aaRS	aminoacyl tRNA synthetases
DNA tRNAome	the DNA encoding tRNA for an organism
GlyRS	i.e., glycine aminoacyl tRNA synthetase
LUCA	The last universal common cellular ancestor

Introduction

We posit that cloverleaf tRNA is the molecular archetype around which translation systems and the genetic code evolved. Evolution of the genetic code was recently comprehensively reviewed, but issues remain unresolved (Koonin, 2017; Koonin & Novozhilov, 2017). We posit that to grasp code evolution requires a focus on tRNA evolution. To make sense of translation systems, for instance, start with tRNA and work out (Pak et al., 2017; Root-Bernstein et al., 2016). Translation systems evolved around cloverleaf tRNA, which has not changed very much since LUCA (the last universal common cellular ancestor), and a tRNA-centric view renders translation a much simpler conceptual problem (Pak et al., 2017; Root-Bernstein et al., 2016). In order to understand evolution of the genetic code, moreover, start with tRNA and work out. As one obvious example, the genetic code is a triplet code because the structure of the tRNA anticodon loop forces a triplet register for two adjacent tRNAs paired to mRNA bound in the ribosome A and P sites at the decoding center. Genetic code evolution, therefore, must have tracked tRNA evolution more closely than mRNA or

---

* The entire chapter was previously published. Transcription. 2018;9(3):137—151. Published online 2018 Feb 6. https://doi.org/10.1080/21541264.2018.1429837. PMC ID: PMC5927645. PMID: 29372672.

The Makings of a Clinical Protocol
ISBN 978-0-323-95749-6
https://doi.org/10.1016/B978-0-323-95749-6.00007-7

ribosome evolution. Furthermore, because of unique features of the tight tRNA 7 nt anticodon loop structure with its specialized U turn (Quigley & Rich, 1976), the anticodon loop of tRNA constrained code evolution much more than the mRNA, which has an extended and partly relaxed conformation on the ribosome. Because of physical limitations of tRNA anticodon wobble sequences immediately following the U turn, the initial expansion of the genetic code in tRNA was limited to 48 anticodons, even though in mRNA all 64 codons are utilized.

Based mostly on analyses of archaeal tRNAs, which are more faithfully conserved from LUCA, a model for evolution of cloverleaf tRNA was proposed (Pak et al., 2017; Root-Bernstein et al., 2016). According to the model, acceptor stems are derived from a GCG repeat and its CGC complement (Pak et al., 2017). The anticodon stem is flanked on both sides by 5 nt relics of acceptor stems, also derived from GCG and CGC repeats. Although largely unpaired in the cloverleaf, in the anticodon loop mini-helix, the last 5 nt of the D loop and the 5 nt V loop was paired as acceptor stems (Pak et al., 2017). The D loop is derived from a UAGCC repeat. The anticodon stem and loop and the T stem and loop are homologous by sequence and structure. Both loops have a U turn (a U-shaped turn) after a U (in the anticodon loop) or a pseudouridine (in the T loop) between loop positions 2 and 3 out of 7 total (Pak et al., 2017). Anticodon loop bases 3–7 are tightly stacked, as if in a helix connecting with the $3'$-anticodon stem, making cloverleaf tRNA a relatively stiff adapter to specify contacts to mRNA in the decoding center of the ribosome smaller subunit. The T loop of tRNA is very similar in structure to the anticodon loop, but differs slightly because of T loop interactions with the D loop. Specifically, intercalation of D loop G19 (G18 in historic numbering), between T loop bases four and five, lifts T loop base 5 to contact the T loop stem and flips loop bases 6 and 7 out of the T loop (Pak et al., 2017). Based on the tRNA-centric view and the tRNA evolution model, we advocate reassessment of the evolution of the genetic code.

It has been suggested that glycine may be a founding amino acid for the genetic code (Bernhardt, 2016; Bernhardt & Patrick, 2014; Bernhardt & Tate, 2008; Trifonov, 2004). Here, we show that archaeal tRNA^{Gly} is very close to the posited root of the tRNA evolutionary tree. We propose therefore the polyglycine hypothesis that the primordial cloverleaf tRNA (tRNA^{Pri}), which most strongly resembles archaeal tRNA^{Gly}, diversified by mutation to include all permitted anticodons. The initial purpose of the 3 nt code may have been, therefore, to synthesize short chains of poly-glycine, used to stabilize protocells for energy generation. Gly$_5$ is the typical length for polyglycine cross-linking in bacterial cell walls (Romaniuk & Cegelski, 2015). In the primitive system, polyglycine chains may have been short in length, because of weak translational processivity and/or mRNA codons lacking a corresponding tRNA anticodon. The polyglycine hypothesis provides a functional root, and the tRNA evolutionary model provides a sequence root to the genetic code. Di Giulio has argued against polyglycine as a founding product of the code, but his arguments are centered on proto-mRNAs encoding multiple peptide products (Di Giulio, 2014). Whether or not Di Giulio is correct about ancient mRNA coding, the genetic code that exists today appears to be evolved around a single primordial cloverleaf tRNA that recruits mRNA (Pak et al., 2017). The nearly universal genetic code, therefore, may be a reinvention of coding that surpassed and suppressed older mRNA-centered systems.

In human tRNA^{Gly}, adenine in the wobble position was shown to be destabilizing for the anticodon loop (Saint-Léger et al., 2016). In bacteria and eukaryotes, adenine in the tRNA wobble position is converted to inosine by a tRNA adenosine deaminase, conferring greater loop stability (Rafels-Ybern et al., 2018; Saint-Léger et al., 2016). In archaea, adenine is rarely or never found in the tRNA wobble position, indicating that, in the RNA-protein world and at LUCA, adenine was negatively selected at the wobble position. This observation shrinks the maximum size of the initial genetic code in tRNA from 64 anticodons to 48 anticodons.

A hierarchy is observed for the importance of the three tRNA anticodon positions for translation (Koonin, 2017; Koonin & Novozhilov, 2009). The middle (second) position of the anticodon is most important for translational fidelity, followed by the third position and then followed by the wobble (first) position. Ambiguity of the wobble position, therefore, describes degeneracy of the code and why only ∼20 amino acids are specified rather than a potentially much larger number (i.e., up to 48). From a structural perspective, when a tRNA binds in the ribosome A site (addition site), the wobble position is ambiguous because the tRNA wobble anticodon base is not fully restrained and can make multiple types of contacts (i.e., Watson-Crick pairs and various non-Watson-Crick wobble pairs) to mRNA. The middle and third positions of the tRNA anticodon, by contrast, are constrained to form accurate Watson-Crick base pairs, and even G∼U wobble pairs, commonly found in RNA stems, are strongly disfavored at these positions (Demeshkina et al., 2012). At the second and third anticodon positions but not the wobble position, the specificity of contacts is checked by a proofreading conformational change in the decoding center of the smaller subunit of the ribosome involving EF-TU and GTP hydrolysis (Ogle et al., 2002). Because pairing of the wobble base involves multiple types of contacts, the ribosome conformational change cannot now be extended to proofread wobble position contacts. For one thing, multiple essential wobble contacts that rely on non–Watson-Crick base pairs would be disallowed. Also, ambiguity at the wobble anticodon position was likely necessary for early stage evolution of the genetic code, and wobble ambiguity remains positively selected (Pak & Burton, 2018). To our knowledge, the structural explanations for degrees of freedom in pairing at the wobble position are not fully known. Also, the central importance of the middle anticodon position has not been completely elucidated. Considering these issues, however, we discuss the evolution and the degeneracy of the code. We argue that, at earlier stages of evolution, as the code grew toward ~ 16 letters, wobble position ambiguity was positively selected.

Archaeal species generally have one tRNA species per permitted anti-codon (excepting adenine in the anticodon wobble position), but there are a few common exceptions. Many archaea have multiple (generally three) $tRNA^{Met}$ (anticodon CAU), including initiator and elongator $tRNA^{Met}$. *Pyrococcus furiosis* (archaea), as a typical example, has two elongator $tRNA^{Met}$ (CAU) and one initiator $tRNA^{iMet}$ (CAU). Interestingly, *Pyrococcus furiosis* has only one $tRNA^{Ile}$ (GAU; in some archaea, a single anti-codon UAU or CAU may be utilized). *Pyrococcus* $tRNA^{Met}$, with three tRNAs (CAU), and $tRNA^{Ile}$, with only one $tRNA^{Ile}$ (generally anticodon GAU), share the same 4-codon sector of the codon-anticodon table (anticodon NAU). From analysis of rooted tRNAomes for ancient archaeal species, it appears that $tRNA^{Met}$ may have been derived from $tRNA^{Ile}$.

Results

Lineages in tRNAs

A DNA tRNAome is defined here as the set of all available coding tRNA DNA sequences from a single strain of a species of organism. Sequences of tRNAs were collected from tRNA databases (Chan & Lowe, 2016; Juhling et al., 2009). Others have used tRNA sequences to indicate phylogenies of species (Widmann et al., 2010). To improve these comparisons, we root tRNAome trees to $tRNA^{Pri}$ and compare tree structures among species. We compare evolutionary trees of rooted DNA tRNAomes from *Pyrococcus furiosis DSM3638, Pyrococcus abyssi GE5, Pyrococcus horikoshii OT3, Staphylothermus marinus F1, Pyrobaculum aerophilum str. IM2, Aeropyrum pernix K1, Sulfolobus solfataricus P2* (archaea) and *Thermus thermophilus HB27* (bacteria). *Pyrococcus, Staphylothermus, Pyrobaculum, Aeropyrum,* and *Sulfolobus* species were selected because their tRNAomes appear to be very similar to the LUCA tRNAome. A *Pyrococcus* typical tRNA for instance shows much stronger conservation than a broader archaeal or bacterial typical tRNA, indicating proximity of *Pyrococcus* to LUCA (Pak et al., 2017). Very strong GCG/CGC (acceptor stem) and UAGCC (D loop) repeats are preserved in *Pyrococcus* tRNAs, indicating that these tRNAomes remain close to the primordial cloverleaf. The *Pyrococcus* typical tRNA sequence is very close to tRNA^{Pri} (the proposed primordial tRNA cloverleaf) sequence (60/79 in-phase identities). Of course, these observations also support our assignment of the tRNA^{Pri} sequence. For consistency of interpretation, tRNAome evolutionary trees were rooted to tRNA^{Pri}. *Thermus* was selected based on inspection of typical tRNA diagrams, which indicated *Thermus* was a bacterial family that was more similar to archaea than others.

In Fig. 1.1, a qualitative interpretation of the evolutionary trees is shown. With some differences, apparent lineages of tRNAs are maintained in different archaeal species and some tRNA lineages are preserved among both archaeal and bacterial species. A sophisticated bioinformatics analysis (not yet complete), therefore, can trace tRNAomes for many species through the archaea and into the bacteria and eukarya. One useful comparison would be an evolutionary tree of tRNAome trees, comparing intact genetic code structures, organism to organism. Rooting trees to tRNA^{Pri} helps to interpret the comparisons.

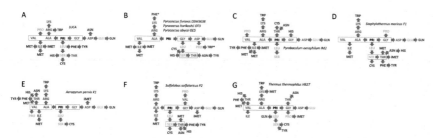

Figure 1.1 Qualitative maps of the radiations of tRNAomes for various species based on interpretation of evolutionary trees. (A) Model for LUCA; (B) Three *Pyrococcus* species; (C) *Pyrobaculum aerophilium str. IM2*; (D) *Staphylothermus marinus F1*; (E) *Aeropyrum pernix K1*; (F) *Sulfolobus solfataricus P2* (archaea); and (G) *Thermus thermophilus HB27* (bacteria). We posit that tRNAs are added to the code in the approximate order cyan → orange → green → purple → red. Asterisks indicate two tRNAs that appear to be reassigned to encode distinct amino acids compared to LUCA and other archaea. (For interpretation of the references to colour in this figure legend, the reader is referred to the web version of this article.)

In Fig. 1.1A, a model for a LUCA tRNAome is shown based on comparison of maps and the relatedness of encoded amino acids. Some of the LUCA tRNAome assignments are based on information in an accompanying paper (Pak & Burton, 2018). The model for LUCA gives an indication of lineages that appear most conserved. Lineages often connect tRNAs for related amino acids, indicating the likelihood of the lineage. A strongly conserved lineage appears to be tRNA^{Pri} → tRNA^{Gly} → tRNA^{Asp} → tRNA^{Glu} → tRNA^{Gln} (Fig. 1.1A−F). Another apparent conserved lineage is tRNA^{Pri}/tRNA^{Gly} → tRNA^{Leu} → tRNA^{Ser} (Fig. 1.1A−G). Because V loop inserts were deleted from tRNA alignments before generating trees, the strong similarity of tRNA^{Leu} and tRNA^{Ser} is due to the tRNA cloverleaf core sequence (1−75) and is not due to alignments of tRNA^{Leu} and tRNA^{Ser} extended V loops. The archaeal species that are most similar to LUCA connect tRNA^{Ser} → tRNA^{Thr}, which appear to radiate to tRNA^{Cys}, tRNA^{His}, possibly to tRNA^{Asn} and possibly to tRNA^{Phe} → tRNA^{Tyr}. Because Phe and Tyr are related to amino acids, a tRNA^{Phe} → tRNA^{Tyr} lineage seems reasonable, whether or not this lineage roots properly to tRNA^{Ser}/tRNA^{Thr}. In *Pyrobaculum aerophilum str. IM2*, for instance, the lineage tRNA^{Ala}/tRNA^{Val} → tRNA^{Ile} → tRNA^{Phe} → tRNA^{Tyr} is indicated. Based on amino acid relatedness and positions in the codon-anticodon table, however, tRNA^{Leu} → tRNA^{Phe} → tRNA^{Tyr} might be a more reasonable model (Pak & Burton, 2018). We posit that tRNA^{Asn} may originally be derived from tRNA^{Asp}, similar to tRNA^{Gln} being apparently derived from tRNA^{Glu}, as might be expected based on amino acid similarity. If this surmise is correct, tRNA^{Asn} was forced to diverge further from tRNA^{Asp} to maintain accuracy of AsnRS and AspRS charging and translation. Another apparent conserved lineage is tRNA^{Ala} → tRNA^{Arg} → tRNA^{Lys} → tRNA^{Trp} (Fig. 1.1A, C−G). In the archaea that appear most closely related to LUCA, tRNA^{Ile} → tRNA^{Met} (one initiator and two elongator) is likely (Fig. 1.1A−C). In more derived species, one tRNA^{Met} and tRNA^{iMet} appear to have specialized and further diverged from tRNA^{Ile} (Fig. 1.1D−F). Some conserved lineage structures appear to extend from archaea → bacteria.

In three *Pyrococcus* species, it appears that two tRNAs may have been reassigned to attach a different amino acid compared to LUCA tRNAs. Partly because of its placement in the map, *Pyrococcus* tRNA^{Phe} appears to be a reassigned tRNA^{Trp}. Also, *Pyrococcus* tRNA^{Trp} appears to be a reassigned tRNA^{Pro} (Fig. 1.1B). Because three *Pyrococcus* species are considered, it is difficult to attribute these apparent tRNA reassignments based on sequencing errors. We posit that these two tRNAs duplicated and evolved to assume new identities. So far as we can judge, the other tRNAs considered for the 8 organisms analyzed here may have maintained their original identities, although divergent and convergent evolution of tRNAs causes tRNAs to move in the lineage maps. Evolution of tRNAs, therefore, can suppress evidence of tRNA reassignments. Migration of tRNAs in the maps tends to make rooting of tRNA lineages ambiguous and causes lineages to appear more shallow (i.e., in bacteria) than in species with tRNAs that are more similar to LUCA tRNAs (i.e., *Pyrococcus* and *Pyrobaculum*).

The polyglycine hypothesis

It appears that the initial purpose of the triplet genetic code may have been to synthesize short chain polyglycine (Bernhardt, 2016; Bernhardt & Patrick, 2014; Bernhardt & Tate, 2008; Trifonov, 2004). A reason to consider this hypothesis is that tRNA^{Pri} most resembles archaeal tRNA^{Gly} (Fig. 1.2).

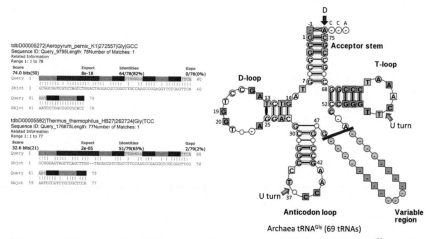

Figure 1.2 The primordial tRNA cloverleaf is most similar to archaeal tRNA^{Gly}. A blast search of the primordial tRNA sequence against the *Aeropyrum pernix* (archaea) DNA tRNAome and the *Thermus thermophilus HB27* DNA tRNAome (bacteria). Coloring of the primordial tRNA sequence: (*green*) acceptor stems; (*magenta*) D loop; (*cyan*) acceptor stem remnants; (*red*) anticodon loop and T loop stems; (*yellow*) anticodon loop and T loop; (*blue*) 3'-ACCA. (*Right image*) A typical tRNA diagram generated from 69 archaeal tRNA^{Gly} sequences is shown. Numbering of the tRNA is based on a 75 nt tRNA core sequence. *Blue arrows* indicate U turns. The *red arrow* indicates the discriminator (D). Only 5 nt of the V loop are considered in the evolutionary model. Longer V loops include inserts (i.e., tRNA^{Leu} and tRNA^{Ser}). (For interpretation of the references to colour in this figure legend, the reader is referred to the web version of this article.)

The GCGGCGG 5′-acceptor stem GCG repeat of tRNA^{Pri} is most similar to an archaeal tRNA^{Gly} acceptor stem. Searching the primordial tRNA sequence against genomic DNA sequences (i.e., all archaea) produces tRNA^{Gly} (GCC) as the top hit (not shown). Searching against the archaeal *Aeropyrum pernix* tRNAome also produces tRNA^{Gly} (anticodon GCC) as the top hit, with an e-value of 8×10^{-18} and 64/78 in-phase identities (Fig. 1.2). Searching against the bacterial *Thermus thermophilus* DNA tRNAome gives tRNA^{Gly} (anticodon UCC) as a top in-phase hit, with an e-value of 2×10^{-5} and 51/79 identities. There is a 2 nt deletion in the D loop of *Thermus* tRNA^{Gly} (anticodon UCC). D loop deletions are found in archaeal tRNAs but are almost universal for bacterial and eukaryotic tRNAs (Pak et al., 2017; Root-Bernstein et al., 2016). Analysis of tRNAomes indicates that tRNA^{Gly} is initially most similar to tRNA^{Pri}.

Radar graphs

Radar graphs (Fig. 1.3) provide a characteristic and identifying tRNAome "fingerprint" that can readily be compared among organisms. Radiations of

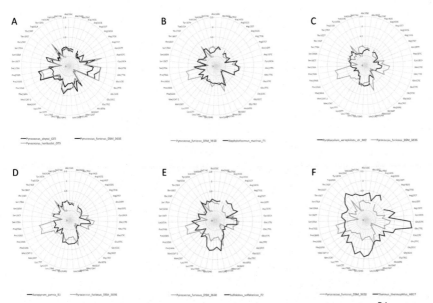

Figure 1.3 Radar graphs of the radiations of DNA tRNAomes from tRNA^{Pri} (at the origin). Evolutionary distances are shown. Comparisons of: (A) three *Pyrococcus* species; (B) *Staphylothermus marinus F1* versus *Pyrococcus furiosis*; (C) *Pyrobaculum aerophilium str. IM2* versus *Pyrococcus furiosis*; (D) *Aeropyrum pernix* versus *Pyrococcus furiosis*; (E) *Sulfolobus solfataricus* versus *Pyrococcus furiosis* (archaea); and (F) *Thermus thermophilus* (bacteria) versus *Pyrococcus furiosis*.

tRNAome sequences from tRNA^{Pri} (at the origin) are compared for *Pyrococcus furiosis DSM3638, Pyrococcus abyssi GE5, Pyrococcus horikoshii OT3, Staphylothermus marinus F1, Pyrobaculum aerophilum str. IM2, Aeropyrum pernix K1, Sulfolobus solfataricus P2* (archaea), and *Thermus thermophilus HB27* (bacteria). We note that radar graphs provide insight into the evolution of species and their relatedness. From the similarity of graphs, three *Pyrococcus* species, *Pyrobaculum, Staphylothermus,* and *Aeropyrum,* appear closely related. In particular, compare radar graphs for tRNAs encoding Asn, Asp, Cys, Gln, Glu, Gly, and His for these organisms to observe the clear similarities in graph shapes. Based on the extent of radiation from tRNA^{Pri}, *Pyrobaculum aerophilum* may be the closest species of those selected to a LUCA tRNAome. In this comparison, the archaeal species were selected to be similar to LUCA by inspection of typical tRNA sequences. In many archaea, tRNA^{Pri} is most similar to tRNA^{Gly}, as expected from the polyglycine hypothesis. Some features of radar graph shapes appear to be conserved to *Thermus* (a bacteria). We conclude that species relatedness and divergence can be determined by analysis of tRNAomes, for instance, as represented in radar graphs. Note that the comparisons shown in radar graphs are also embedded in evolutionary trees although, using trees, the information is more difficult to visually compare species to species.

In archaea, tRNA^{Gly} is generally most similar to tRNA^{Pri} (Figs. 1.2 and 1.3A–D). Although others have also posited that glycine is the founding amino acid for evolution of the code (Bernhardt, 2016; Bernhardt & Patrick, 2014; Bernhardt & Tate, 2008; Trifonov, 2004), we believe we are the first to posit the polyglycine hypothesis, which simplifies the understanding of genetic code evolution. Also, archaeal tRNAs are more closely related to tRNA^{Pri} than bacterial tRNAs (Figs. 1.2 and 1.3F), indicating that archaeal tRNAomes are more similar to LUCA tRNAomes than bacterial tRNAomes (Pak et al., 2017; Root-Bernstein et al., 2016).

tRNA mutagenesis and divergence

Anticodon loop positions 1 and 7 weakly interact (i.e., a C~A wobble pair), restricting the sequences that appear at these positions (Giege & Eriani, n.d.). Loop position 2 is always U in archaea and bacteria. Loop position 6 is always A or G. Generally, the anticodon of tRNA can mutate without much effect on the cloverleaf fold, so the anticodon is expected to be the fastest sequence in tRNA to change in evolution. Many other substitutions in the cloverleaf, however, are likely to be disruptive and may require a compensatory mutation or a tRNA modification to rescue the fold. Substitutions in stems, for instance, generally require a compensating mutation in the complementary base. We posit, therefore, consistent with the polyglycine hypothesis, that essentially all allowed anticodons became available on tRNAs with acceptor stems that are expected to attach glycine. So populating the codon-anticodon table with a primordial tRNA^{Gly} may have occurred before additional amino acids were added to the code. Polyglycine (Gly$_5$) is a component of bacterial cell walls (Romaniuk & Cegelski, 2015). We imagine that, in the evolving RNA-protein world, polyglycine cross-linking stabilized protocells to facilitate energy generation, which requires a proton gradient across a membrane. Polyglycine, therefore, had evolutionary value from early times, and the genetic code may have initially evolved as an improved mechanism to generate short-chain polyglycine to stabilize protocells. If the polyglycine hypothesis is correct, different tRNA species within an organism are expected to have lineages derived from tRNA^{Pri}/tRNA^{Gly} (Figs. 1.1 and 1.3) Initially, tRNA^{Pri} and its earliest radiations appear to have attached glycine and then evolved to attach other amino acids.

Adenine in the anticodon wobble position

In bacteria and archaea, adenine is rarely found in the anticodon wobble position of tRNAs (Fig. 1.4) (Rafels-Ybern et al., 2018; Saint-Léger et al., 2016). In 6368 bacterial tRNAs (shown as DNA), adenine is underrepresented at the wobble position (180/6368 → 2.8%). In most bacterial species, tRNA^{Arg} (anticodon ACG) is the only tRNA with adenine encoded in the anticodon wobble position. In 1088 archaeal tRNAs (Juhling et al., 2009), remarkably, adenine is never found at the wobble position. In bacteria and eukaryotes, adenine in the anticodon wobble position is converted to inosine by a tRNA adenosine deaminase, a modification that may stabilize the anticodon loop and that expands wobble position contacts to mRNA (Rafels-Ybern et al., 2018; Saint-Léger et al., 2016). Archaea lack the tRNA adenosine deaminase to convert wobble adenine to inosine (Saint-Léger et al., 2016). We conclude that, at LUCA, adenine in the anticodon wobble position was under strong negative selection, probably, for two reasons. Wobble adenine can have destabilizing effects on the anticodon loop (Saint-Léger et al., 2016). Also, adenine can only pair strongly with uridine, whereas inosine can pair with adenine, cytidine, or uridine (Agris et al., 2017, 2018), indicating positive selection for ambiguity

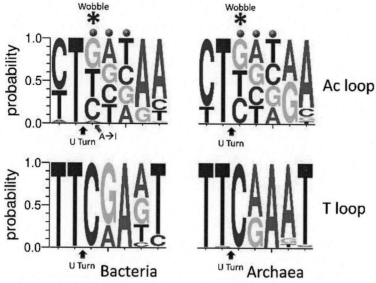

Figure 1.4 A strong negative selection against adenine in the anticodon wobble position. The homologous T loop is shown below the anticodon (Ac) loop for comparison. DNA sequence logos of the 7 nt anticodon and T loops are shown. Right panels) In archaea (1088 tRNAs), no A is detected at the anticodon wobble position. Left panels) In bacteria (6368 tRNAs), adenine (A) is rarely used, except in tRNA^{Arg} (anticodon ACG), and adenine is converted to inosine (I) by tRNA adenosine deaminase. Blue dots indicate the anticodon positions of the loop. The asterisk indicates the wobble position of the anticodon loop. (For interpretation of the references to colour in this figure legend, the reader is referred to the web version of this article.)

at the wobble anticodon position. Unmodified adenine, therefore, in the tRNA anticodon wobble position specifies U in the mRNA codon wobble position. Because of these restrictions, the initial genetic code included at most 48 and not 64 anticodons, as has generally been believed. Interestingly, 46 tRNAs (44 unique anticodons) are found in many archaeal and bacterial species, allowing for 3 stop codons. In prokaryotes, there are generally three tRNA^{Met} (CAU anticodon), including one initiator and two elongator tRNA^{Met} (CAU anticodon), and commonly absent tRNA^{Ile} (generally a GAU anticodon is preferred and UAU is not utilized) (Juhling et al., 2009).

Evolution of the standard genetic code

Not all of the utilized 48 anticodons specify distinct amino acids, so the genetic code is considered to be degenerate, and structural ambiguity in the reading of the wobble anticodon position causes degeneracy. Because of loop destabilization and the potential for wobble position overspecification, in the initial code, adenine never occupies the anticodon wobble position. Others have noted that the tRNA second (middle) base of the anticodon is most important for translational accuracy, followed by the third base, and then followed by the first (wobble) base, which is recognized with ambiguity (Koonin, 2017; Koonin & Novozhilov, 2009).

These observations suggest an order in which amino acids may have been added to the genetic code (Fig. 1.5) (Novozhilov & Koonin, 2009; Sengupta & Higgs, 2015). We propose the following approximate pathway for sectoring the code. Initially, essentially all permitted anticodons specify Gly to synthesize polyglycine. Then, the code divides into sectors according to the middle position of the anticodon, which is most important for translational accuracy, to specify Val, Ala, Asp, and Gly (Sengupta & Higgs, 2015). The second most important position for translational accuracy is the third anticodon position, but this position sectors with difficulty. We posit that the third position initially sectors between purines and pyrimidines to add Leu, Pro, Glu, and Ser. Subsequently, the third anticodon position is

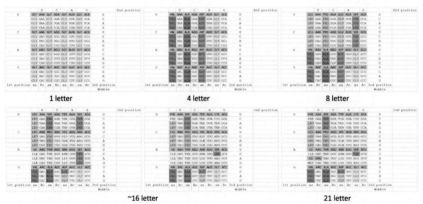

Figure 1.5 Sectoring of the genetic code. A codon-anticodon (Ac) table is shown. The code is posited to sector from a 1 → 4 → 8 → ~16 → 21 letter code (20 aa + Ter (Stop)). Approximate intermediates are shown. Red 1-codon sectors are not utilized in archaea and are rare in bacteria because adenine in the anticodon wobble position is negatively selected. tRNA^{Ile} (UAU) is rarely utilized as the single tRNA^{Ile} in archaea and bacteria. (For interpretation of the references to colour in this figure legend, the reader is referred to the web version of this article.)

utilized to divide the code into 4-codon sectors adding Ile, Ser, Thr, Lys, Ter (Stop), Arg, and Ser. We consider that Leu may have continued to hold two 4-codon sectors. Potentially, full sectoring of the second and third anticodon positions might correlate with the EF-TU mediated conformational tightening of the decoding center of the ribosome smaller subunit, in order to verify the accuracy of Watson-Crick base pairing of the tRNA second and third anticodon positions to mRNA (Demeshkina et al., 2012). At this stage, tRNA^{Ser} occupies three 4-codon sectors of the codon-anticodon table, explaining how tRNA^{Ser} alone of all tRNAs ends up occupying two separated and disconnected 4-codon sectors. We cannot currently explain why serine was of so much apparent importance at this stage of evolution.

Next, Asn and Gln may have been added to the code. As shown in Fig. 1.1A–F, tRNA^{Asn} is more diverged from tRNA^{Asp} than tRNA^{Gln} is diverged from tRNA^{Glu}. Perhaps, AspRS (class IIB aaRS) requires greater divergence of tRNA^{Asp} and tRNA^{Asn} for accurate discrimination than GluRS (class IB aaRS) requires for tRNA^{Glu} and tRNA^{Gln}. Comparing relevant structures (PDBs 1ASY, 4WJ3, 1OOB and 1ZJW) (Bullock et al., 2003; Gruic-Sovulj et al., 2005; Ruff et al., 1991; Suzuki et al., 2015), dimeric class IIB aaRS enzymes such as AspRS and AsnRS appear to make weaker determining acceptor stem contacts than monomeric class IB enzymes such as GluRS and GlnRS, and, therefore, may require greater divergence of tRNAs for discrimination.

Because the code initially sectors to encode a small set of amino acids (i.e., up to ~16), but the code is forced by tRNA anticodon loop geometry to be in a register of 3 nt, wobble position ambiguity is likely to be an early advantage. If, from the onset, the code had evolved with accurate wobble specification, for instance, too many amino acids would be specified at too early a stage of evolution. Such an inflexible code puts heavy pressures on mRNA to also adopt a highly complex complementary code, slowing the pace of code evolution. So, in early stages, too few amino acids were available to justify a 3 nt code, and the barriers to code evolution were too high to allow for an inflexible 3 nt code. Also, if the wobble position were strongly specified in coding, the codon-anticodon table would likely not have broken so completely into 4-codon sectors. As the intermediate code becomes established, evolutionary pressures change to add additional chemistry to the code, and innovation at the wobble position becomes a more viable strategy. As an example, adenine in the anticodon wobble position might only pair with uridine in mRNA, supporting an inflexible

3 nt code, whereas guanine in the wobble position pairs with either cytidine or uridine. Restrictions against an overly inflexible 3 nt code may partly explain negative selection against adenine in the anticodon wobble position. Converting tRNA wobble adenine → inosine allows pairing with mRNA codon adenine, cytidine and uridine, increasing ambiguity of mRNA interpretation (Rafels-Ybern et al., 2018; Saint-Léger et al., 2016) Evolution of the adenine → inosine modification, therefore, provides evidence for positive selection for ambiguity at the tRNA wobble anticodon position, in order to match tRNAs with a larger set of cognate synonymous codons in mRNAs (Pak & Burton, 2018).

We posit that the last ~5−6 amino acids to be added to the code may include Met, His, Cys, Phe, Tyr, and Trp. The upper right 4-codon sector no longer encodes Ser but rather Cys, Ter (stop), and Trp. Arg invades another Ser 4-codon sector. Met appears to invade the Ile sector, which, judging from the unutilized tRNA^{Ile} anticodons in archaea and bacteria, may never have been fully occupied by Ile. In the archaea that are most similar to LUCA, tRNA^{Met} appears to be derived from tRNA^{Ile} (Fig. 1.1A−C), as might be expected from its position in the table (Bhattacharyya & Varshney, 2016). In archaea and bacteria, tRNA^{Ile} generally utilizes only a single GAU, UAU, or CAU anticodon from its 4-codon sector (generally GAU).

A bacterial and eukaryotic tRNA anticodon modification

Bacteria modified the genetic code in tRNA by utilizing tRNA^{Arg} (ACG), which, for the most part, is the only anticodon in bacteria with adenine encoded in the wobble position (Rafels-Ybern et al., 2018; Saint-Léger et al., 2016). Bacteria utilize a tRNA adenosine deaminase to convert adenine to inosine at the wobble position. The advantage to bacteria of the ACG → ICG modification, which is missing in archaea, may be to protect the Arg (ACG, GCG, UCG, CCG) 4-codon sector from dividing into two 2-codon sectors, adding a new amino acid not encoded in archaea to the bacterial code. Inosine can interact with mRNA codons ending in wobble A, C, and U. Because of this ambiguity in reading mRNA codons, it becomes difficult to subdivide this Arg 4-codon sector. Eukaryotes (and a few bacteria) have altered the genetic code further to include other tRNAs with adenine → inosine in the anticodon wobble position. In eukaryotes, tRNA^{Leu} (AAG), tRNA^{Ile} (AAU), tRNA^{Val} (AAC), tRNA^{Ser} (AGA), tRNA^{Pro} (AGG), tRNA^{Thr} (AGU), tRNA^{Ala} (AGC), and tRNA^{Arg} (ACG) with adenine converted to inosine in the wobble position are utilized (Rafels-Ybern et al., 2018).

Discussion

A model for tRNA evolution

A model for evolution of the tRNA cloverleaf has been proposed and strongly supported using statistical tests (Pak et al., 2017). Essentially, all predictions of the model have been verified for archaeal and bacterial tRNAs. The model is based on ligation of three 31 nt minihelices followed by two internal, symmetrical 9 nt deletions to yield a 75 nt cloverleaf core (1–75), with the attached discriminator base (76) and 3'-CCA (77–79). By contrast, historical tRNA numbering utilizes a 72 nt core, which is based on eukaryotic tRNAs with 3 nt deleted in the D loop relative to tRNA^{Pri}. In cloverleaf evolution, one of the three ligated minihelices became the D loop, one the anticodon loop and one the T loop. 9 nt deletions are within ligated acceptor stem sequences, leaving two 5 nt relics of what were initially complementary acceptor stems surrounding the anticodon stem. The anticodon stem and loop and the T stem and loop are homologous, and obviously so, particularly for archaeal tRNAs, and homology is starkly evident from inspection of typical tRNA diagrams (i.e., of *Pyrococcus* tRNAs) (Pak et al., 2017).

Two minihelix tRNA evolution models

In a competing two minihelix model for tRNA evolution, proposed by others (Di Giulio, 2009, 2012; Widmann et al., 2010), the cloverleaf sequence is essentially divided through the anticodon loop, and the halves are expected to be homologous, even though, in the cloverleaf, the halves are expected to be complementary. In the two minihelix model, because, for the comparison, the anticodon stem and loop were bisected, the anticodon loop and the T loop cannot be homologs, although they clearly are, both from inspection of archaeal tRNAs and using statistical tests (Pak et al., 2017). In the two minihelix model, the D loop and the T loop ought to be homologs, although they clearly are not (in any alignment register). By contrast, the tRNA evolution model utilized here is predictive and apparently accurate and competing models are falsified. Identification of tRNA^{Pri} based on the tRNA evolution model is highly predictive for the evolution of the genetic code (Figs. 1.1, 1.2 and 1.3).

tRNA and rugged evolution

A tightly folded RNA such as the tRNA cloverleaf is subject to rugged evolution in which many or most substitutions are catastrophic for folding (Curtis & Bartel, 2013; Novozhilov et al., 2007). For instance, most substitutions in a tRNA stem are expected to require rescue by a complementary mutation (except for many C → U substitutions in stems, which allow G~U pairing). In our model for tRNA evolution from tRNA[Pri], very few substitutions (if any) are required to obtain a folded cloverleaf. By contrast, in a two minihelix model for tRNA evolution, many substitutions are necessary to obtain a cloverleaf. Because of rugged RNA evolution and the required number of compensating substitutions, a two minihelix model is untenable. Furthermore, a two minihelix model requires unimaginable convergent evolution of the T stem and loop and the anticodon stem and loop to apparent structural and sequence homology. Because cloverleaf tRNA is subject to rugged evolution (Curtis & Bartel, 2013; Novozhilov et al., 2007; Pak et al., 2017), many disqualifying criticisms are generated for a two minihelix model. Other tRNA evolution models also appear to be inconsistent with rugged evolution of RNA (Caetano-Anollés et al., 2013; Rodin et al., 2011).

A root for the tRNA evolutionary tree

The model for tRNA evolution indicates a sequence for tRNA[Pri] (Pak et al., 2017), which is most similar to archaeal tRNA[Gly], indicating that Gly may be the founding amino acid of the code (Fig. 1.2) (Bernhardt, 2016; Bernhardt & Patrick, 2014). The polyglycine hypothesis is posited, that tRNA initially evolved to synthesize short-chain polyglycine to stabilize protocells. Very rapidly, every permitted anticodon was initially assigned as tRNA[Gly] before reassignment to specify other amino acids (Fig. 1.5). Cloverleaf tRNA and the genetic code appear to be prerequisites for cellular and DNA genome-based life, which originate at LUCA. In the RNA-protein world, genes were more independent than they subsequently became, in compact, streamlined and rapidly replicating DNA genomes encapsulated in cells. We propose, therefore, that colonies of independently replicating tRNA genes in an RNA-polymer world quickly diversified to include all permitted anticodon sequences, which, initially, encoded glycine (i.e., based on acceptor stem sequence, discriminator A (as in tRNA[Pri] and archaeal tRNA[Gly] (Fig. 1.2)) and typical tRNA sequences. Of course, specification of glycine attachment by tRNA[Pri] need not have been highly accurate. It appears that errors in tRNA charging drove code evolution (Koonin, 2017; Koonin & Novozhilov, 2009; Novozhilov & Koonin, 2009).

Degeneracy and sectoring

We favor a simple stepwise model for evolution and sectoring of the genetic code (Fig. 1.5). The model describes why the code specifies ∼20 amino acids and is degenerate. As we argue here, the initial genetic code probably consisted of 48 and not 64 permitted anticodons, because adenine in the wobble position of the anticodon loop is destabilizing and would be expected to interact awkwardly with mRNA (Saint-Léger et al., 2016). Furthermore, adenine in the anticodon wobble position probably supports a genetic code that is overly inflexible during initial code evolution, because adenine too strongly specifies uridine in the mRNA wobble codon position. Because of early positive selection for ambiguity in reading the anticodon wobble position, the genetic code should be considered initially to be primarily a 2 nt code encoding at most 16 amino acids (or 15 amino acids + Ter (stop)) in a register of 3 nt. Discrimination using the wobble anticodon position is only achieved with difficulty and, because of the ambiguity of tRNA anticodon-mRNA codon interactions in the ribosome decoding center (Rozov et al., 2016), recognition at the wobble anticodon base is not strongly constrained by Watson-Crick base pairing. Despite early selection for ambiguity reading the mRNA wobble position, the tRNA anticodon wobble position was later innovated to add an additional ∼5−6 letters to the code ($16 + 5 = 21$ letters total, including stops).

Wobble pairing: the importance of being ambiguous

Negative selection against adenine in the anticodon wobble position indicates that tRNA-mRNA wobble A∼C pairing is negatively selected when A is the tRNA anticodon wobble base (Pak & Burton, 2018). We note, however, that G∼U and U∼G wobble pairings are allowed. This raises the question of whether C∼A pairing might have been allowed, if C was the tRNA anticodon wobble base. Modifications of tRNA wobble C improve C∼A base pairing, including agmatidine (archaea), 2-lysidine (bacteria) and 5-formylcytidine (mitochondria, eukarya) (Machnicka et al., 2013). Many tRNAs have a weak C∼A hydrogen bonding interaction between the 7 nt anticodon loop base position 1 (i.e., 2′-O-methyl-C

(C=O or N)) and loop base position 7 (i.e., A (NH$_2$)). From PDB 4TRA, it appears that the weak $1 \rightarrow 7$ C$\sim$A interaction is modulated by Mg^{2+}, and elevated Mg^{2+} is reported to induce translation errors (Agafonov & Spirin, 2004; Scarlat et al., 1969). During the early stages of code evolution, therefore, ambiguous wobble base pair interactions appear to have been positively selected. We posit that, for translation, a wobble tRNA base C (or modified C) may pair mRNA base A more efficiently than a wobble tRNA base A will pair mRNA base C, partly explaining the strong negative selection of A in the tRNA anticodon wobble position. It appears that tRNA anticodon wobble C is not as strongly negatively selected as wobble A. We note the possibility that tRNA anticodon wobble C modification to pair mRNA codon A may have occurred very early in evolution to compensate for an otherwise overly restrictive code. Also, there may be a selected preference for G and C over A and U during early evolution of the code. The genetic code initially evolved to be a $\sim$16 letter code before innovating the wobble position to expand to a 21 letter code.

Covalent modifications of tRNAs are common. In Figure S10, archaeal tRNA modifications determined for *Haloferax volcanii* tRNAs from the Modomics database (Machnicka et al., 2013) are displayed on a *Pyrococcus* typical tRNA. In concept, tRNA modifications could be used as determinants for aaRS enzymes to discriminate different tRNAs (i.e., tRNA^{Phe} in bacteria, which requires tRNA^{Phe} modifications for accurate charging by PheRS) (Perona & Gruic-Sovulj, 2014), although, to our knowledge, such a mechanism has not yet been clearly demonstrated for any archaeal tRNA. In archaea, many covalent modifications are found in the anticodon loop particularly at loop positions 1 and 3 (wobble). Modifications in the anticodon loop may: (1) help stabilize the tight U turn structure; (2) affect anticodon readout; and/or (3) modify weak anticodon loop positions $1 \rightarrow 7$ interactions. Contacts between loop positions 1 and 7 affect loop dynamics and modify wobble position readout (Agris et al., 2017, 2018). Modifications of the D loop, T loop, and V loop may stabilize loop and stem conformations, D loop-T loop interactions and/or stability of the overall cloverleaf fold. Of course, for bacteria and eukaryotes, tRNA modifications allow expansions of the anticodon repertoire, as seen for the enzymatic conversion of wobble position adenine $\rightarrow$ inosine (Rafels-Ybern et al., 2018; Saint-Léger et al., 2016).

Cloverleaf tRNA as an evolutionary archetype

In ancient evolution from about 3.8 to 4 billion years ago, cloverleaf tRNA was the defining innovation that made possible the RNA-protein world and then cellular life (Pak et al., 2017). Essentially, without cloverleaf tRNA, the genetic code was impossible, and the RNA-protein world and cellular life were, therefore, impossible. 17 nt microhelices and 31 nt minihelices (17 nt microhelices with 2 × 7 nt acceptor stems) may have supported polyglycine synthesis, but there is little evidence that much more complex products were possible based on minihelix adapters (Pak et al., 2017). For one thing, from the cloverleaf tRNA^{Pri} sequence, the 31 nt minihelix posited to have given rise to the D loop appears to have had glycine-specifying acceptor stems, indicating that, because at least two distinct minihelices (D loop and anticodon loop/T loop) appeared to have specified glycine, few products, if any, other than polyglycine were made.

In a minihelix world, the D loop minihelix could not have supported a 3 nt genetic code register, because the D loop minihelix cannot form a 7 nt U turn. By contrast, the minihelices that gave rise to the anticodon loop and the T loop form the tight 7 nt U turn loop. The anticodon loop and the T loop are homologous to each other and distinct in sequence from the D loop minihelix, except in the acceptor stems, which appear initially to be identical (GCG and CGC repeats) (Pak et al., 2017). We posit, therefore, that polypeptide synthesis based on primitive minihelix adapters was chaotic, limited and inefficient.

Based on cloverleaf tRNA sequence, structure and evolution, we posit a strange polymer world that included acceptor stems (GCG and CGC repeats), D loop minihelices (UAGCC repeats with acceptor stems) and anticodon and T loop minihelices (∼GGCCCUUCAAAAGGGCC with acceptor stems) (Pak et al., 2017). Replication of minihelices is expected to involve ligation and an unknown mechanism of complementary replication (i.e., ribozyme-based replication) producing complementary sequences. Ligation of 3 minihelices and symmetrical RNA processing is sufficient to generate the tRNA cloverleaf, indicating that the minihelix-polymer world quickly gave way to a world dominated by cloverleaf tRNA. We posit that cloverleaf tRNA was quickly adopted as an improved mechanism to synthesize polyglycine, which stabilized protocells to support an unknown mechanism of energy generation (i.e., ribozyme-based). As described in this paper, cloverleaf tRNA rapidly diversified to encode 20 amino acids and stop codons, sufficient to support the RNA-protein world and leading subsequently to DNA genome-based cellular life at LUCA with a relatively modern translation system and genetic code. Cloverleaf tRNA, therefore, is proposed to be the most essential and central molecular archetype that made the RNA-protein world and cellular life possible. Remarkably, particularly in archaea (i.e., *Pyrococcus* and *Staphylothermus*), cloverleaf tRNA is little changed since LUCA (Pak et al., 2017).

Explosive evolution

Francis Crick referred to the evolution of tRNA and the genetic code as a "frozen accident," in which, very rapidly, tRNA, the code, and aaRS enzymes coevolved into existence (Bernhardt & Tate, 2008; Koonin, 2017; Rafels-Ybern et al., 2018; Rodin et al., 2011). Logically, tRNA must initially evolve and diversify, indicating that the first aminoacyl transferase functions were ribozymes that later were replaced by aaRS enzymes. We consider that a separate and robust (i.e., ribozyme-based) genetic code is unlikely to have existed prior to cloverleaf tRNA. Indeed, evolution to the mature and nearly universal genetic code must have been rapid. The mechanisms that brought closure to code evolution, which we begin to address here, now particularly need to be explained. We posit that the code evolved mostly in two stages. In the initial stage up to ~16 letters, ambiguity of the wobble anticodon position was of positive value so that more mRNA wobble sequences could be tolerated and so an initial code could be established to encode proteins using a limited set of available amino acids. As we show here, rejection of inflexible code evolution can partly explain the negative selection of adenine at the tRNA anticodon wobble position. In the later stage of evolution, innovation at the wobble position was utilized to complete sectoring of the code. According to this view, closure occurs to balance initial positive selection for wobble position ambiguity and the growing requirement to accurately encode robust proteins with sufficient chemistry.

Computational approaches

We simplify and shrink the problem of evolution of the genetic code. We show that adenine is generally not utilized in the tRNA anticodon wobble position, unless adenine can be converted to inosine by a tRNA adenosine deaminase, which is missing in archaea and was probably, therefore, missing at LUCA (Saint-Léger et al., 2016). Because adenine at the wobble position is destabilizing for the anticodon loop, and because the tRNA wobble position is selected to be ambiguous, the primordial genetic code reduces to 44 unique tRNA anticodons +3 stop codons. The archaeal species analyzed here have 46 tRNAs, including a missing tRNA^{Ile} (UAU) and a total of 3 tRNA^{Met} (CAU), matching the expectation of 44 unique tRNA antico-dons. We propose orderly mechanisms by which the code might have sectored to produce the current genetic code and suggest mechanisms by which the code progressed to universality and closure at ∼20 amino acids + stops. These observations render the evolution of the genetic code more reasonably accessible for computational approaches, such as machine learning and artificial intelligence. Because each organism solves the prob-lem of tRNA evolution somewhat differently, we advocate for computa-tional methods that relate tRNAomes, compared organism to organism, as we begin here. The collection of tRNAs for each organism is a set with bound limits (i.e., because of cloverleaf structure and rugged evolution) in which tRNAs are powerfully coevolved. Clearly, machine learning can be applied to the comprehensive comparison of tRNAome evolutionary trees (i.e., used as "fingerprints" to discriminate species). We note that advances in tRNA modifications found in bacteria and eukaryotes but missing in archaea expand the possibilities for the tRNA anticodon repertoire in ways that can be predicted and/or identified, for instance, as observed for adenine→ inosine conversion at the anticodon wobble position.

Methods

Cloverleaf tRNA evolution

The model for tRNA evolution (Pak et al., 2017; Root-Bernstein et al., 2016) was developed by inspection of typical tRNA sequences (Juhling et al., 2009), sequence logos (Schneider & Stephens, 1990) and sequences obtained from tRNA databases (Chan & Lowe, 2016; Juhling et al., 2009). It is clear (i.e., from inspection of typical tRNA diagrams and sequence logos) that archaeal tRNAs are more similar to LUCA tRNAs than bacterial tRNAs (Pak et al., 2017), so archaeal tRNAs are mostly used in the analyses shown here. Because the tRNA database often gives genomic DNA sequences for archaeal tRNAs, DNA sequences and RNA sequences are presented here according to convenience. We use a numbering system for tRNAs based on a 75 nt core sequence (Pak et al., 2017). Traditional tRNA numbering is based on a 72 nt core sequence, which is determined from eukaryotic tRNAs with 3 nt deleted from the D loop (Pak et al., 2017; Root-Bernstein et al., 2016). The traditional numbering system is confusing.

Sequence logos

Sequence logos were prepared using Weblogo 3.6 (http://weblogo.threeplusone.com/create.cgi) (Schneider & Stephens, 1990).

NCBI blast searches

The tRNA^{Pri} sequence (GCGGCGGTAGCCTAGCCTGGCCTAGGCG GCCGGGTTNNNAACCCGGCCGCCCCGGGTTCAAATCCCGGC CGCCGCACCA) (Pak et al., 2017) was searched using NCBI (National Center for Biotechnology Information; https://www.ncbi.nlm.nih.gov/) nucleotide blast versus genomic archaeal and bacterial sequences and against collections of tRNA sequences from different organisms (Fig. 1.2). A top hit in these searches is tRNA^{Gly}. Nucleotides 19 (A → G) and 67 (A → T) were adjusted in tRNA^{Pri} because these sequences are invariant in archaea and conserved in bacteria. To allow for different anticodon sequences, the anti-codon is represented by NNN. The typical *Pyrococcus* tRNA sequence is very similar to tRNA^{Pri} and typical archaeal tRNA^{Gly} (GCGGCGGTAGCN-TAGCCTGGTNNAGNGCGCCGGNCTNNNGANCCGGNGGTCC CGGGTTCAAATCCCGGCCGCNGCACC) (Fig. 1.2) (Pak et al., 2017).

Evolutionary trees

Evolutionary trees for tRNAs were generated using PASTA (https://github.com/smirarab/pasta) (Mirarab et al., 2015). All available tRNAs for an organism were collected and annotated by hand. V loop inserts were removed in alignments to eliminate detection of possible false similarity comparing tRNA^{Leu} and tRNA^{Ser}, which both have V loop inserts and which test as closely related tRNAs whether or not V loop inserts are included in the alignment (the comparison is not shown). To root trees, the tRNA^{Pri} sequence was included. Evolutionary distances were determined by adding the distances of related branches. PASTA output was analyzed using FigTree (http://tree.bio.ed.ac.uk/software/figtree/). Alignments of tRNAs were checked against the tRNA databases. If PASTA is allowed to align bacterial tRNAs, gap errors are sometimes generated aligning the 5′-acceptor stem and the D loop. In some cases, apparent errors in tRNA databases (i.e., unlikely mispairing of the 5′- and 3′-acceptor stems) were detected from analysis of tRNAomes and apparent misplacing of a tRNA in the tree. Adjustment of the tRNA sequence resulted in more appropriate placement of the tRNA in the tree. Finding likely sequence errors based on tRNAome trees indicates the probable reliability of trees. *Pyrococcus* tRNAomes appear to have two reassigned tRNAs (tRNA^{Phe} (from tRNA^{Trp}) and tRNA^{Trp} (from tRNA^{Pro})) (Fig. 1.1).

Funding Statement

This work was partly supported by National Science Foundation CAREER [Grant number DBI-0953738 to Y.S].

Acknowledgments

We thank Kevin Liu (Computer Science and Engineering, Michigan State University), Robert Root-Bernstein (Physiology, MSU), and Bruce Kowiatek (Blue Ridge Community and Technical College Martinsburg, WV) for helpful discussions and advice. This work was partly supported by National Science Foundation CAREER Grant DBI-0953738 to Y.S.

References

Agafonov, D. E., & Spirin, A. S. (2004). The ribosome-associated inhibitor a reduces translation errors. *Biochemical and Biophysical Research Communications, 320*(2), 354—358. https://doi.org/10.1016/j.bbrc.2004.05.171

Agris, P. F., Eruysal, E. R., Narendran, A., Väre, V. Y. P., Vangaveti, S., & Ranganathan, S. V. (2018). Celebrating wobble decoding: Half a century and still much is new. *RNA Biology, 15*(4—5), 537—553. https://doi.org/10.1080/15476286.2017.1356562

Agris, P. F., Narendran, A., Sarachan, K., Väre, V. Y. P., & Eruysal, E. (2017). The importance of being modified: The role of RNA modifications in translational fidelity. *Enzymes, 41*, 1—50. https://doi.org/10.1016/bs.enz.2017.03.005

Bernhardt, H. S. (2016). Clues to tRNA evolution from the distribution of class II tRNAs and serine codons in the genetic code. *Life, 6*(1). https://doi.org/10.3390/life6010010

Bernhardt, H. S., & Patrick, W. M. (2014). Genetic code evolution started with the incorporation of glycine, followed by other small hydrophilic amino acids. *Journal of Molecular Evolution, 78*(6), 307—309. https://doi.org/10.1007/s00239-014-9627-y

Bernhardt, H. S., & Tate, W. P. (2008). Evidence from glycine transfer RNA of a frozen accident at the dawn of the genetic code. *Biology Direct, 3*. https://doi.org/10.1186/1745-6150-3-53

Bhattacharyya, S., & Varshney, U. (2016). Evolution of initiator tRNAs and selection of methionine as the initiating amino acid. *RNA Biology, 13*(9), 810—819. https://doi.org/10.1080/15476286.2016.1195943

Bullock, T. L., Uter, N., Nissan, T. A., & Perona, J. J. (2003). Amino acid discrimination by a class I aminoacyl-tRNA synthetase specified by negative determinants. *Journal of Molecular Biology, 328*(2), 395—408. https://doi.org/10.1016/S0022-2836(03)00305-X

Caetano-Anollés, G., Wang, M., Caetano-Anollés, D., & Maga, G. (2013). Structural phylogenomics retrodicts the origin of the genetic code and uncovers the evolutionary impact of protein flexibility. *PLoS ONE, 8*(8), e72225. https://doi.org/10.1371/journal.pone.0072225

Chan, P. P., & Lowe, T. M. (2016). GtRNAdb 2.0: An expanded database of transfer RNA genes identified in complete and draft genomes. *Nucleic Acids Research, 44*(1), D184–D189. https://doi.org/10.1093/nar/gkv1309

Curtis, E. A., & Bartel, D. P. (2013). Synthetic shuffling and in vitro selection reveal the rugged adaptive fitness landscape of a kinase ribozyme. *RNA, 19*(8), 1116–1128. https://doi.org/10.1261/rna.037572.112

Demeshkina, N., Jenner, L., Westhof, E., Yusupov, M., & Yusupova, G. (2012). A new understanding of the decoding principle on the ribosome. *Nature, 484*(7393), 256–259. https://doi.org/10.1038/nature10913

Di Giulio, M. (2009). A comparison among the models proposed to explain the origin of the tRNA molecule: A synthesis. *Journal of Molecular Evolution, 69*(1), 1–9. https://doi.org/10.1007/s00239-009-9248-z

Di Giulio, M. (2012). The origin of the tRNA molecule: Independent data favor a specific model of its evolution. *Biochimie, 94*(7), 1464–1466. https://doi.org/10.1016/j.biochi.2012.01.014

Di Giulio, M. (2014). The genetic code did not originate from an mRNA codifying polyglycine because the proto-mRNAs already codified for an amino acid number greater than one. *Journal of Theoretical Biology, 361*, 204–205. https://doi.org/10.1016/j.jtbi.2014.09.006

Giege, & Eriani, G. (n.d.). Transfer RNA recognition and aminoacylation by synthetases. John Wiley & Sons.

Gruic-Sovulj, I., Uter, N., Bullock, T., & Perona, J. J. (2005). tRNA-dependent aminoacyl-adenylate hydrolysis by a nonediting class I aminoacyl-tRNA synthetase. *Journal of Biological Chemistry, 280*(25), 23978–23986. https://doi.org/10.1074/jbc.M414260200

Juhling, F., Morl, M., Hartmann, R. K., Sprinzl, M., Stadler, P. F., & Putz, J. (2009). tRNAdb 2009: Compilation of tRNA sequences and tRNA genes. *Nucleic Acids Research, 37*, D159–D162. https://doi.org/10.1093/nar/gkn772. Database.

Koonin, E. V. (2017). Frozen accident pushing 50: Stereochemistry, expansion, and chance in the evolution of the genetic code. *Life, 7*(2). https://doi.org/10.3390/life7020022

Koonin, E. V., & Novozhilov, A. S. (2009). Origin and evolution of the genetic code: The universal enigma. *IUBMB Life, 61*(2), 99–111. https://doi.org/10.1002/iub.146

Koonin, E. V., & Novozhilov, A. S. (2017). Origin and evolution of the universal genetic code. *Annual Review of Genetics, 51*, 45–62. https://doi.org/10.1146/annurev-genet-120116-024713

Machnicka, M. A., Milanowska, K., Oglou, O. O., Purta, E., Kurkowska, M., Olchowik, A., Januszewski, W., Kalinowski, S., Dunin-Horkawicz, S., Rother, K. M., Helm, M., Bujnicki, J. M., & Grosjean, H. (2013). MODOMICS: A database of RNA modification pathways—2013 update. *Nucleic Acids Research, 41*(1), D262–D267. https://doi.org/10.1093/nar/gks1007

Mirarab, S., Nguyen, N., Guo, S., Wang, L. S., Kim, J., & Warnow, T. (2015). PASTA: Ultra-large multiple sequence alignment for nucleotide and amino-acid sequences. *Journal of Computational Biology, 22*(5), 377–386. https://doi.org/10.1089/cmb.2014.0156

Novozhilov, A. S., & Koonin, E. V. (2009). Exceptional error minimization in putative primordial genetic codes. *Biology Direct, 4.* https://doi.org/10.1186/1745-6150-4-44

Novozhilov, A. S., Wolf, Y. I., & Koonin, E. V. (2007). Evolution of the genetic code: Partial optimization of a random code for robustness to translation error in a rugged fitness landscape. *Biology Direct, 2.* https://doi.org/10.1186/1745-6150-2-24

Ogle, J. M., Murphy IV, F. V., Tarry, M. J., & Ramakrishnan, V. (2002). Selection of tRNA by the ribosome requires a transition from an open to a closed form. *Cell, 111*(5), 721–732. https://doi.org/10.1016/S0092-8674(02)01086-3

Pak, & Burton, Z. F. (2018). *Aminoacyl-tRNA synthetase proofreading, anticodon wobble preference and sectoring of the genetic code via tRNA charging errors.* Transcription.

Pak, D., Root-Bernstein, R., & Burton, Z. F. (2017). tRNA structure and evolution and standardization to the three nucleotide genetic code. *Transcription, 8*(4), 205–219. https://doi.org/10.1080/21541264.2017.1318811

Perona, J. J., & Gruic-Sovulj, I. (2014). Synthetic and editing mechanisms of aminoacyl-tRNA synthetases. *Topics in Current Chemistry, 344*, 1–41. https://doi.org/10.1007/128_2013_456

Quigley, G. J., & Rich, A. (1976). Structural domains of transfer RNA molecules. *Science, 194*(4267), 796–806. https://doi.org/10.1126/science.790568

Rafels-Ybern, À., Torres, A. G., Grau-Bove, X., Ruiz-Trillo, I., & Ribas de Pouplana, L. (2018). Codon adaptation to tRNAs with Inosine modification at position 34 is widespread among Eukaryotes and present in two Bacterial phyla. *RNA Biology, 15*(4–5), 500–507. https://doi.org/10.1080/15476286.2017.1358348

Rodin, A. S., Szathmáry, E., & Rodin, S. N. (2011). On origin of genetic code and tRNA before translation. *Biology Direct, 6*. https://doi.org/10.1186/1745-6150-6-14

Romaniuk, J. A. H., & Cegelski, L. (2015). Bacterial cell wall composition and the influence of antibiotics by cell-wall and whole-cell NMR. *Philosophical Transactions of the Royal Society B: Biological Sciences, 370*(1679). https://doi.org/10.1098/rstb.2015.0024

Root-Bernstein, R., Kim, Y., Sanjay, A., & Burton, Z. F. (2016). tRNA evolution from the proto-tRNA minihelix world. *Transcription, 7*(5), 153–163. https://doi.org/10.1080/21541264.2016.1235527

Rozov, A., Demeshkina, N., Westhof, E., Yusupov, M., & Yusupova, G. (2016). New structural insights into translational miscoding. *Trends in Biochemical Sciences, 41*(9), 798–814. https://doi.org/10.1016/j.tibs.2016.06.001

Ruff, M., Krishnaswamy, S., Boeglin, M., Poterszman, A., Mitschler, A., Podjarny, A., Rees, B., Thierry, J. C., & Moras, D. (1991). Class II aminoacyl transfer RNA synthetases: Crystal structure of yeast aspartyl-tRNA synthetase complexed with tRNAAsp. *Science, 252*(5013), 1682–1689. https://doi.org/10.1126/science.2047877

Saint-Léger, A., Bello, C., Dans, P. D., Torres, A. G., Novoa, E. M., Camacho, N., Orozco, M., Kondrashov, F. A., & De Pouplana, L. R. (2016). Saturation of recognition elements blocks evolution of new tRNA identities. *Science Advances, 2*(4). https://doi.org/10.1126/sciadv.1501860

Scarlat, I. V., Rutkovskaya, N. O., Ginevskaya, V. A., & Agol, V. I. (1969). Magnesium-induced errors of translation in a cell-free system from krebs-II ascites carcinoma cells. *FEBS Letters, 5*(3), 231–232. https://doi.org/10.1016/0014-5793(69)80340-6

Schneider, T. D., & Stephens, R. M. (1990). Sequence logos: A new way to display consensus sequences. *Nucleic Acids Research, 18*(20), 6097–6100. https://doi.org/10.1093/nar/18.20.6097

Sengupta, S., & Higgs, P. G. (2015). Pathways of genetic code evolution in ancient and modern organisms. *Journal of Molecular Evolution, 80*(5–6), 229–243. https://doi.org/10.1007/s00239-015-9686-8

Suzuki, T., Nakamura, A., Kato, K., Söll, D., Tanaka, I., Sheppard, K., & Yao, M. (2015). Structure of the *Pseudomonas aeruginosa* transamidosome reveals unique aspects of bacterial tRNA-dependent asparagine biosynthesis. *Proceedings of the National Academy of Sciences, 112*(2), 382–387. https://doi.org/10.1073/pnas.1423314112

Trifonov, E. N. (2004). The triplet code from first principles. *Journal of Biomolecular Structure and Dynamics, 22*(1), 1–11. https://doi.org/10.1080/07391102.2004.10506975

Widmann, J., Harris, J. K., Lozupone, C., Wolfson, A., & Knight, R. (2010). Stable tRNA-based phylogenies using only 76 nucleotides. *RNA, 16*(8), 1469–1477. https://doi.org/10.1261/rna.726010

CHAPTER 2

Aminoacyl-tRNA synthetase evolution and sectoring of the genetic code[*]

Daewoo Pak[1], Yunsoo Kim[2] and Zachary F. Burton[3]
[1]Center for Statistical Training and Consulting, Michigan State University, East Lansing, MI, United States; [2]IMC Trading, Chicago, IL, United States; [3]Department of Biochemistry and Molecular Biology, Michigan State University, East Lansing, MI, United States

Abbreviations

aminoacyl–tRNA synthetase enzymes (aaRS)	(i.e., glycine aminoacyl–tRNA synthetase (GlyRS))
Homo sapiens	(*Hs*)
inosine	(I)
Pyrobaculum aerophilum	(*Pae*)
Pyrococcus furiosis	(*Pfu*)
Staphylothermus marinus	(*Sma*)
Sulfolobus solfataricus	(*Sso*)
The last universal common cellular ancestor	(LUCA)
Thermus thermophilus	(*Tth*)

Introduction

Aminoacyl-tRNA synthetases (aaRS; i.e., GlyRS) accurately add amino acids to the 3′-CCA ends of tRNAs. Two distinct protein folds are identified for aaRS, described as class I and class II (Giege & Eriani, n.d.; Perona & Gruic-Sovulj, 2014). Class, I aaRS have an active site that adenylates an amino acid at a "Rossmannoid" fold of parallel β-sheets before transferring the amino acid to tRNA. Class II aaRS, by strong contrast, have an active site of antiparallel β-sheets. The evolutionary relationship of class I and class II enzymes has not been clearly demonstrated, although the interesting suggestion has been made that class I and class II aaRS enzymes were encoded on opposing strands of a bi-directional ancestral gene (Carter, 2017; Carter & Wills, 2018; Pak et al., 2018; Pham et al., 2007; Rodin &

[*] The entire chapter was previously published. Transcription. 2018;9(4):205-224. https://doi.org/10.1080/21541264.2018.1467718. Epub 2018 May 30.

The Makings of a Clinical Protocol
ISBN 978-0-323-95749-6
https://doi.org/10.1016/B978-0-323-95749-6.00002-8

27

Ohno, 1995; Rodin et al., 2009; Wills & Carter, 2018). We provide a simpler explanation. We show amino acid sequence similarity in archaea that indicates that class I and class II aaRS enzymes arose from unidirectional in-frame translation starting from different N-termini. The longer N-terminal region of class I aaRS enzymes forces class I to fold and prevents the class II fold. To detect class I and class II aaRS sequence similarity, one only has to gaze toward LUCA (the last universal common cellular ancestor; ~3.85 billion years ago) by comparing sequences in ancient archaea (Pak et al., 2018).

All 64 codons are utilized in mRNA, but only a subset of matching anticodons is utilized in tRNA. A subset of tRNA anticodons is possible because of the degeneracy of the genetic code and ambiguity in tRNA anticodon wobble bases reading mRNA codons, allowing (and limiting) a tRNA anticodon to read multiple synonymous mRNA codons. Potentially, therefore, ambiguity in tRNA anticodons reading mRNA codons could be positively selected in evolution, which might be reflected in anticodon wobble base preferences. It appears that tRNAomes (the collection of tRNAs for an organism) are generally selected to be small (even in complex eukaryotes) without skipping the recognition of mRNA codons (Pak et al., 2018). Specialization of tRNAs may occur (Percudani, 2001), but this was not the major driving force in early evolution.

In a recent review, it was suggested that evidence was sparse for the error minimization hypothesis in standard genetic code evolution (Koonin & Novozhilov, 2017). The error minimization theory describes the sectoring of the code to minimize the impacts of random mutations in tRNAs and of tRNA charging errors. Clustering similar amino acids in the codon-anticodon table might be selected in order to reduce the impact of translation errors. Massey has argued, however, that the code likely did not sector strongly to minimize errors in translation and coding, but, rather, that clustering of similar amino acids occurred through the evolutionary sectoring mechanism (Massey, 2008). Here, we also argue against error minimization as a strong selection pressure in building the genetic code. Rather, we argue that the sectoring of the code was largely driven by tRNA charging errors, and, therefore, error minimization resulted from the pathway of code evolution, essentially as proposed by Massey. Specifically, we show that minimization of translation errors via aaRS proofreading appears to have limited sectoring of the genetic code, indicating that tRNA charging errors led to reassignments of tRNAs during early code evolution. Reassignments of tRNAs could result in subdividing a 4-codon sector of the codon-anticodon table into two 2-codon sectors and adding a newly encoded amino acid. Mutations in the anticodon loop of tRNAs can also initiate an invasion of neighboring genetic code sectors, but this process moves amino acids in the table without introducing new amino acids into the code. Because tRNA charging errors drove code evolution, mechanisms ensuring tRNA charging accuracy brought the code to closure and universality. The dominant model to analyze genetic code evolution, therefore, should be that tRNA charging errors induced sectoring of the code, and the evolution of accuracy mechanisms brought the code to universality and closure. The coevolution hypothesis posits that tRNAs, amino acids, the genetic code, and aaRS enzymes are coevolved, an idea that we support in this paper.

In recent work, we describe how a cloverleaf tRNA evolution model (Pak et al., 2017; Root-Bernstein et al., 2016) is highly predictive for models of genetic code evolution (Pak et al., 2018). Further, we show that the evolution of the genetic code is centered more on tRNA than on mRNA or the ribosome. Primitive archaea have 46 tRNAs and 3 stop codons. Translation termination signals are recognized by proteins (not tRNA) that bind to an mRNA stop codon in the ribosome decoding center and reach into the ribosome peptidyl transferase center to terminate translation (Bertram et al., 2001). Included in sets of 46 tRNAs are encoded 44 unique tRNA anticodons. There are 3 tRNA^{Met} (CAU anticodon) including 1 initiator tRNA^{iMet} and 2 elongator tRNA^{Met} (Chan & Lowe, 2016; Juhling & Hartmann, 2009). Generally, in ancient archaea and bacteria, only a single tRNA^{Ile} (GAU) is utilized. All other permitted anticodons are found in tRNAs except for three potential anticodon sequences corresponding to stop codons. Because only 44 unique anticodons and 3 stop codons need to be considered in early code evolution, but all 64 codons are utilized in mRNA, tRNA anticodon structure and presentation appear to have placed the greatest restrictions on expansions of the genetic code (Pak et al., 2018).

In archaea, little or no tRNA wobble position adenine is found (Pak et al., 2018; Saint-Léger et al., 2016). In bacteria, only tRNA^{Arg} (ACG) generally has adenine encoded in the anticodon wobble position (Rafels-Ybern et al., 2017; Saint-Léger et al., 2016). In bacteria and eukarya, tRNA wobble adenine is modified to inosine (A → I) by a tRNA adenosine deaminase. Wobble A in tRNA specifies U in mRNA codons, but wobble inosine pairs A, C, and U, indicating that increasing ambiguity in mRNA codon interpretation was positively selected as long as the specificity of coding remained unchanged (Rafels-Ybern et al., 2017; Saint-Léger et al., 2016). Because tRNA wobble A is negatively selected, according to a tRNA-centric view, only 44 unique anticodons and 3 stop codons need to be considered in the earliest standard genetic code evolution rather than 64 (Pak et al., 2018). Because of tRNA wobble ambiguity reading mRNA, however, the maximum number of amino acids that can be encoded by a genetic code read by tRNA is 31 aas with stops.

Although the initial evolution of the genetic code may have involved ribozyme-catalyzed tRNA aminoacylation (Lee et al., 2000; Rodin & Rodin, 2006; Xiao et al., 2008), at later stages, tRNAs coevolved with aaRS enzymes that attach amino acids at the tRNA 3′-CCA ends (Giege & Eriani, n.d.; Perona & Gruic-Sovulj, 2014). Some aaRS enzymes have the capability to proofread tRNA-aa attachments by moving an improperly joined amino acid from the aaRS synthetic site, where the amino acid is linked to the tRNA 3′-CCA end, to a separate aaRS editing or proofreading site, where the noncognate amino acid is removed (Giege & Eriani, n.d.; Perona & Gruic-Sovulj, 2014). We make observations about aaRS editing that are not noted in reviews nor, as far as we can discover, in the literature. We make the observation that aaRS editing appears to inhibit continued sectoring of the code utilizing the anticodon wobble position, giving insight into the roles of tRNA charging errors in the evolution of the code. Furthermore, in eukaryotes, the left half and mostly 4-codon sectors of the genetic code, for which the aaRS enzymes have proofreading capacity, are also the sectors that have introduced the adenine → inosine anticodon wobble position modification. A → I modification blocks subdivision of a 4-codon sector to two 2-codon sectors because sectoring would result in translation errors. A → I conversions and U versus C wobble preference to increase the ambiguity of the tRNA wobble base to allow broader sequence contacts to synonymous mRNA codons.

Results
Evolution and homology of class I and class II aaRS enzymes

The evolution of *Pyrococcus furiosis* (*Pfu*) aaRS enzymes is described in Fig. 2.1. Interestingly, the apparent pathways for *Pfu* aaRS divergence show similarities to the proposed pathways for LUCA tRNA evolution (Pak et al., 2018). AaRS enzymes, amino acids, and tRNAs are coevolved, as predicted by the coevolution hypothesis (Koonin & Novozhilov, 2017). To construct the pathway for aaRS evolution, NCBI (National Center for Biotechnology Information) Blast tools were used with relaxed search metrics to identify the closest apparent relationships between aaRS proteins. *Pfu* was selected as an example of ancient archaea with a similar translation

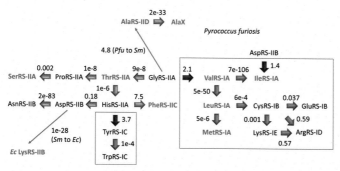

Figure 2.1 *Pyrococcus furiosis* (*Pfu*) aaRS enzymes were searched using NCBI Blast tools for nearest homologs in *Pfu*. In some cases, *Staphylothermus marinus* (*Sma*) (archaea) and *Escherichia coli* (*Eco*) (bacteria) homologs are identified. AlaX is one of a set of tRNA^Ala editing enzymes in *Pfu*.

system to LUCA (Pak et al., 2018). Separate structural comparisons (structural dendograms) of class I and class II aaRS enzymes have been published (O'Donoghue & Luthey-Schulten, 2003; Smith & Hartman, 2015; Valencia-Sánchez et al., 2016), and our analysis is consistent with these. Surprisingly, however, we identify similarities in protein sequences comparing class I and class II *Pfu* aaRS enzymes, which, to our knowledge, have not previously been reported (however, see Fig. 13 of reference O'Donoghue & Luthey-Schulten, 2003). Specifically, GlyRS-IIA and ValRS-IA are similar in amino acid sequence (e-value = 2.1). AspRS-IIB and IleRS-IA are similar (e-value = 1.4 or 1.5; depending on the alignment). HisRS-IIA and TyrRS-IC are also similar (e-value = 3.7). TyrRS-1C is more similar in sequence to HisRS-IIA (e-value = 3.7) than it is to other class I aaRS enzymes, with the sole exception of closely related TrpRS-IC (e-value = 1e−4). ThrRS-IIA is similar in sequence to IleRS-IA (e-value = 4.2). The e-value scores are for the best local alignments, but *Pfu* GlyRS-IIA and ValRS-IA are similar in sequence over nearly the entire length of GlyRS-IIA.

Interestingly, *Pfu* ValRS-IA, LeuRS-IA, IleRS-IA, and MetRS-IA are all very similar enzymes by aaRS structural class (IA) and e-value, and Val, Leu, Ile, and Met are similar neutral and hydrophobic amino acids within the first column of the codon–anticodon table. For *Pfu*, therefore, amino acids, tRNAs, and aaRS enzymes are coevolved for the first column of the code, as expected from the coevolution hypothesis. ThrRS-IIA, ProRS-IIA, and SerRS-IIA are found in the second column of the code, and these are related enzymes by aaRS class, e-value and apparent lineage. Gly,

Asp, Val, and Ala have been proposed to be the first four amino acids in the code (Pak et al., 2018). Interestingly, GlyRS-IIA, AspRS-IIB, ValRS-IA, and AlaRS-IID are very different enzymes, indicating that, at the base of code evolution across rows, discrimination of tRNAs by distinct aaRS enzymes was strongly selected. Apparently, there is a greater tendency for amino acid, tRNA, and aaRS coevolution within columns than across rows of the genetic code, particularly at the base of the code and at the earliest stage of code evolution. These observations appear to partly explain the distributions of similar amino acids within codon–anticodon table columns.

Fig. 2.2 shows the alignment of GlyRS-IIA, ValRS-IA, and IleRS-IA enzymes from ancient archaea. The alignment in Fig. 2.2 includes a Zn-binding motif that is shared among class I and class II aaRS enzymes. Some other features of class I and class II aaRS enzymes also appear to be

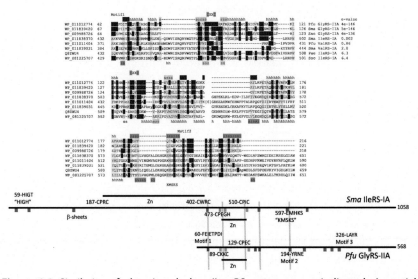

Figure 2.2 Similarity of class I and class II aaRS enzymes are indicated. A partial sequence alignment of GlyRS-IIA, ValRS-IA, and IleRS-IAs enzymes is shown demonstrating sequence similarity of a shared Zn-binding motif and GlyRS-IIA Motif 2 with IleRS-IA KMSKS motif. *Red shading* indicates identity comparing class I and class II aaRS. *Yellow shading* indicates similarity comparing class I and class II aaRS. *Green shading* is used to highlight Zn-binding motifs. *Cyan shading* indicates active site β-sheets (sss). *Magenta shading* indicates 3 β-sheets in GlyRS-IIAs expected to block class I folding by a class II aaRS. *Gray shading* indicates β1–β3 of the shared Zn-binding domain. The schematic diagram shows how *Pfu* GlyRS-IIA and *Sma* IleRS-IA align (*gray lines* highlight some similarities). *Pae, Pyrobaculum aerophilum; Sso, Sulfolobus solfataricus.* (For interpretation of the references to colour in this figure legend, the reader is referred to the web version of this article.)

conserved. A summary of the alignment data is shown in the schematic in Fig. 2.2. Relative to *Pfu* GlyRS-IIA, *Staphylothermus marinus* (*Sma*) IleRS-IA has an N-terminal extension that includes essential active site β-sheets, the HIGH active site motif, and a Zn-binding motif, all of which are missing in GlyRS-IIA. These unique N-terminal determinants of class I aaRS enzymes are likely to ensure the class I fold and to block the C-terminus of the protein from assuming the class II fold. The shorter GlyRS-IIA aligns to IleRS-IA and ValRS-IA over its entire length, and the C-terminus of these proteins is reasonably conserved across aaRS classes. GlyRS-IIA, IleRS-IA, and ValRS-IA share: (1) a β-sheet in the Motif 1 region of GlyRS-IIA aligning with an active site β-sheet in IleRS-IA; (2) a Zn-binding domain including 3 similarly positioned β-sheets; (3) a β-sheet of GlyRS-IIA just C-terminal to the shared Zn-binding domain aligns with an active site β-sheet of IleRS-IA; and (4) the active site Motif 2 of GlyRS-IIA and the active site KMSKS of IleRS-IA align, including a shared active site β-sheet and loop. The quality of the amino acid sequence alignment is probably sufficient to demonstrate GlyRS-IIA, ValRS-IA, and IleRS-IA homology. The structural similarities, such as the shared Zn-binding motif, strongly reinforce this conclusion. Local alignments with e-values as low as 0.001−0.002 have been obtained for GlyRS-IIA and IleRS-IA (i.e., a 1:500−1:1000 chance that the alignment is due to a random event).

In order to analyze the shared Zn-binding motif, a homology model for *Pfu* GlyRS-IIA was generated (Fig. 2.3). The closest structure identified using the Phyre2 server was human GlyRS-IIA (i.e., PDB 4KQE and 4QEI) (Deng et al., 2016; Qin et al., 2016). Although human GlyRS-IIA lacks cysteine ligands for Zn binding, the fold of the homologous region of the protein is maintained, so a model of the conserved *Pfu* GlyRS-IIA Zn-binding domain was obtained. *Thermus thermophilus* (*Tth*) ValRS-IA (PDB 1GAX) includes a shortened version of the shared Zn-binding domain (Fukai et al., 2000). In Fig. 2.4, the shared Zn-binding regions are compared. Within the Zn-binding region, three similarly arranged β-sheets are identified (β1−β3) comparing *Tth* ValRS-IA, *Pfu* GlyRS-IIA, and human GlyRS-IIA structures. We conclude from this structural comparison that class I and class II aaRS enzymes are homologous.

Incompatibility of class I and class II aaRS folds

ValRS-IA and GlyRS-IIA folds are incompatible (Fig. 2.5). The N-terminal extension of ValRS-IA helps to form the class I aaRS active site

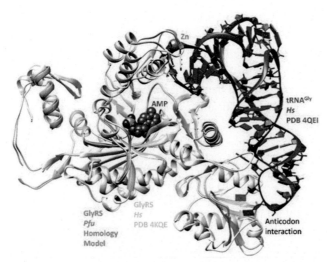

Figure 2.3 A homology model of *Pyrococcus furiosis* GlyRS-IIA was constructed by homology threading to human GlyRS-IIA (PDB 4KQE). The homology model (*powder blue*), PDB 4KQE (*white*) and related PDB 4QEI (*magenta*) were overlaid. Although human (*Hs*) GlyRS-IIA lacks Zn binding, the shape of the loops is maintained. (For interpretation of the references to colour in this figure legend, the reader is referred to the web version of this article.)

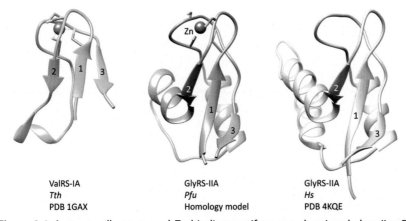

Figure 2.4 A structurally conserved Zn-binding motif among class I and class II aaRS enzymes. Similar orientations of *Tth* ValRS-IA (*green*), *Pfu* GlyRS-IIA (*magenta*), and human GlyRS-IIA (*white*) are shown. (For interpretation of the references to colour in this figure legend, the reader is referred to the web version of this article.)

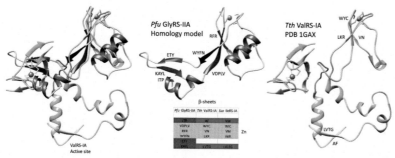

Figure 2.5 Incompatibility of class I and class II aaRS folding patterns. An overlay of the shared Zn-binding motif of GlyRS-IIA (secondary structure representation) and ValRS-IA (*green*) demonstrates a clash by three antiparallel GlyRS-IIA β-sheets with the N-terminal ValRS-IA Zn-binding domain. A ValRS-IA active site β-sheet (LVLEG) is *yellow*. LVLEG corresponds to *Pfu* GlyRS-IIA β-sheet KAYL in the 3 antiparallel β-sheet cluster surrounding the shared Zn-binding motif. (For interpretation of the references to colour in this figure legend, the reader is referred to the web version of this article.)

(i.e., the HIGH motif and essential active site β–sheets), and the N–terminal Zn-binding region of class I aaRS enzymes blocks class II aaRS folding. By comparison of structures, three antiparallel β–sheets in GlyRS–IIA that surround the shared Zn-binding motif establish a clash with the ValRS–IA N-terminal Zn-binding domain. The most C–terminal β–sheet of the *Pfu* GlyRS–IIA β–sheets (157–KAYL) corresponds to an active site β–sheet in *Tth* ValRS–IA (485–LVTG; *Sma* IleRS–IA 553–FIVEG), so the formation of the β–sheets in GlyRS–IIA is incompatible with the formation of the ValRS–IA active site. Formation of the ValRS–IA active site, therefore, is dependent on the N-terminal domain of *Tth* ValRS–IA, which includes the "HIGH" active site motif, parts of the active site parallel β–sheets (β–sheets 35–PFVIF, 73–EAVWL(P)GT, 137–DWSREAF) and the class I–specific Zn-binding domain. The more N–terminal class I–specific Zn-binding domain blocks class II aaRS folding.

The standard genetic code

The initial standard genetic code, which is found in many ancient archaea, is shown in Fig. 2.6 as a codon-anticodon table (top table, for archaea). Because of the central importance of tRNA in genetic code evolution, codon-anticodon tables are more informative than simpler representations. When the standard genetic code was established (i.e., in the RNA-protein world before LUCA), the anticodons shaded in red in the top chart were disallowed, because adenine was negatively selected in the tRNA anticodon wobble position (Pak et al., 2018; Saint-Léger et al., 2016). Adenine in the wobble position can destabilize the anticodon loop. Also, because wobble A pairs with U much better than with C in mRNA, adenine in the tRNA wobble position supports an inflexible code that was negatively selected (Pak et al., 2018). In addition, only one tRNA[Ile] (GAU) is generally utilized. Therefore, only 44 unique tRNA anticodons and 3 stop codons need to be considered in early genetic code evolution (Pak et al., 2018).

AaRS enzymes that proofread

In the course of studies on genetic code evolution, we analyzed archaeal, bacterial, and eukaryotic aaRS enzymes with proofreading active sites versus the standard genetic code (Fig. 2.6) (Giege & Eriani, n.d.; Perona & Gruic-Sovulj, 2014). The figure also accounts for the tRNA wobble adenine → inosine modification lacking in archaea but found in bacteria (tRNA[Arg] (ACG)) and eukarya (tRNA[Leu] (AAG), tRNA[Ile] (AAU), tRNA[Val] (AAC), tRNA[Ser] (AGA), tRNA[Pro] (AGG), tRNA[Thr] (AGU), tRNA[Ala] (AGC) and tRNA[Arg] (ACG)) (Rafels-Ybern et al., 2017; Saint-Léger et al., 2016). Remarkably, the aaRS enzymes that proofread in archaea are restricted to the left half of the codon-anticodon table, and, in eukaryotes, aaRS enzymes that proofread correlate strongly with the wobble A → I modification. SerRS-IIA proofreads, but Ser is split between the left and right halves of the table. In bacteria and eukaryotes, LysRS-IIB proofreads, but, in archaea, LysRS-IE does not (Giege & Eriani, n.d.; Perona & Gruic-Sovulj, 2014). To our knowledge, near restriction of aaRS editing to the left half of the codon-anticodon chart is not recorded in recent reviews or in the literature on aaRS enzymes, tRNAs, or the genetic code, although this observation is informative about code structure and evolution. Because the A → I wobble modification strongly correlates with aaRS enzymes that are proofread, this added structure of the code requires explanation.

Synonymous anticodon preferences
I >> G >> A

In part because of strong G >> A anticodon wobble preference in archaea (Fig. 2.6), we considered wobble preference more generally in tRNA (Carter & Wills, 2018; Pak et al., 2018; Wills & Carter, 2018). Unlike codon preference, anticodon wobble preference does not appear to be largely driven by gene regulation (i.e., to match codon bias). Inspection of anticodon wobble base frequencies indicated that, for each synonymous

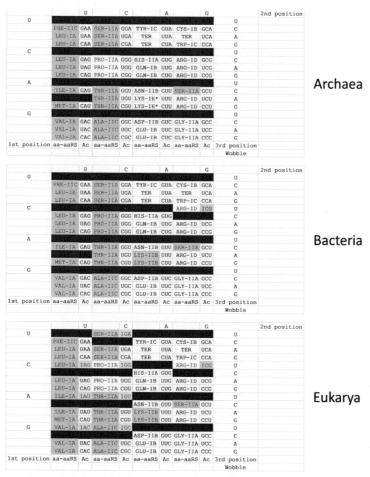

Figure 2.6 Codon-anticodon tables. Proofreading by aaRS enzymes in archaea is confined to the left half of the codon-anticodon table. *Gray shading* indicates editing by aaRS enzymes. *Red shading* indicates anticodons that are disallowed or strongly underrepresented. *Green shading* indicates adenosine → inosine conversion in bacteria and eukaryotes (tRNA^{Arg} (ACG → ICG)). *Yellow shading* indicates adenosine → inosine conversion in eukaryotes (very rarely, these modifications are found in some bacteria).

ANN and GNN pair (encoding the same amino acid) (Fig. 2.7), there was a strong preference for wobble G >> A, unless A was deaminated to inosine, in which case, interestingly, the preference was strongly I >> G. In archaea, A is largely excluded in the wobble position. In bacteria, only tRNA^{Arg} (ACG → ICG) is strongly favored over tRNA^{Arg} (GCG). In all other cases in bacteria, G is strongly favored over A, as in archaea. In eukaryotes, tRNA^{Leu} (AAG vs. GAG), tRNA^{Ile} (AAU vs. GAU), tRNA^{Val} (AAC vs. GAC), tRNA^{Ser} (AGA vs. GGA), tRNA^{Pro} (AGG vs. GGG), tRNA^{Ala} (AGC vs. GGC) and tRNA^{Arg} (ACG vs. GCG), for which wobble anticodons encoding A modify A $\rightarrow$ I, inosine is strongly favored over G. The A $\rightarrow$ I conversion is expected to increase the encoding of tRNAs with ANN anticodons but is not necessarily expected to so strongly suppress the use of synonymous GNN anticodons, which are functional in archaea. In this regard, tRNA wobble G can pair with mRNA codon C or U, but tRNA wobble I can pair with A, C, or U. Apparently, I >> G preference in the anticodon wobble position reflects strong positive selection for broader recognition by tRNA of synonymous mRNA codons.

We note that in eukaryotes tRNA^{Ser} (ACU << GCU) shows G preference over A in the anticodon wobble position. For tRNA^{Ser} (ACU), A is not converted to I. If tRNA^{Ser} (ACU) converted A $\rightarrow$ I, this would cause recognition by tRNA^{Ser} (ICU) of AGA Arg codons in mRNA, causing translation errors (Ser replacement of Arg in proteins). The A $\rightarrow$ I conversion, therefore, only occurs in 4-codon sectors to prevent spillover of tRNA specificity to a 2-codon sector encoding a different amino acid. In eukaryotes, the only 4-codon sector of the genetic code for which there is no A $\rightarrow$ I conversion is tRNA^{Gly} (ACC) (Saint-Léger et al., 2016).

U > C

Generally, U is preferred over C. In archaea, U is slightly preferred over C, for all synonymous anticodon pairs. Trp (CCA) is a special case because anticodon UCA represents a UGA stop codon that is read in mRNA by a protein. In bacteria, U is generally preferred over C, except for Leu2 (UAG < CAG) and Arg1 (UCG << CCG). For Val (UAC >> CAC), Ser (UGA >> CGA), Pro (UGG >> CGG), Thr (UGU >> CGG), Ala (UGC >> CGC), Gln (UUG >> CUG), Lys (UUU >> CUU), Glu (UUC >> CUC) and Gly (UCC >> CCC), wobble U is strongly preferred over wobble C. In eukaryotes, the U > C tRNA anticodon wobble preference is apparent for sets of synonymous anticodons for Ser (UGA > CGA), Pro (UGG >> CGG), Thr (UGU >> CGU), Ala (UGC >> CGC), Gln (UUG > CUG), Glu (UUC >> CUC), Arg1 (UCG >> CCG) and Arg2 (UCU > CCU). Interestingly, U versus C bias is the opposite for eukaryotic Arg1 (UCG >> CCG) and bacterial Arg1 (UCG << CCG). Also, U < C anticodon wobble preference is observed in eukaryotes for Leu1 (UAA < CAA), Leu2 (UAG < CAG), and Val (UAC << CAC), Lys (UUU << CUU), and Gly (UCC < CCC).

Ile, Met, and Trp are special cases. Ile and Met occupy the same 4-codon sector in the codon-anticodon table. For archaea and bacteria, tRNA^{Ile} (GAU) is highly used and tRNA^{Ile} (AAU and UAU) is very rarely used. Generally, archaea and bacteria utilize a single tRNA^{Ile} (GAU). In eukaryotes, tRNA^{Ile} (AAU → IAU) is strongly favored and tRNA^{Ile} (GAU) is suppressed, as expected. Interestingly, tRNA^{Ile} (UAU) is commonly utilized in eukaryotes, although tRNA^{Ile} (UAU) can potentially be ambiguous with tRNA^{Met} (CAU). Trp (CCA) shares a 2-codon sector with a stop codon, which is read in mRNA by a protein rather than a tRNA (Bertram et al., 2001).

Synonymous anticodon wobble preference in mitochondria

To maintain a small organelle genome, mitochondria encode a subset of tRNAs, potentially limiting available anticodons. Mitochondrial anticodon wobble preference, indeed, is strange and limited in coding capacity. In particular, Leu, Val, Pro, Thr, Ala, Arg, and Gly tRNAs have scant or no mitochondria-encoded wobble anticodon A or G. In terms of codon usage and preference, however, mitochondria utilize mRNA codons with wobble C and U encoding these amino acids. Because mitochondria import cytosolic tRNAs (Salinas et al., 2012; Salinas-Giegé et al., 2015; Schneider, 2011), deficiencies in mitochondrial coding can be compensated, and, perhaps, all of the apparent mitochondrial anticodon wobble deficiencies are compensated by imported cytosolic tRNAs. We note that the import of tRNAs with inosine in the anticodon wobble position (encoding Leu, Val, Pro, Thr, Ala, and Arg) would be almost sufficient to compensate for limiting mitochondrial tRNAs. The import of tRNA^{Ile} (IAU) is less important because tRNA^{Ile} (GAU) is encoded in the mitochondria and can suffice to read mRNA codons AUC and AUU. Import of a cytosolic tRNA^{Gly} (GCC) also appears necessary, and cytosolic tRNA^{Gly} (GCC) is imported into mitochondria (Salinas et al., 2012). Furthermore, mitochondrial-encoded tRNAs are heavily biased toward wobble U rather than C. Interestingly, tRNA^{Trp} (UCA >> CCA) in mitochondria utilizes the UCA anticodon corresponding to the UGA stop codon in place of the CCA anticodon, which is utilized to encode Trp in archaea, bacteria, and eukaryotes. Of course, tRNA wobble U reads a broader spectrum of synonymous mRNA codons than tRNA wobble C because wobble U can pair with mRNA wobble A or G but wobble C strongly prefers to pair with mRNA wobble G.

Arg coding

Interesting features of tRNA^{Arg} distributions in eukarya and bacteria: notably, tRNA^{Arg} (CCG) is somewhat limiting in eukaryotes and tRNA^{Arg} (UCG) is limiting or absent in bacteria. In Fig. 2.7, the consequences of these tRNA limitations are reviewed. It appears that eukaryotes primarily use tRNA^{Arg} (UCG) to read CGG codons. Bacteria primarily use tRNA^{Arg} (CCG) to read CGG codons. Eukaryotes read CGA codons using tRNA^{Arg} (ICG and UCG). Bacteria read CGA codons primarily using tRNA^{Arg} (ICG). The absence of tRNA^{Arg} (UCG) in some bacteria, therefore, appears to explain the evolution of the A → I tRNA wobble modification, which, in the case of missing tRNA^{Arg} (UCG), is required to read CGA codons.

The editing hypothesis

Based on the genetic code table left side biased distribution of 4-codon sectors correlating with proofreading aaRS enzymes (Fig. 2.6), sectoring of the genetic code utilizing the tRNA anticodon wobble position was likely inhibited by aaRS proofreading. Furthermore, editing appears to be limited to hydrophobic and neutral amino acids with limited charge and absent or limited side-chain hydrogen bonding potential. Proofreading generally occurs for amino acids that are smaller than, or very similar to, the cognate amino acid (Giege & Eriani, n.d.; Perona & Gruic-Sovulj, 2014). Smaller amino acids may attach to a noncognate tRNA and require editing because they can fit the synthetic aaRS active site and, therefore, can be linked to a noncognate tRNA (Giege & Eriani, n.d.; Perona & Gruic-Sovulj, 2014). Aminoacylation errors are less likely for amino acids with charged side chains and/or with more hydrogen bonding groups because more readily distinguished amino acids are more fully specified in their cognate aaRS synthetic active site. Interestingly, the densest sectoring employing the tRNA anticodon wobble position is observed for the third column (Glu, Asp, Lys, Asn, Gln, His, Ter (stop), Tyr) and the uppermost 4-codon sector of the fourth column (Trp, Ter, Cys). None of the

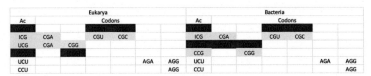

Figure 2.7 Consequences of apparent limiting of tRNA^{Arg} (CCG) in eukarya and tRNA^{Arg} (UCG) in bacteria. Ac for anticodon.

corresponding aaRS enzymes are proofread in archaea (Fig. 2.6). In bacteria, LysRS-IIB proofreads to reject amino acids that are from outside the code (homocysteine, homoserine, and ornithine). Lys and Arg are readily discriminated in the LysRS and ArgRS active sites. Lys has a flexible side chain with a localized positive charge. Arg, by contrast, has a much stiffer side chain with a distributed positive charge and hydrogen bonding potential. We posit that, as the code evolved, right half tRNAs initially did not require editing by an aaRS because encoded amino acids with more identifying functional groups were easier to specify through interactions in the aaRS synthetic active site. Because accurate specification of a cognate amino acid in the aaRS synthetic site limits tRNA charging errors, it is likely that the right half of the genetic code is sectored more completely to 2-codon sectors prior to full evolution of accurate amino acid selectivity by aaRS enzymes. A slightly different but related view might be that amino acids with more identifying characteristics were more aggressive at invading 4-codon sectors compared to neutral amino acids with limited hydrogen bond-forming potential. The evolution of aaRS anticodon recognition domains (in all aaRS except for AlaRS and SerRS) also enhanced the accuracy of tRNA charging and brought the code to universality. Fidelity mechanisms, therefore, continued to evolve and potentially take precedence over one another as the code continued to the sector.

Because aaRS proofreading appears to inhibit sectoring, and because archaeal aaRS enzymes from the right half of the code do not edit, sectoring is more innovative on the right half of the codon-anticodon table than the left half. We posit the editing hypothesis that aaRS proofreading inhibited genetic code sectoring. Invasion to reassign a 4-codon sector encoding a single amino acid to two 2-codon sectors, each encoding a distinct amino acid, for instance, was initiated by aminoacylation errors on existing tRNAs. During code evolution, the invasion could be by an amino acid that was not yet encoded, resulting in an increase in the complexity of the code. Note that accuracy of translation and tRNA charging continued to improve as the code evolved, and editing and specificity, therefore, became ever more important later in evolution as additional amino acids became encoded. Also, metabolism generates amino acids that are not encoded but could be charged to tRNAs in error and could be removed by aaRS proofreading. Alternatively, amino acids could be attached to tRNAs in error, and, then, through selection, could be added to the genetic code by dividing a 4-codon sector into two 2-codon sectors. With the exceptions of Met and Trp, only a stop codon (UGA), which is recognized in mRNA by a protein, not a tRNA, can occupy a 1-codon sector. Only in eukaryotes does Ile strongly occupy the UAU anticodon (adjacent to Met (CAU)), by which time in evolution, mechanisms were developed for tRNA modifications to support accurate tRNA^{Ile} (UAU) and tRNA^{Met} (CAU) discrimination (Agris et al., 2007, 2017).

Met, which appears to have invaded a partially occupied 4-codon Ile sector, may be an apparent exception to the rule that proofreading aaRS enzymes resist sectoring around the wobble anticodon position. MetRS proofreads to remove homocysteine, which is not part of the genetic code (Giege & Eriani, n.d.; Perona & Gruic-Sovulj, 2014). The Ile–Met 4-codon sector, however, appears to be a special case (see Discussion). In evolution, Phe may have invaded a 4-codon Leu sector (AAA, GAA, UAA, CAA), perhaps being recruited from outside the code. Arg appears to have invaded a 4-codon Ser sector (ACU, GCU, UCU, CCU) (Fig. 2.6), apparently demonstrating movement of amino acids within the code.

Coevolution of aaRS enzymes and tRNAs and the editing hypothesis

Fig. 2.8 shows how aaRS enzymes may have coevolved with tRNAs (Giege & Eriani, n.d.; Perona & Gruic-Sovulj, 2014). Before the code was substantially evolved, it is difficult to imagine tRNA recognition by proteins, and initial aminoacyl transfers may have been catalyzed by ribozymes

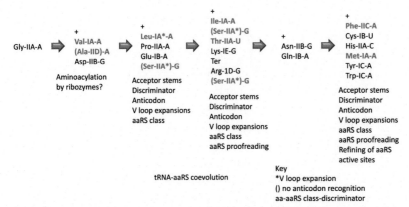

Figure 2.8 An approximate sequence of events for the requirement of different mechanisms for discrimination of tRNA identities by aaRS enzymes and the evolution of ribosome fidelity. *Green text* indicates aaRS proofreading (in archaea). (For interpretation of the references to colour in this figure legend, the reader is referred to the web version of this article.)

(Lee et al., 2000; Turk et al., 2010; Xiao et al., 2008). A proposed sequence of events was developed according to the aaRS mechanisms now used to discriminate different archaeal tRNAs. As aaRS enzymes evolved, acceptor stems of tRNAs and the discriminator base (position 76; 73 in historic numbering) (Pak et al., 2017, 2018) may have been the most important initial determinants for discrimination. In archaea, most discriminators are A, so the discriminator base is only used for a subset of amino acids (i.e., generally A in archaea except (G) Asp, Ser, Arg, Asn; (U) Thr, Cys; and (C) His) (Juhling & Hartmann, 2009). We posit that recognition of the anti-codon of tRNAs by aaRS enzymes subsequently became a mechanism for tRNA specification that restricted further sectoring of the code. Only AlaRS and SerRS lack anticodon recognition domains. Later, the aaRS enzyme class (i.e., class I vs. class II aaRS) became a determinant (Giege & Eriani, n.d.; Perona & Gruic-Sovulj, 2014). Without knowing the exact order of events, at some stage, longer V loops in tRNA^{Leu} and tRNA^{Ser} became important as determinants and anti-determinants for tRNA charging. From Fig. 2.1 of a recent paper (Pak et al., 2018), many tRNAs appear to be derived from tRNA^{Leu} and tRNA^{Ser}, which could have driven tRNA^{Leu} and tRNA^{Ser} V loop expansions in the evolving code in order to discriminate partially radiated tRNAs that may attach related amino acids. In archaea, generally, other tRNAs do not have V loop expansions (i.e., tRNA^{Tyr} and tRNA^{Sec} (Sec for selenocysteine) in bacteria). Along the pathway, active sites of aaRS enzymes continued to evolve to exclude the attachment of incorrect amino acids. This exclusion is more difficult for amino acid side chains that are uncharged and that form only one hydrogen bond (i.e., the left half of the codon-anticodon table), explaining why aaRS proofreading became such a dominant mechanism for the left half of the code. At a late stage, therefore, proofreading by aaRS enzymes is posited to have been recruited as a mechanism for discrimination mostly restricted to the left half of the codon-anticodon table (Fig. 2.6), consistent with the editing hypothesis that aaRS proofreading maintained 4-codon sectors in the genetic code by suppressing further sectoring. For bacteria, only LysRS-IIB from the right half of the table is capable of proofreading. AaRS enzyme classes are structurally related enzymes for archaea, bacteria, and eukaryotes, except for LysRS, which is typically structural class IE in archaea and class IIB in bacteria and eukarya (Giege & Eriani, n.d.; Perona & Gruic-Sovulj, 2014). GlyRS is class IIA in archaea and eukaryotes but class IID (or historically classified IIC) in many bacteria (Turk et al., 2010).

Proofreading is not utilized for GlyRS, ArgRS, and archaeal/eukaryotic ProRS (Fig. 2.6). Glycine is the smallest amino acid, so the GlyRS synthetic active site is constrained to block the loading of larger amino acids (PDB 4KR2) (Qin et al., 2014). Arginine is a large amino acid that is much less flexible than lysine. As with lysine, arginine is charged (+1), and arginine has significant hydrogen bonding potential. These distinguishing features of arginine are utilized in the ArgRS synthetic active site to exclude incorrect amino acids (PDB 1F7U) (Delagoutte et al., 2000). Proline is the only encoded imino acid, so proline is readily distinguished in the ProRS active site from other encoded amino acids. ProRS proofreads in bacteria because of the addition of a bacterial-specific editing domain to ProRS-IIA that is missing in ProRS-IIA of archaea and eukaryotes. Of course, aaRS editing and accurate cognate amino acid specification also suppress inaccurate charging of tRNAs with amino acids that are generated from metabolism but are not encoded. When the code was evolving, aaRS enzymes were likely more error-prone in attaching amino acids, supporting sectoring of the code via tRNA charging errors.

Interestingly, the tRNAs added by eukaryotes (compared to bacteria) with adenine $\rightarrow$ inosine in the anticodon wobble position are mostly proofread by aaRS enzymes and, also, generally occupy 4-codon sectors on the left half of the genetic code table (Fig. 2.6) (Rafels-Ybern et al., 2017; Saint-Léger et al., 2016). In eukaryotes, all tRNAs on the left half of the chart utilize the A $\rightarrow$ I conversion except tRNA^{Met}, which lacks an anticodon with encoded wobble A, and tRNA^{Phe}, which occupies a 2-codon sector and, therefore, cannot adopt the A $\rightarrow$ I modification without substituting Phe for Leu in proteins. It appears that eukaryotes adopted a mechanism evolved in bacteria for tRNA^{Arg} (ACG $\rightarrow$ ICG) in order to modify and stabilize the left half of the eukaryotic genetic code table. Perhaps, most interestingly, when wobble inosine is utilized, the synonymous GNN anticodon is suppressed, indicating that the broader mRNA synonymous codon recognition of inosine compared to G is positively selected. The adenine $\rightarrow$ inosine modification has only invaded sectors with 4 codons because inosine pairs A, C, and U in mRNA codons. In eukaryotes, the only 4-codon sector that is not altered with the adenine $\rightarrow$ inosine modification is the Gly sector (Saint-Léger et al., 2016).

In mitochondria, it appears that tRNAs encoding Leu, Val, Pro, Thr, Ala, and Arg, which all convert wobble A $\rightarrow$ I in eukaryotes, must be imported from the cytosol, indicating a strong preference for utilizing tRNAs encoded in the cell nucleus with inosine in the wobble anticodon position. Because the mitochondrion was derived from an α-proteo-bacterial endosymbiont, mitochondria would encode tRNAs with wobble G to specify these amino acids. It appears that the mitochondria prefer to import nuclear-encoded eukaryotic tRNAs with inosine in the wobble position rather than to utilize mitochondrial tRNAs with G in the wobble position, indicating once again the importance of increasing ambiguity in tRNA reading synonymous mRNA codons. The import of tRNAs is a fascinating process supporting the mitochondria-eukaryote symbiotic relationship. Without a eukaryotic host to supply missing tRNAs, mitochondria would not be able to translate mitochondria-encoded mRNA.

Discussion

Alternate class I and class II aaRS folding

Class I and class II aaRS enzymes are related by amino acid sequence homology identified in archaeal species (Figs. 2.1 and 2.2). A shared Zn-binding domain in GlyRS-IIA and ValRS-IA is identified (Fig. 2.4). Class, I aaRS enzymes have an N-terminal extension that can include a second Zn-binding domain, which may have been a determinant in distinct class I aaRS folding. Additionally, class I aaRS active site β-sheets and the active site "HIGH" motif are found within the class I-specific N-terminus, so the class I aaRS active site cannot be assembled without the N-terminal domain. Because class I and class II enzymes have largely incompatible protein folds and bind to opposite faces of tRNA, we posit that an ancestral aaRS enzyme folded in distinct class I and class II conformations for three reasons. First, the N-terminal extension in class I aaRS enzymes that includes the HIGH active site motif and active site β-sheets and that can include a Zn-binding domain helped to enforce the class I fold. Second, a set of three antiparallel β-sheets in class II aaRS enzymes would clash with the N-terminal Zn motif found in class I aaRS and block assembly of the class I aaRS active site (Fig. 2.5). Third, opposite faces of cloverleaf tRNA bind class I and class II aaRS enzymes, and tRNA binding may have helped direct the alternate aaRS folds. As domains evolved to take on the appropriate fold, Zn-binding disappears from some domains that initially evolved around Zn binding (Fig. 2.4). The more complex model that class I and class II aaRS enzymes arose from transcription and translation of an ancestral bidirectional gene (Carter, 2017; Pham et al., 2007; Rodin et al., 2009; Rodin & Ohno, 1995) we find less likely. Early in evolution, Zn-binding appears to have directed the stability and folding conformations of large proteins such as aaRS enzymes and RNA polymerase. Over time, some Zn domains hardened in conformation so that Zn binding was no longer necessary (Fig. 2.4). Based on the determinants for class, I and class II aaRS folding identified here, domain swap experiments can likely switch the folding of the two aaRS structural forms.

The maximal size of the genetic code

The standard genetic code is generally considered to potentially encode 64 amino acids. Because adenine is not utilized in the anticodon wobble position in archaea, however, this reduces the number of utilized anticodons to 48 at the base of code evolution (Pak et al., 2018). When wobble A is encoded in tRNA, A $\rightarrow$ I modification occurs, and wobble G is suppressed in the synonymous anticodon. The most heavily divided 4-codon sectors of the standard genetic code that encode amino acids, and do not stop codons, or Met, are divided into two 2-codon sectors. The reason that 2-codon sectors resisted further subdivision into 1–codon sectors encoding two different amino acids is that tRNA anticodons with wobble U and wobble C are read ambiguously to recognize mRNA codons with both wobble A and G. Anticodon wobble C is thought to mostly recognize codon wobble G but may have recognized mRNA codon A well enough to have supported ambiguous reading of mRNAs during the early evolution of the code. Interestingly, anticodon wobble C was not excluded from tRNA as strictly as anticodon wobble A, and this observation requires further explanation. Because of ambiguous coding, and reading tRNAs, the largest number of amino acids that could be encoded using a triplet tRNA code is 32 (or 31 aas with stops). Because the division of 4-codon sectors was limited by aaRS proofreading, evolutionary refinement of aaRS active sites, aaRS anticodon recognition, and the A $\rightarrow$ I modification, the standard genetic code has only 20 aas with stops, and, with minor partial exceptions, the code has remained universal in the three domains of life.

Coevolution of aaRS accuracy and genetic code universality

There is a "chicken and egg" problem to consider in terms of aaRS evolution. Notably, there is no known mechanism to generate aaRS proteins until the code has evolved, and, to our knowledge, there is no clear model for making functional proteins with subsets of amino acids. At this time, we offer no simple solution to this problem. Ribozymes as small as 5 nt created in vitro can aminoacylate tRNAs (Lee et al., 2000; Turk et al., 2010;

Xiao et al., 2008), but these ribozyme functions appear now to be fully replaced by aaRS enzymes, so a natural record of aminoacylating ribozymes may not now exist.

Because tRNA and aaRS enzymes must be coevolved (Giege & Eriani, n.d.; Perona & Gruic-Sovulj, 2014), however, aaRS enzymes and proofreading by aaRS enzymes are considered with regard to the evolution of the code. We note that aaRS proofreading, in archaea, is limited to the left half of the codon-anticodon table, which encodes only hydrophobic and neutral amino acids (Fig. 2.6). ProRS, from the left half of the code, does not edit in archaea and eukarya, but ProRS edits in bacteria (Fig. 2.6). Another partial exception to the left half rule is tRNA^{Ser} (GCU). Ser is the only amino acid that is split into both the left and right halves of the table and SerRS-IIA edits. We posit that a 4-codon sector encoding Ser (anticodons ACU, GCU, UCU, CCU) may have been invaded by Arg probably before SerRS proofreading evolved to adequately resist sectoring. Also, because Ser is encoded within separated genetic code sectors, SerRS did not recognize the tRNA^{Ser} anticodon for discrimination in accurate Ser attachment, which may have increased tRNA^{Ser} charging errors, leading to Ser sensitivity to invasion by Arg (Giege & Eriani, n.d.; Perona & Gruic-Sovulj, 2014). In bacteria, LysRS (class IIB (editing) in most bacteria; class IE (nonediting) in most archaea) is also a partial exception (Fig. 2.6) (Perona & Gruic-Sovulj, 2014). we observe that the third column and also the uppermost 4-codon sector of the fourth column of the codon-anticodon table are the most heavily innovated, indicating that the evolution of aaRS proofreading inhibited code sectoring, limiting the expansion of the code. Based on this observation, we posit that errors in amino acid attachment to tRNA were important to continue sectoring the code by utilizing the wobble anticodon position. As errors become more difficult to make or to sustain, i.e., because of aaRS synthetic site-specificity (mostly the right half of the code) or because of aaRS editing (left half of the code), the code evolved toward closure and universality. Also, the tRNA cloverleaf structure and rugged RNA evolution may limit the potential size of the code. The advent of specific tRNA modifications (i.e., in bacteria and eukaryotes) can be assessed by expanding permitted anticodon contacts to mRNA (Agris & Narendran, 2017; Agris et al., 2017; Väre et al., 2017). Rugged evolution occurs when many or most substitutions are disruptive to structure, as expected for tRNA (Curtis & Bartel, 2013; Kun & Szathmáry, 2015; Novozhilov et al., 2007). Expanding the code beyond 20 amino acids, therefore, may strain the capacity of tRNAs and aaRS enzymes to coevolve for adequate accuracy and discrimination. The evo-lution of tRNA covalent modifications supported innovation and refine-ment of the code (i.e., discrimination of tRNA^{Ile} (UAU) and tRNA^{Met} (CAU) in eukaryotes) (Agris & Narendran, 2017; Agris et al., 2017; Ven-deix et al., 2008).

Positive selection of tRNA anticodon wobble ambiguity

Because tRNA wobble bases make ambiguous contacts with mRNA, a single tRNA can recognize multiple synonymous mRNA codons, but tRNA wobble ambiguity also limited the capacity for code expansions to encode new amino acids. Although there may be selection for tRNAs with specific purposes, particularly in complex eukaryotes (Percudani, 2001), generally, the selection was for increased ambiguity in reading tRNA an-ticodons. In the evolution of the genetic code, tRNA anticodon wobble inosine is strongly preferred to guanine, which is strongly preferred to adenine. Anticodon wobble inosine recognizes A, C, and U in mRNA. Anticodon wobble G recognizes C and U. Anticodon wobble A recognizes U, but tRNA wobble A recognizes mRNA wobble C poorly. We posit that the I >> G >> A preference reflects a positive selection of increasing ambiguity in the tRNA anticodon wobble position without affecting the reading of synonymous mRNA codons. Because inosine recognizes A, C, and U in mRNA codons, the A → I substitution strongly selects for, and can only occur in, 4-codon sectors. Similarly, anticodon wobble U can pair with both mRNA codon A and G. Anticodon wobble C pairs much more strongly with G than with A. We posit that U is generally preferred to C in the anticodon wobble position because tRNA wobble U recognizes syn-onymous mRNA wobble A and G more rapidly and readily than tRNA wobble C recognizes mRNA wobble A. It is also possible that G = C wobble pairs are (or were) too stable to be optimal for translation (i.e., gave slow tRNA release on the ribosome). The selection pressures at the inception of the code were different than subsequent selection pressures.

Resistance to forming 1-codon sectors

The reason that 4-codon sectors of the genetic code split into two 2-codon sectors around purine and pyrimidine wobble bases is that tRNA wobble bases are read ambiguously. The only 1-codon sectors are for tRNA^{Met} (CAU) and tRNA^{Trp} (CCA). The Ile-Met 4-codon sector is a special case (see below). In the Cys-Ter-Trp 4-codon sector, tRNA^{Trp} (CCA) shares a

2-codon sector with a stop codon UGA (anticodon UCA), which is recognized in mRNA by a protein, not a tRNA. In mitochondria, however, anticodon UCA (corresponding to stop codon UGA) is utilized to encode Trp. Because of tRNA wobble ambiguity, the maximum coding potential of the standard genetic code is for 32 letters: 31 aas + stops.

The Ile–Met sector

Questions remain with regard to the early evolution of the Ile–Met 4-codon sector of the standard genetic code. In archaea, typically only a single $tRNA^{Ile}$ (GAU), two elongator $tRNA^{Met}$ (CAU), and one $tRNA^{iMet}$ (CAU) are found. The sectoring and early proliferation of $tRNA^{Met}$ (CAU) are unusual and near the base of code evolution and require explanation. From the analysis of archaeal tRNA radiations from the primordial cloverleaf $tRNA^{Pri}$, it appears that $tRNA^{Met}$ and $tRNA^{iMet}$ may be derived from $tRNA^{Ile}$, as might be expected from code structure (Pak et al., 2018). Furthermore, one $tRNA^{Met}$ and $tRNA^{iMet}$ appear to radiate further and further from $tRNA^{Ile}$ in more derived archaeal species. Perhaps the 4-codon Ile–Met sector can be viewed as a partially occupied 4-codon Ile sector, partly invaded by Met. Invasion of the Ile 4-codon sector by Met probably involved the recruitment of Met from outside the code via inaccurate $tRNA^{Ile}$ (i.e., CAU) charging. Met invasion of Ile and $tRNA^{Met}$ proliferation were partly driven to establish the start signal for translation. Because Met (CAU) evolved at LUCA to discriminate three $tRNA^{Met}$ (CAU; 2 elongator and 1 initiator), at eukaryogenesis, discrimination of potentially synonymous Met (CAU) and Ile (UAU) could be supported by previously evolved tRNA modifications (Agris & Narendran, 2017; Agris et al., 2017; Vendeix et al., 2008).

Evolution of the standard genetic code

Three main hypotheses for the evolution of the standard genetic code include (1) variations on the Gamow hypothesis (the stereochemical hypothesis: that amino acids interact directly with RNAs, i.e., codons or anticodons, leading to the matching of codons and anticodons with amino acids and evolution of the code); (2) the coevolution theory (that code complexity coevolved with advances in amino acid metabolism); and (3) the error minimization theory (that the code evolved to minimize tRNA charging and translation errors) (Koonin & Novozhilov, 2017). Recently, it was pointed out that these long-standing hypotheses may have limitations for furthering our understanding of code evolution (Koonin & Novozhilov, 2017). Here, we give a simple hypothesis partly relating to, and slightly at odds with, the error minimization theory. We posit that the standard genetic code evolved through mechanisms of inaccurate tRNA charging, tRNA anticodon mutation, and tRNA diversification. Mechanisms that enforced tRNA charging accuracy, therefore, brought the code to universality. We posit that similar amino acids are encoded in neighboring sectors and often in the same column of the codon-anticodon table because sectoring was driven by two mechanisms. First, errors in aaRS-catalyzed amino acid attachments to tRNAs induced the division of sectors, generally involving the recruitment of similar amino acids, from outside the code, that attached to initially similar tRNAs. Secondly, tRNA anticodon mutations could result in local migrations to a neighboring sector, moving similar amino acids to nearby positions within the code. Selection for incorporation of a new amino acid into proteins drove tRNAs to diverge and discriminate amino acid attachments, leading to a more complex code with an increased number of sectors encoding different amino acids. We posit that the code was built by sectoring in a series of stages described in a recent paper (Pak et al., 2018).

Koonin and Novozhilov ask why the code is a triplet code (Koonin & Novozhilov, 2017). The code is triplet because of the structure of the tRNA anticodon loop, which forces a triplet register for two adjacent tRNAs bound to adjacent mRNA codons (Pak et al., 2017). In strict terms of coding, however, the code is almost a 2-nucleotide code, because of degeneracy in the anticodon wobble position, explaining why there are 20 amino acids + stops in the standard genetic code rather than a larger number (up to 31 aas + stops).

Koonin and Novozhilov suggest that translation systems should be analyzed to understand code evolution (Koonin & Novozhilov, 2017). We identify two features of translation systems that are relevant. First of all, in the decoding center of the ribosome, proofreading of anticodon base-pair attachments to mRNA codons, involving small ribosomal subunit conformational closure enabling EF-Tu and GTP hydrolysis, applies to the second and third anticodon positions only, not the first (wobble) position (Demeshkina et al., 2012). For most amino acids, the tRNA wobble position was selected to broaden recognition of mRNA codons, supporting code degeneracy and making tRNAs more readily available for insertion of the encoded amino acid. Secondly, translation systems evolved around tRNA, so a focus on tRNA evolution helps to interpret genetic code evolution. The tRNA-centric view significantly simplifies the problem of standard genetic code evolution, i.e., by shrinking the relevant number of anticodons. Because the genetic code is degenerate, analyzing code evolution from the point of view of mRNA is deceptive, because all 64 codons are utilized in mRNA, but only 44 unique tRNA anticodons and 3 stop codons were utilized at the inception of the standard genetic code (LUCA and ancient archaea). Furthermore, because of tRNA wobble ambiguity, the maximal capacity of the genetic code only expands to 31 amino acids + stops, but aaRS proofreading, accurate aaRS synthetic site specification of amino acid substrates, aaRS anticodon recognition, ribosome conformational proofreading of the anticodon–codon interaction and perhaps the A $\rightarrow$ I modification limited code expansions to 20 amino acids by preserving 4-codon sectors. Evolving 1-codon sectors of the genetic code were strongly resisted particularly via aaRS and ribosome fidelity mechanisms.

Ribosome proofreading of the anticodon-codon interaction

In a recent paper, we posit that the genetic code sectored from a $1 \rightarrow 4 \rightarrow 8 \rightarrow 16 \rightarrow 21$ letter code (20 aas + stops) (Pak et al., 2018). The initial code evolved to utilize any mRNA sequence to synthesize polyglycine, used to stabilize protocells. According to this view, conformational tightening and EF-Tu and GTP proofreading of Watson-Crick base pairing between the anticodon and the codon in the second and third anticodon positions (Demeshkina et al., 2012) became necessary at the $8 \rightarrow 16$ letter stage. The 8-letter stage is characterized by resolution of purines and pyrimidines only, but not individual bases, in the first mRNA codon position and the corresponding third tRNA anticodon position. At the 8-letter stage of code evolution, reading the third anticodon position is similar to the sectoring of the wobble position of the standard code, indicating that ribosome proofreading was not yet evolved at this stage. In order to fully resolve A, G, C, and U in the first codon position and the corresponding third anticodon position, conformational tightening and EF-Tu and GTP hydrolysis proofreading were essential. The model for sectoring of the genetic code, therefore, makes a prediction about the evolution of translational fidelity mechanisms that brought the code to universality.

Correlation of aaRS proofreading and A $\rightarrow$ I modification

In eukaryotes, there is a strong correlation between aaRS editing and tRNA wobble A $\rightarrow$ I modification (Fig. 2.6; bottom panel). Primarily, we attribute this correlation to 4-codon sectors. Proofreading by aaRS enzymes maintains 4-codon sectors by inhibiting tRNA charging errors that could lead to further sectoring. A → I modification is most utilized by eukaryotes, which are about 2.2 billion years old. The standard genetic code, by vast contrast, is probably >3.8 billion years old. Because the code is ancient and universal, eukaryotic innovations do not bear on the birth of the code, although eukaryotic innovations may have stabilized the eukaryotic code to prevent further sectoring and a possible escape by eukaryotes from code universality. A → I conversion is limited to 4-codon sectors, because tRNA wobble inosine recognizes mRNA wobble A, C, and U. A → I modification in a 2-codon sector, therefore, spills into a neighboring 2-codon sector, causing translation errors. Much earlier in code evolution, tRNA charging errors induced sectoring, adding amino acids to the code. Now such errors are lethal because they induce translation errors. In bacteria, the Arg (ACG, GCG, UCG, CCG) 4-codon sector was protected by the A → I modification, but, in archaea, the Arg 4-codon sector was faithfully preserved without the A → I modification, perhaps because of the high specificity of the ArgRS synthetic active site, ArgRS anticodon recognition and EF-TU proofreading on the ribosome.

Because Gly occupies a 4-codon sector of the code, this raises the question of why tRNA^{Gly} (ACC) is not modified A → I in eukaryotes (Saint-Léger et al., 2016). GlyRS resists charging errors because of the small size of the synthetic active site, which made the Gly (ACC, GCC, UCC, CCC) sector resistant to the subdivision. A similar argument can be made for the Arg (ACG, GCG, UCG, CCG) sector. ArgRS does not have a proofreading active site. The ArgRS synthetic active site accurately specifies Arg, however, because of the distinctive Arg side chain. The specificity of charging is enhanced because ArgRS recognizes the tRNA^{Arg} anticodon. The Arg 4-codon sector resists further division in bacteria and eukarya because ArgRS charging is accurate and because the A → I modification limits sectoring. It is possible that the standard genetic code is universal (i.e., in archaea, bacteria, and eukaryotes), in part, because aaRS proofreading, high aaRS synthetic site-specificity, anticodon recognition by aaRS, EF-TU proofreading, and A → I modification prevented the introduction of new 2-codon sectors in the bacterial and eukaryotic genetic codes. Because tRNA charging errors resulted in code sectoring, evolving mechanisms that enhanced the accuracy of amino acid attachments to tRNAs led to the closure and universality of the genetic code.

The tRNA-centric view

We advocate a tRNA-centric view of genetic code and ribosome evolution (Pak et al., 2017, 2018). The complexity of the genetic code was limited by tRNA anticodon loop structure and tRNA wobble degeneracy reading mRNA. The primitive ribosome might have been a decoding scaffold and a mobile peptidyl transferase center. According to our view, cloverleaf tRNA was the essential biological intellectual property leading to the evolution of the code and to the encoding of proteins including aaRS enzymes. According to this view, cloverleaf tRNA^{Pri} was a prerequisite to the coevolution of tRNAomes, aaRS enzymes, ribosomes, and the genetic code. It appears to us that a small collection of ribozymes, most of which have been generated in vitro, is sufficient to convert a strange polymer and minihelix world into a cloverleaf tRNA world that leads inevitably to an RNA-protein world and cellular life. As described previously, tRNA^{Pri} evolved initially as an improved mechanism to synthesize polyglycine to stabilize protocells (as in bacterial cell walls) before the coevolution of tRNAomes, aaRS enzymes, ribosomes, and the genetic code. Alternate views have been expressed by others (Carter, 2017; Carter & Wills, 2018; Pettersen et al., 2004; Rodin et al., 2009).

Methods

NCBI Blast

NCBI Blast tools (https://blast.ncbi.nlm.nih.gov/Blast.cgi#alnHdr_ 317113484) were used to analyze the relatedness of *Pfu* aaRS enzymes (Fig. 2.1) and to obtain alignments (Fig. 2.2).

Anticodon wobble preference

Sequences for tRNAs were collected from the tRNA database (http://trna. bioinf.uni-leipzig.de/) and the genomic tRNA database (http://gtrnadb. ucsc.edu/) (Chan & Lowe, 2016; Juhling & Hartmann, 2009). Anticodon wobble position preference was analyzed for synonymous anticodons with A and G (ANN vs. GNN) or U and C (UNN vs. CNN).

Homology modeling

Pfu GlyRS-IIA was modeled to human GlyRS-IIA (PDB 4KQE) (Deng et al., 2016) using the program Phyre2 (Kelley et al., 2015; Kim et al., 2018). Atomic coordinates were refined using the YASARA energy minimization server (http://www.yasara.org/minimizationserver.htm). UCSF Chimera was used to visualizing molecules (Pettersen et al., 2004; Yang et al., 2012). Zn was oriented to ligands as previously described (Kim et al., 2018). Because of low sequence similarity in shared Zn fingers, *Pfu* GlyRS-IIA and *Tth* ValRS-IA Zn fingers were aligned manually using Chimera.

Statistical methods

Anticodon wobble preference data sets were analyzed using chi-square goodness of fit test (http://www.stat.yale.edu/Courses/1997-98/101/chigf.htm). Because of the large datasets used and the differences observed, all comparisons were judged to be significant (P-value < .0001).

Acknowledgments

We thank Bruce Kowiatek (Blue Ridge Community and Technical College, WV) and Robert Root-Bernstein, Michigan State University, MI) for encouragement and helpful suggestions. Kristopher Opron (University of Michigan, Bioinformatics Core) helped with sequence alignments.

Disclosure of potential conflicts of interest:

The authors have no potential conflicts of interest in the publication of this work.

References

Agris, E., & Narendran. (2017). Celebrating wobble decoding: Half a century and still much is new. n.d. *RNA Biology*, 1—17

Agris, P. F., Narendran, A., Sarachan, K., Väre, V. Y. P., & Eruysal, E. (2017). The importance of being modified: The role of RNA modifications in translational fidelity. *Enzymes, 41*, 1—50. https://doi.org/10.1016/bs.enz.2017.03.005

Agris, P. F., Vendeix, F. A. P., & Graham, W. D. (2007). tRNA's wobble decoding of the genome: 40 years of modification. *Journal of Molecular Biology, 366*(1), 1—13. https://doi.org/10.1016/j.jmb.2006.11.046

Bertram, G., Innes, S., Minella, O., Richardson, J. P., & Stansfield, I. (2001). Endless possibilities: Translation termination and stop codon recognition. *Microbiology, 147*(2), 255—269. https://doi.org/10.1099/00221287-147-2-255

Carter, C. W. (2017). Coding of class I and II aminoacyl-tRNA synthetases. *Advances in Experimental Medicine and Biology, 966*, 103—148. https://doi.org/10.1007/5584_2017_93

Carter, C. W., & Wills, P. R. (2018). Interdependence, reflexivity, fidelity, impedance matching, and the evolution of genetic coding. *Molecular Biology and Evolution, 35*(2), 269—286. https://doi.org/10.1093/molbev/msx265

Chan, P. P., & Lowe, T. M. (2016). GtRNAdb 2.0: An expanded database of transfer RNA genes identified in complete and draft genomes. *Nucleic Acids Research, 44*(1), D184—D189. https://doi.org/10.1093/nar/gkv1309

Curtis, E. A., & Bartel, D. P. (2013). Synthetic shuffling and in vitro selection reveal the rugged adaptive fitness landscape of a kinase ribozyme. *RNA, 19*(8), 1116—1128. https://doi.org/10.1261/rna.037572.112

Delagoutte, B., Moras, D., & Cavarelli, J. (2000). tRNA aminoacylation by arginyl-tRNA synthetase: Induced conformations during substrates binding. *EMBO Journal, 19*(21), 5599—5610. https://doi.org/10.1093/emboj/19.21.5599

Demeshkina, N., Jenner, L., Westhof, E., Yusupov, M., & Yusupova, G. (2012). A new understanding of the decoding principle on the ribosome. *Nature, 484*(7393), 256—259. https://doi.org/10.1038/nature10913

Deng, X., Qin, X., Chen, L., Jia, Q., Zhang, Y., Zhang, Z., Lei, D., Ren, G., Zhou, Z., Wang, Z., Li, Q., & Xie, W. (2016). Large conformational changes of insertion 3 in human glycyl-tRNA synthetase (hGlyRS) during catalysis. *Journal of Biological Chemistry, 291*(11), 5740—5752. https://doi.org/10.1074/jbc.M115.679126

Fukai, S., Nureki, O., Sekine, S.i., Shimada, A., Tao, J., Vassylyev, D. G., & Yokoyama, S. (2000). Structural basis for double-sieve discrimination of L-valine from L-isoleucine and L-threonine by the complex of tRNA(Val) and valyl-tRNA synthetase. *Cell, 103*(5), 793—803. https://doi.org/10.1016/S0092-8674(00)00182-3

Giege, & Eriani, G. (n.d.). Transfer RNA recognition and aminoacylation by synthetases. John Wiley & Sons.

Juhling, M., & Hartmann, R. (2009). Compilation of tRNA sequences and tRNA genes. *Nucleic Acids Research, 37*. https://doi.org/10.1093/nar/gkn772. PMID:18957446 [Crossref].

Kelley, L. A., Mezulis, S., Yates, C. M., Wass, M. N., & Sternberg, M. J. E. (2015). The Phyre2 web portal for protein modeling, prediction and analysis. *Nature Protocols, 10*(6), 845—858. https://doi.org/10.1038/nprot.2015.053

Kim, Y., Benning, N., Pham, K., Baghdadi, N., Caruso, G., Colligan, M., Grayson, A., Hurley, A., Ignatoski, N., Mcclure, S., Mckaig, K., Neag, E., Showers, C., Tangalos, A., Vanells, J., Padmanabhan, K., & Burton, Z. F. (2018). Homology threading to generate RNA polymerase structures. *Protein Expression and Purification, 147*, 13—16. https://doi.org/10.1016/j.pep.2018.02.002

Koonin, E. V., & Novozhilov, A. S. (2017). Origin and evolution of the universal genetic code. *Annual Review of Genetics, 51*, 45—62. https://doi.org/10.1146/annurev-genet-120116-024713

Kun, Á., & Szathmáry, E. (2015). Fitness landscapes of functional RNAs. *Life, 5*(3), 1497—1517. https://doi.org/10.3390/life5031497

Lee, N., Bessho, Y., Wei, K., Szostak, J. W., & Suga, H. (2000). Ribozyme-catalyzed tRNA aminoacylation. *Nature Structural Biology, 7*(1), 28—33. https://doi.org/10.1038/71225

Massey, S. E. (2008). A neutral origin for error minimization in the genetic code. *Journal of Molecular Evolution, 67*(5), 510—516. https://doi.org/10.1007/s00239-008-9167-4

Novozhilov, A. S., Wolf, Y. I., & Koonin, E. V. (2007). Evolution of the genetic code: Partial optimization of a random code for robustness to translation error in a rugged fitness landscape. *Biology Direct, 2*. https://doi.org/10.1186/1745-6150-2-24

O'Donoghue, P., & Luthey-Schulten, Z. (2003). On the evolution of structure in aminoacyl-tRNA synthetases. *Microbiology and Molecular Biology Reviews, 67*(4), 550—573. https://doi.org/10.1128/mmbr.67.4.550-573.2003

Pak, D., Du, N., Kim, Y., Sun, Y., & Burton, Z. F. (2018). Rooted tRNAomes and evolution of the genetic code. *Transcription, 9*(3), 137—151. https://doi.org/10.1080/21541264.2018.1429837

Pak, D., Root-Bernstein, R., & Burton, Z. F. (2017). tRNA structure and evolution and standardization to the three nucleotide genetic code. *Transcription, 8*(4), 205—219. https://doi.org/10.1080/21541264.2017.1318811

Percudani, R. (2001). Restricted wobble rules for eukaryotic genomes [1]. *Trends in Genetics, 17*(3), 133—135. https://doi.org/10.1016/S0168-9525(00)02208-3

Perona, J. J., & Gruic-Sovulj, I. (2014). Synthetic and editing mechanisms of aminoacyl-tRNA synthetases. *Topics in Current Chemistry, 344,* 1—41. https://doi.org/10.1007/128_2013_456

Pettersen, E. F., Goddard, T. D., Huang, C. C., Couch, G. S., Greenblatt, D. M., Meng, E. C., & Ferrin, T. E. (2004). UCSF Chimera—a visualization system for exploratory research and analysis. *Journal of Computational Chemistry, 25*(13), 1605—1612. https://doi.org/10.1002/jcc.20084

Pham, Y., Li, L., Kim, A., Erdogan, O., Weinreb, V., Butterfoss, G. L., Kuhlman, B., & Carter, C. W. (2007). A minimal TrpRS catalytic domain supports sense/antisense ancestry of class I and II aminoacyl-tRNA synthetases. *Molecular Cell, 25*(6), 851—862. https://doi.org/10.1016/j.molcel.2007.02.010

Qin, X., Deng, X., Chen, L., & Xie, W. (2016). Crystal structure of the wild-type human GlyRS bound with tRNAGly in a productive conformation. *Journal of Molecular Biology, 428*(18), 3603—3614. https://doi.org/10.1016/j.jmb.2016.05.018

Qin, X., Hao, Z., Tian, Q., Zhang, Z., Zhou, C., & Xie, W. (2014). Cocrystal structures of glycyl-tRNA synthetase in complex with tRNA suggest multiple conformational states in glycylation. *Journal of Biological Chemistry, 289*(29), 20359—20369. https://doi.org/10.1074/jbc.M114.557249

Rafels-Ybern, Torres, A., & Grau-Bove, X. (2017). Codon adaptation to tRNAs with Inosine modification at position 34 is widespread among Eukaryotes and present in two bacterial phyla. n.d. *RNA Biology,* 1—8

Rodin, S. N., & Ohno, S. (1995). Two types of aminoacyl-trna synthetases could be originally encoded by complementary strands of the same nucleic ACID. *Origins of Life and Evolution of the Biosphere, 25*(6), 565—589. https://doi.org/10.1007/BF01582025

Rodin, S. N., & Rodin, A. S. (2006). Origin of the genetic code: First aminoacyl-tRNA synthetases could replace isofunctional ribozymes when only the second base of codons was established. *DNA and Cell Biology, 25*(6), 365—375. https://doi.org/10.1089/dna.2006.25.365

Rodin, A. S., Rodin, S. N., & Carter, C. W. (2009). On primordial sense-antisense coding. *Journal of Molecular Evolution, 69*(5), 555—567. https://doi.org/10.1007/s00239-009-9288-4

Root-Bernstein, R., Kim, Y., Sanjay, A., & Burton, Z. F. (2016). tRNA evolution from the proto-tRNA minihelix world. *Transcription, 7*(5), 153—163. https://doi.org/10.1080/21541264.2016.1235527

Saint-Léger, A., Bello, C., Dans, P. D., Torres, A. G., Novoa, E. M., Camacho, N., Orozco, M., Kondrashov, F. A., & De Pouplana, L. R. (2016). Saturation of recognition elements blocks evolution of new tRNA identities. *Science Advances, 2*(4). https://doi.org/10.1126/sciadv.1501860

Salinas-Giegé, T., Giegé, R., & Giegé, P. (2015). TRNA biology in mitochondria. *International Journal of Molecular Sciences, 16*(3), 4518—4559. https://doi.org/10.3390/ijms16034518

Salinas, T., Duby, F., Larosa, V., Coosemans, N., Bonnefoy, N., Motte, P., Maréchal-Drouard, L., Remacle, C., & Zhang, J. (2012). Co-evolution of mitochondrial tRNA import and codon usage determines translational efficiency in the green alga chlamydomonas. *PLoS Genetics, 8*(9), e1002946. https://doi.org/10.1371/journal.pgen.1002946

Schneider, A. (2011). Mitochondrial tRNA import and its consequences for mitochondrial translation. *Annual Review of Biochemistry, 80*, 1033–1053. https://doi.org/10.1146/annurev-biochem-060109-092838

Smith, T. F., & Hartman, H. (2015). The evolution of Class II Aminoacyl-tRNA synthetases and the first code. *FEBS Letters, 589*(23), 3499–3507. https://doi.org/10.1016/j.febslet.2015.10.006

Turk, R. M., Chumachenko, N. V., & Yarus, M. (2010). Multiple translational products from a five-nucleotide ribozyme. *Proceedings of the National Academy of Sciences of the United States of America, 107*(10), 4585–4589. https://doi.org/10.1073/pnas.0912895107

Valencia-Sánchez, M. I., Rodríguez-Hernández, A., Ferreira, R., Santamaría-Suárez, H. A., Arciniega, M., Dock-Bregeon, A. C., Moras, D., Beinsteiner, B., Mertens, H., Svergun, D., Brieba, L. G., Grøtli, M., & Torres-Larios, A. (2016). Structural insights into the polyphyletic origins of glycyl tRNA synthetases. *Journal of Biological Chemistry, 291*(28), 14430–14446. https://doi.org/10.1074/jbc.M116.730382

Väre, V. Y. P., Eruysal, E. R., Narendran, A., Sarachan, K. L., & Agris, P. F. (2017). Chemical and conformational diversity of modified nucleosides affects tRNA structure and function. *Biomolecules, 7*(1). https://doi.org/10.3390/biom7010029

Vendeix, F. A. P., Dziergowska, A., Gustilo, E. M., Graham, W. D., Sproat, B., Malkiewicz, A., & Agris, P. F. (2008). Anticodon domain modifications contribute order to tRNA for ribosome-mediated codon binding. *Biochemistry, 47*(23), 6117–6129. https://doi.org/10.1021/bi702356j

Wills, P. R., & Carter, C. W. (2018). Insuperable problems of the genetic code initially emerging in an RNA world. *BioSystems, 164*, 155–166. https://doi.org/10.1016/j.biosystems.2017.09.006

Xiao, H., Murakami, H., Suga, H., & Ferré-D'Amaré, A. R. (2008). Structural basis of specific tRNA aminoacylation by a small in vitro selected ribozyme. *Nature, 454*(7202), 358–361. https://doi.org/10.1038/nature07033

Yang, Z., Lasker, K., Schneidman-Duhovny, D., Webb, B., Huang, C. C., Pettersen, E. F., Goddard, T. D., Meng, E. C., Sali, A., & Ferrin, T. E. (2012). UCSF Chimera, MODELLER, and IMP: An integrated modeling system. *Journal of Structural Biology, 179*(3), 269–278. https://doi.org/10.1016/j.jsb.2011.09.006

CHAPTER 3

Evolution of life on earth: tRNA, aminoacyl-tRNA synthetases and the genetic code

Lei Lei[1] and Zachary F. Burton[2]

[1]School of Biological Sciences, University of New England, Biddeford, ME, United States; [2]Department of Biochemistry and Molecular Biology, Michigan State University, East Lansing, MI, United States

Introduction

Fig. 3.1 shows a schematic model for evolution of life on Earth (Kim et al., 2019). Life evolved around tRNA and the tRNA anticodon. Notably, without tRNA, no complex life supported by genetic coding could evolve. Abiogenesis is the physical and chemical process by which prelife led to cellular life. We refer to a late stage of abiogenesis, characterized by rapidly evolving translation systems, as the DNA/RNA-protein world. LUCA (the

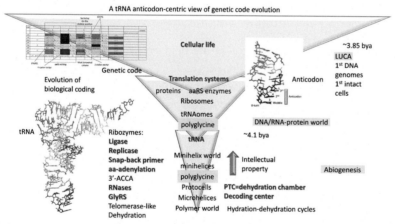

Figure 3.1 A model for evolution of life on Earth. Evolution of tRNA and translation systems leads to the first cells. A small number of ribozymes appears sufficient to generate tRNA from pre-tRNA sequences that are known because they are conserved in tRNAs. Red (dark gray in print version) type indicates ribozymes generated in vitro. Evolution of tRNA leads to evolution of translation systems and the genetic code. Triangles indicate the increases in biological potential associated with advances in coding. Abbreviations: *LUCA*, last universal common cellular ancestor; *PTC*, peptidyl transferase center.

The Makings of a Clinical Protocol
ISBN 978-0-323-95749-6
https://doi.org/10.1016/B978-0-323-95749-6.00008-9

59

last universal common cellular ancestor) indicates the first cells and the first intact DNA genomes. We posit that tRNA was the essential biological intellectual property that drove evolution. We posit that the first tRNA was a tRNA^{Gly} charged by a ribozyme GlyRS (Glycine aminoacyl-tRNA synthetase) (Kim et al., 2019; Pak et al., 2017; Pak, Du, et al., 2018). The initial purpose of tRNA was to generate polyglycine as a cross-linking agent to stabilize protocells. From tRNA arose tRNAomes (all of the tRNAs for an organism), ribosomes, the genetic code, and protein synthesis. Before tRNA, a minihelix world and a polymer world prevailed. Evidence for these more ancient worlds is conserved in tRNA sequences (Burton, 2020). A dominant purpose of polymers and minihelices was also the synthesis of polyglycine, so initially tRNA evolved as an improved mechanism for polyglycine synthesis. Once the genetic code evolved, synthesis of RNA-encoded proteins was possible, and complex life on Earth became inevitable. Triangles in the figure indicate the expansions in biological capacity with innovations in coding, and Darwinian selections became more stringent with successive enhancements in biological information processing. In this report, we return to this figure as we add additional detail and support to the model.

This report

The purpose of this report is to focus the broader origin-of-life field on its central story, which is evolution of tRNA and the genetic code. Once tRNA evolves, evolution of tRNAomes and the genetic code became inevitable. The evolution of genomes, cells, replication systems, and biological complexity then became inevitable. Recently, a review described the origin-of-life field partly in terms of conflicting approaches and philosophies (Mariscal et al., 2019). In this report, we consider a lack of focus in origin studies to be an issue. Another recent paper made a bold attempt to describe the advent of biological coding (Chatterjee & Yadav, 2019), which we also consider.

The basic approach advocated here is to start at the center of a system of coevolutionary partners and to work out. tRNA is the central feature of the evolution of life (Fig. 3.1). tRNAomes, the genetic code, mRNA, and ribosomes evolved around tRNA (Kim et al., 2019); Opron & Burton, 2019). A strategy of beginning with the core was previously applied by us to analyze the evolution of transcription systems, including RNA polymerase, general transcription factors, and promoters (Burton et al., 2016; Burton, 2014; Burton & Burton, 2014). Although transcription was then a more familiar problem to us, transcription systems are more derived evolution-arily than translation systems, making translation systems in some ways a simpler problem. By applying lessons from transcription systems, we rapidly gained insight into translation systems. We found that focus on tRNA was the key to understanding translation. Focusing on the central issue was also the strategy underlying the RNA world hypothesis, although we find the hypothesis is enriched by focus on tRNA. This report, therefore, focuses the RNA world approach on generation of tRNA and the genetic code. RNA, of course, and tRNA form the core of the genetic coding system. From RNA, DNA can be generated via reverse transcription. From RNA, protein can be generated via translation. As coding of protein enzymes evolved, in order to supply needed enzymes, ribozymes can function as catalysts. Here, we advocate a very similar approach to that underlying the RNA world hypothesis. The advantage of the inside-out approach is that we provide a means of thinking critically and holistically about ancient evolution problems, and we make potential sense out of otherwise bewildering complexity.

Investigators have advocated top-down and bottom-up approaches to analysis of ancient evolutionary events (Mariscal et al., 2019; Ma, 2017; Kunnev & Gospodinov, 2018). Here, we start with a top-down strategy of inferring prelife from molecules such as tRNA that were conserved in the prelife→cellular life transition. Studying tRNA evolution, the tRNA molecule suggests both top-down and bottom-up strategies to understand evolution of repeating polymers, inverted repeats, minihelices, tRNA, tRNAomes, ribosomes, and the genetic code (Fig. 3.1) (Kim et al., 2019; Opron & Burton, 2019). Significantly, tRNA was generated from highly patterned sequences: repeats and inverted repeats (Kim et al., 2019; Pak, Du, et al., 2018; Burton, 2020; Kim et al., 2018). Therefore, tRNA in-dicates features of a minihelix world and polymer world that preceded tRNA. Understanding tRNA sequence and structure from existing se-quences helps to model how tRNAomes and the genetic code evolved. Centering the focus on tRNA, ribosome evolution becomes easier to understand (Opron & Burton, 2019). Analysis of tRNA, therefore, suggests both top-down and bottom-up strategies to model and understand the central story in evolution of life on Earth.

A working model for evolution of the genetic code

Anticodon preference rules govern filling the genetic code

Amino acids appear to enter the genetic code according to ordered rules for tRNA anticodon preference. In the 2nd and 3rd anticodon positions, the preference appears to be C > G > U >> A (Kim et al., 2019). Therefore, 2nd and 3rd anticodon position C is favored. We posit that C is favored in the anticodon because it is a pyrimidine (a smaller base), and C forms three hydrogen bonds to G in mRNA. G and U are approximately equally favored, but amino acids with anticodon 3rd position G enter the code before amino acids with 3rd position U, indicating a preference for anti-codon G over U (see below). The explanation we offer is that G forms three hydrogen bonds, but is a purine (a larger base), and U forms two hydrogen bonds but is a pyrimidine (a smaller base). A is strongly dis-favored, particularly in the 3rd anticodon position. A is disfavored because it forms two hydrogen bonds and is a purine (a larger base). Of course, despite preferences, single base recognition (A, G, C, and U) was essential in the 2nd and 3rd anticodon positions to evolve a complex code. The preferences we identify for tRNA anticodons could potentially be preferences for the complementary bases in mRNA. Perhaps insight could be obtained from molecular dynamics or quantum mechanical simulations of codon-anticodon pairs on the ribosome to help judge the precise chemistry un-derlying these base selections.

Preferences in the anticodon wobble position appear to be G > (U/C) >>>> A. Only purine versus pyrimidine resolution is achieved for the tRNA wobble position, limiting the size of the code (Kim et al., 2019). This is why the genetic code is divided into mostly 4-codon sectors and 2-codon sectors, with few 1-codon sectors. In the wobble position, anticodon G reads mRNA codon C, as a Watson—Crick pair, and U, as a wobble pair. Anticodon U reads codon A, as a Watson—Crick pair, and G, as a wobble pair. Without modification, anticodon C reads mRNA codon G with reasonable fidelity. In evolution, anticodon wobble U and C are rarely or never split in coding two amino acids, as we discuss below. Anticodon wobble A is essentially never used in Archaea and Bacteria (Pak, Kim, & Burton, 2018; Saint-Léger et al., 2016). When A is modified to inosine, however, A may be encoded in tRNA in Bacteria and Eukarya.

Guidelines for placements of amino acids into the code

A highly detailed description of genetic code evolution is provided in this report. The working model is built on a simple set of ideas: (1) glycine was the first encoded amino acid, and at an early stage, all anticodons and all codons encoded glycine; (2) amino acids were added to the code via invasion of occupied sectors; (3) the genetic code evolved around the tRNA anticodon; (4) the final sectoring structure of the code depended strongly on the order of additions of amino acids; and (5) the code was highly structured according to its evolutionary history, and code structure was largely maintained through subsequent evolution. That glycine was the first encoded amino acid has also been hypothesized by others (Bernhardt, 2016; Bernhardt & Patrick, 2014; Bernhardt & Tate, 2008). We posit this idea mostly based on the observation that tRNA^{Pri} (a primordial tRNA) is closest in sequence to tRNA^{Gly} in ancient Archaea (Kim et al., 2019; Pak, Du, et al., 2018). Also, Gly occupies what appears to be the most favored anticodon positions in the code (tRNA^{Gly} (GCC, UCC, CCC); favored column 4 (2nd anticodon position C); favored row 4 (3rd anticodon position C).

We posit that Gly, Asp, Glu, Ala, and Val are among the first encoded amino acids. All of these amino acids are found in favored row 4 of the code (3rd anticodon position C). Furthermore, Phe, Tyr, Trp, and Cys appear to be among the last amino acids encoded (disfavored 3rd anticodon position A) (Brooks et al., 2002; Fournier & Alm, 2015). Also, stop codons are recognized in mRNA by protein translation release factors (Burroughs & Aravind, 2019), so stop codons do not follow rules for tRNA anticodons. Phe, Tyr, Trp, Cys, and stop codons are all located in row 1 of the genetic code, indicating that row 1 (3rd anticodon position A) is disfavored. We conclude that the position of amino acids in the code reflects a Darwinian tRNA anticodon preference selection and also the order of amino acid additions to the code. The genetic code is highly structured according to its evolutionary history, and structure is demonstrated best by evolution of aminoacyl-tRNA synthetases (aaRS; i.e., GlyRS) (Kim et al., 2019; Pak, Kim, & Burton, 2018). We discuss the evidence below.

Darwinian selection in a prelife world

There has been controversy about whether Darwinian selection can function before living systems evolve (Mariscal et al., 2019; Chatterjee & Yadav, 2019; Ma, 2017; Kunnev & Gospodinov, 2018). For instance, some have argued against evolution of the genetic code based on Darwinian selection. In this paper, we posit Darwinian selections for minihelices, tRNA, tRNAomes, ribosomes, and the genetic code. As indicated in Fig. 3.1, we do recognize that Darwinian selections become more stringent as biological coding evolved. To us, there is no mystery about how to evolve the most central prelife systems.

The archaeal domain is the most ancient

Although this point has been argued, we posit that, for translation functions, the archaeal domain is the most similar to LUCA and to prelife (Kim et al., 2019;Pak, Du, et al., 2018; Pak, Kim, & Burton, 2018; Battistuzzi et al., 2004). We have previously argued this point based on evolution of transcription systems (Burton et al., 2016; Burton, 2014; Burton & Burton, 2014). In compelling support of this hypothesis, archaeal tRNAs are more ancient than bacterial tRNAs (Kim et al., 2019; Pak, Du, et al., 2018; Burton, 2020). This point can be demonstrated by inspection of archaeal and bacterial tRNAomes. Archaeal tRNAs are much closer to tRNA[Pri] (a primordial tRNA). Bacterial tRNAs are more derived. Therefore, to understand evolution of the genetic code, we focus first on archaeal systems before determining how the genetic code differs in the bacterial system. Eukarya are a more complex problem, because eukaryotes arose as an archaeal and bacterial fusion.

Evolution of tRNA

Because life evolved around tRNA, solving the evolution of tRNA is a core issue in the evolution of life (Kim et al., 2019; Pak et al., 2017; Pak, Du, et al., 2018; Burton, 2020; Kim et al., 2018). Fortunately, the evolution of tRNA is a simple problem. tRNA arose from ordered sequences: repeats (GCG, CGC, and UAGCC) and inverted repeats (~CCGGGUUAAAAACCCGG). Despite some controversy (Di Giulio, 2019; Demongeot & Seligmann, 2020; Di Giulio, 2020), the evolution of tRNA is a known and solved problem (Burton, 2020). Specifically, tRNA evolved from ligation of three 31-nt minihelices followed by 9-nt internal deletion(s). A single internal 9-nt deletion generated a type II tRNA (initially 84-nt) with an expanded variable loop. Two internal 9-nt deletions generated a type I tRNA (initially 75-nt). There are other opinions about tRNA evolution (Di Giulio, 2019; Demongeot & Seligmann, 2020; Di Giulio, 2020; Di Giulio, 2012; Branciamore & Di Giulio, 2011;

Demongeot & Seligmann, 2019; Demongeot et al., 2009; Demongeot & Moreira, 2007) that are inconsistent with tRNA sequences (Burton, 2020). The three minihelix model for tRNA evolution, by contrast, fully describes genesis of tRNA from the beginning of the 5′-acceptor stem to the end of the 3′-acceptor stem (Burton, 2020). The 3′-ACCA (initial sequence), to which the amino acid is bound, may have attached by ligation. The evolution of tRNA was solved as a puzzle, and this is a puzzle that anyone can solve. Because tRNA evolved from minihelices of known sequence, information is obtained about a minihelix world and a polymer world that preceded tRNA (Fig. 3.1).

The minihelix and polymer worlds

We posit that polyglycine was generated in a prelife world, as a cross-linking agent to stabilize protocells. Polyglycine (i.e., Gly_5) is a component of bacterial peptidoglycan cell walls (Pinho et al., 2013; Zapun et al., 2008; Scheffers & Pinho, 2005). Gly_5 is a cross-linker connecting short peptide chains (i.e., L-Ala-D-Glu-L-Lys-D-Ala) that are linked to glycan chains (i.e., (N-acetylglucosamine-N-acetylmuramic acid)$_n$). Gly_5 cross-links L-Lys on one peptide chain to D-Ala on another. We posit that, in the prelife world, an analogous peptidoglycan coat was synthesized using ribozyme catalysts and assembled on protocells. One prediction of our model is that peptidoglycan coats enhance the functions of protocells for prebiotic chemistry.

Because aspects of the more ancient minihelix and polymer worlds are known from current tRNA sequences (Kim et al., 2019; Pak, Du, et al., 2018; Burton, 2020; Kim et al., 2018), what were the functions of these ancient repeats and inverted repeats? We posit that a large set of 31-nt minihelices was utilized with RNA templates of diverse sequences to synthesize polyglycine, which stabilizes cells or protocells as a component of peptidoglycan (Pinho et al., 2013; Zapun et al., 2008; Scheffers & Pinho, 2005). In tRNA sequences, only two minihelices survive: a D loop minihelix (GCGGCGGUAGC-CUAGCCUAGCCUACCGCCGC) and two copies of an anticodon/T loop minihelix (~GCGGCGGCCGGGUUAAAAACCCGGCCGCCGC). The only sequence ambiguity in the anticodon and T loop minihelices is in the ~UUAAAAA loops. There is no ambiguity in the 5′- (GCGGCGG) and 3′-acceptor stems (CCGCCGC) or in the 5-nt stems (5′ CCGGG and 3′ CCCGG). We posit that minihelices evolved to synthesize polyglycine using mixed mRNA templates, and mixed 31-nt minihelices, including some no longer known. A D loop minihelix would be expected to project a single C to recognize G in a mRNA (1-nt code). An anticodon/T loop minihelix would be expected to project ~ AAA to recognize ~ UUU in a mRNA (3-nt code). Although only two minihelix sequences survived in tRNAs, it is expected that many minihelices with different core sequences than those described here helped a primitive translation system to recognize different mRNA sequences. In order to attach glycine using a ribozyme GlyRS (GlyRS-RBZ), we posit that ACCA was ligated to 31-nt minihelices. A primitive decoding center scaffold and a mobile peptidyl transferase center (PTC) appear sufficient to serve as an ancient ribosome (Opron & Burton, 2019).

Because minihelices were generated from GCG, CGC, and UAGCC repeating polymers, a mechanism must have existed to generate these repeats. Although there may be other possibilities, we posit that RNA repeats may have been synthesized via a telomerase-like ribozyme with a guide RNA template. These and other RNAs may have been replicated via a template-dependent ribozyme replicase (Fig. 3.1). A mechanism must have existed to generate RNA fragments of defined lengths (i.e., 7-nt (acceptor stems) and 17-nt (minihelix cores)). To attach fragments of RNA, a ribozyme ligase was required. A ribozyme replicase must have existed to utilize and replicate inverted repeats. A 31-nt minihelix ligated to an RNA can be utilized as a snap-back primer for replication of the complementary strand. A ribozyme GlyRS must have existed, but no other aminoacyl transferases were initially required. The system requires RNA scaffolds to act as a primitive decoding center. The system further requires a mobile peptidyl-transferase center (Zhang & Cech, 1998). We posit that the primitive ribosome was approximately built on this model (Opron & Burton, 2019). Such an ancient minihelix world could be reconstructed in a laboratory and/or generated computationally to challenge these ideas. All of the requisite ribozymes have been generated or approximated through selection in vitro (Zhang & Cech, 1998; Pressman et al., 2019; Illangasekare & Yarus, 2012; Yarus, 2011; Turk et al., 2011; Turk et al., 2010; Chumachenko et al., 2009; Kim & Joyce, 2004; Paul & Joyce, 2002; Rogers & Joyce, 2001; Jaeger et al., 1999). Although not discussed here, similar hypotheses can be posited for the more ancient polymer world. It was initially a surprise to us that the central functions of life appear to have been generated from ordered, rather than random polymer sequences. Existing tRNA sequences show that life evolved, at least in part, from ordered sequences (repeats and inverted repeats) (Caetano-Anollés & Caetano-Anollés, 2016).

So, we posit that the Darwinian selection for the minihelix world was primarily to synthesize polyglycine to stabilize protocells (Fig. 3.1). Protocells were necessary to generate membrane potentials to harness redox energy. Initially, tRNA evolved from minihelices as an improved mechanism to generate polyglycine. Evolution of tRNA eventually established the 3-nt genetic code. From this modest beginning, the genetic code evolved, driven by Darwinian selection.

The genetic code has order

A number of models have been advanced to describe the evolution of the genetic code (Chatterjee & Yadav, 2019; Demongeot & Seligmann, 2019; Koonin, 2017; Koonin, 2017; Koonin & Novozhilov, 2009; Caetano-Anollés & Caetano-Anollés, 2016; Rogers, 2019; Marcello Barbieri, 2019a, b; Barbieri, 2019a, b; Hartman & Smith, 2019). With the exception of our model, we find these alternate opinions to be potentially flawed. Specifically, alternate models show the genetic code in terms of mRNA codons, rather than tRNA anticodons. As we have shown, however, the genetic code complexity was determined by how the tRNA anticodon was read on the ribosome. Reading of the tRNA anticodon, therefore, limited the size and complexity of the genetic code (Kim et al., 2019; Pak, Du, et al., 2018; Kim et al., 2018; Pak, Kim, & Burton, 2018). Sixty-four codons are recognized in mRNA ($4 \times 4 \times 4$), but, in tRNA, the maximal complexity of the genetic code is 32 anticodons ($2 \times 4 \times 4$), explaining why only 20 amino acids and stop codons were encoded. Most notably, only pyrimidine versus purine discrimination was initially permitted at the anticodon wobble position, limiting the size of the code in tRNA.

Because tRNA limits the size of the genetic code, we show the code as a 32-assignment, codon-anticodon table (Figs. 3.2 and 3.3). Fig. 3.2 is annotated to describe important features of the code (i.e., start and stop codons). Because of the history of code evolution, the genetic code is primarily broken into 2-codon and 4-codon sectors. Because of tRNA anticodon wobble ambiguity, separating 1-codon sectors to encode two amino acids poses a challenge. Trp (CCA) is a special case because stop codon (UGA; anticodon UCA) is recognized only as a codon in mRNA by a protein translation release factor (Burroughs & Aravind, 2019), which does not obey tRNA anticodon rules. In Archaea, Ile (CAU) and Met (CAU) anticodons cooccupy a sector and are discriminated by distinct

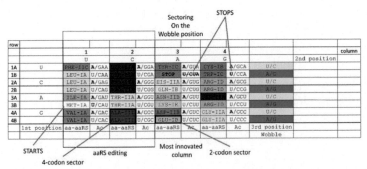

Figure 3.2 The genetic code is highly structured. A 32-assignment codon-anticodon (Ac) table is shown. Codon sequences are shown on the outside (1st position, 2nd position, and 3rd wobble position). aaRS enzymes are indicated by their structural subclass (i.e., GlyRS-IIA). aaRS enzymes that edit inaccurately attached amino acids (aa) are found in columns 1 and 2. The color shading scheme reflects how amino acids were added to the code and is described in future figures. Red (dark gray in print version) letters indicate very rarely used tRNAs and stop codons (strike-through).

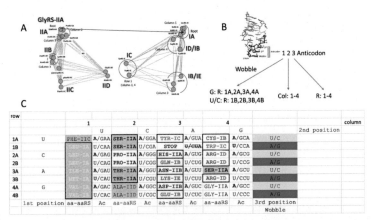

Figure 3.3 Evolution of the genetic code occurred mostly in columns. (A) Evolution of aaRS enzymes. Distances represent evolutionary differences. The red (dark gray in print version) arrow indicates that ValRS-IA is derived from its sequence homolog GlyRS-IIA. (B) The relationship of the tRNA anticodon to the genetic code columns (Col) and rows (R). (C) The genetic code in Archaea. Gray shading indicates aaRS that possess a separate active site to edit inappropriately attached amino acids. Colors highlight genetically similar aaRS enzymes demonstrating evolution primarily in columns.

anticodon wobble C modifications (Satpati et al., 2014; Voorhees et al., 2013; Mandal et al., 2010). In order to reduce translation errors, tRNA[Ile] (UUA) anticodons are rarely used in prokaryotes. Ile (CAU) and Met (CAU) anticodons provide insight into how amino acids invaded the code.

Therefore, apparent 1-codon sectors are special cases that can be ignored in the initial establishment of the code. Because the code does not divide easily into 1-codon sectors, the initial maximal code complexity must be 32-assignments as shown in Figs. 3.2 and 3.3.

History of genetic code evolution is largely preserved in aaRS evolution patterns

Aminoacyl-tRNA synthetases (aaRS; i.e., GlyRS) belong to two structural classes (class I and class II; i.e., GlyRS-IIA) with many structural subclasses (A—E) (Figs. 3.2 and 3.3). Class I and class II aaRS have incompatible folds. Sometimes, structural subclasses do not adequately reflect evolutionary relatedness of aaRS enzymes, but these relationships have recently been clarified (Fig. 3.3A) (Kim et al., 2019). Class II aaRS are older than class I aaRS. GlyRS-IIA is the primordial aaRS from which all aaRS enzymes were derived. As we have shown, GlyRS-IIA is a simple sequence homolog of ValRS-IA and IleRS-IA (Kim et al., 2019; Pak, Kim, & Burton, 2018). GlyRS-IIA was refolded into (probably) ValRS-IA from which other class I aaRS enzymes radiated (Kim et al., 2019). Much of the structure of the genetic code and the history of genetic code evolution can be inferred from the relatedness of aaRS enzymes. The distributions of related aaRS enzymes indicate that much of the evolution of the genetic code occurred within code columns (Fig. 3.3C). Code columns relate to the middle position of the tRNA anticodon, which is the most important position for translational accuracy (Fig. 3.3B). Because the genetic code has a highly conserved structure, a history of genetic code evolution is preserved, particularly in aaRS evolution patterns.

The "frozen accident"

The genetic code is highly structured, indicating that a history of its evolution is recorded and preserved (Figs. 3.2 and 3.3). Conservation of structure in the code is reminiscent of Francis Crick's "frozen accident" (Kim et al., 2019; Koonin, 2017; Koonin & Novozhilov, 2017; Doig, 2017; Kun & Radványi, 2018; Ribas de Pouplana et al., 2017), which we interpret as a rapid evolution to an enduring form with very little subsequent movement or replacement of amino acids within the code. Our approach to describe the sectoring of the code has been explained in some detail, but in this report, we improve and extend our previous model. Notably, making a small number of simplifying assumptions, the entire sectoring of the code can be described via Darwinian selection. Although we relate events from ~4 billion years ago, we consider these detailed working models to be very useful to strengthen our understanding of the evolution of the code.

Establishment of the genetic code requires consideration of a challenging "chicken and egg" problem: that is, how do you establish the functions and structural sub-classes of aaRS enzymes before the code evolves to provide complex proteins (Kim et al., 2019; Kunnev & Gospodinov, 2018)? We posit that this is part of the frozen accident. We posit that the code was sectored initially by ribozyme aaRS enzymes. Once the code was sufficiently established, protein aaRS enzymes replaced their ribozyme analogues. aaRS enzyme divergence was driven by the pressures caused by the increasing requirements for accurate coding. Such a model requires very rapid and almost irreversible establishment of the genetic code, consistent with conserved code structure (Figs. 3.2 and 3.3). We reiterate that there is very little evidence for migration of amino acids in the code. Ser and Arg may be exceptions.

Analysis of tRNAomes indicates much higher order in more ancient species such as ancient Archaea (Pak et al., 2017). By contrast, bacterial tRNAomes are much more derived. We posit, therefore, that tRNAomes are highly dynamic in evolution, and that much of the initial patterning of tRNAomes has been lost, as tRNAs mutate, in order for, for instance, aaRS enzymes to distinguish one tRNA from another. tRNAomes occupy a limited evolutionary space. Because tRNAs are small RNAs that must maintain their characteristic structure, tRNAomes undergo both divergent and convergent evolution within a constrained Darwinian space. Therefore, the frozen accident, which was initially established in the tRNAome, is now more apparent in aaRS sequences and structures than in tRNAomes. We posit that during genetic code sectoring, tRNAomes were more highly structured, and that, when proteins gained sufficient complexity, aaRS evolution tracked the highly ordered tRNAome. After sectoring of the code and aaRS enzymes was mostly complete, Darwinian selection drove diversification of tRNAomes to enhance translational accuracy. Consistent with this view, we observe that archaeal tRNAomes are more highly structured (and more similar to LUCA) than bacterial tRNAomes, which are more derived (Pak, Du, et al., 2018). We posit that the ~8 → ~16 amino acid transition (see below) provides proteins of sufficient complexity to evolve the patterned code.

The tRNAome is an enclosed evolutionary space because tRNAs must maintain their shape and structure. Analysis of tRNAomes indicates both convergent and divergent evolution of tRNAomes. For instance, for the most part, the *Pyrococcus furiosis* (an ancient Archaea) tRNAome resembles a highly ordered LUCA tRNAome. By contrast, *Escherichia coli* (a highly derived Bacterium) has a much more scrambled tRNAome with little apparent evolutionary order. As evidence of continuing tRNA evolution, we have presented evidence in Archaea for recruitment of a duplicated tRNA^{Pro} to become reassigned as a tRNA^{Trp} and a duplicated tRNA^{Trp} to become reassigned as a tRNA^{Phe}, indicating how tRNAs diverge, converge, and jump within a tRNAome (Pak, Du, et al., 2018). To avoid errors in coding, tRNAs sometimes jump to a neighboring column within a genetic code row, in contrast to aaRS enzymes that tend to evolve within columns (Fig. 3.3). We posit that initially, tRNAomes mostly evolved within columns, followed by tRNAome divergence that includes tRNA jumping to a neighboring column but often maintaining the original row.

Polyglycine world

We posit that the initial selection to evolve the 3-nt genetic code was to synthesize polyglycine to stabilize protocells (Fig. 3.1). Polyglycine (i.e., Gly$_5$) is a cross-linking agent in the peptidoglycan layer of bacterial cell walls (Pinho et al., 2013; Scheffers & Pinho, 2005; Zapun et al., 2008). We posit that, before the advent of true cells at LUCA, protocells may have been stabilized by a very similar peptidoglycan structure: that is, glycan chains with short polypeptide linkages (i.e., added by ribozymes) including polyglycine cross-links. Such a scenario makes polyglycine of selectable value in the primordial world. We posit that the prebiotic world generated peptidoglycan supports for protocells. This hypothesis suggests that, for prebiotic chemistry, protocells with a peptidoglycan cage would have advantages over naked protocells. Others have also proposed that glycine was the first encoded amino acid (Bernhardt, 2016; Bernhardt & Patrick, 2014; Bernhardt & Tate, 2008). So far as we know, we are the first to propose that the initial entire genetic code (all anticodons and all codons) encoded polyglycine. This simple assumption enables a detailed model for evolution of life on Earth.

Evolution of the genetic code (a working model)

We posit that the genetic code initially evolved primarily to synthesize polyglycine used as a cross-linking agent to stabilize protocells (Fig. 3.1). What this indicates is that, after evolution of tRNA, the tRNA anticodon mutated rapidly to all possible sequences. We imagine a primordial world in which tRNA can replicate and mutate rapidly. tRNA sequences other than the anticodon posed difficulties to mutate because alterations in stems or in the "elbow" (where the D loop, V loop, and T loop interact; the bend of the tRNA "L" shape) likely require compensatory mutations to stabilize the tRNA fold. After generating many anticodon mutations, all anticodons and all mRNA codons encoded glycine to synthesize polyglycine. We posit that a GlyRS ribozyme charged these tRNAs (mostly) with glycine. Mistakes in aminoacyl-tRNA charging and modifications of amino acids linked to tRNAs drove the evolution of the code.

Because tRNA has a specialized 7-nt anticodon loop, the 3-nt tRNA anticodon projects to recognize a 3-nt mRNA codon. On a primitive ribosome, reading 3-nt codons, however, presented some difficulties that can still be recognized in wobble anticodon-codon interactions (Fig. 3.3B). On the ribosome, the accuracy of the anticodon-codon interaction is monitored by the EF-Tu latch (Loveland et al., 2017; Rozov, Demeshkina, et al., 2016; Rozov et al., 2018; Rozov, Westhof, et al., 2016). EF-Tu is a GTPase that enforces Watson–Crick geometry on the anticodon-codon interaction at the 2nd and 3rd anticodon positions. By contrast, the 1st anticodon wobble position is monitored by the latch, but, at the wobble position, only pyrimidine-purine recognition is initially achieved. The 2nd anticodon position is recognized with the highest accuracy. The 3rd anticodon position also requires Watson–Crick geometry. Rarely, in translation errors, G ∼ U wobble pairs occur at the 3rd anticodon position, but G ∼ U wobble pairs require tautomerization of either the anticodon or codon base to allow Watson–Crick geometry. Because of the EF-Tu latch, generally, unstable wobble pairs in the 3rd anticodon position are rejected before amino acid misincorporation.

The genetic code evolved according to these tRNA anticodon recognition rules (2nd > 3rd > 1st (wobble) anticodon positions). To generate polyglycine, little translational accuracy in reading the anticodon is initially required because all anticodons encoded Gly. As amino acids invaded the code, we posit the following. The code generally sectored first around the 2nd anticodon position (most important), followed mostly by the 3rd position (next most important), followed by the wobble position (generally least important) (Kim et al., 2019). Therefore, the genetic code evolved mostly around the tRNA anticodon, generating a clear working model for evolution of the code. In the model advocated here, the entire code is first populated with Gly. As other amino acids enter the code, Gly is displaced.

As Gly is displaced, however, Gly retains the most-favored anticodons, based on a clear set of selection rules, and Gly gives up less favored anticodons to incoming amino acids (Kim et al., 2019). According to such a model, the placements of all amino acids in the genetic code can be readily described. Alternate models for populating the genetic code are more problematic. Any model that attempts to populate the code from a single sector followed by migrating into other sectors appears unlikely (Błażej et al., 2019; Paweł Błażej et al., 2018; Demongeot & Seligmann, 2019). For such an alternate model, how can a Darwinian selection be imagined for matching assignments of codons and anticodons? In our model, codon and anticodon specifications coevolve, and evolution of the tRNA anticodon drives evolution of codons. We find alternate models to make few useful predictions. By contrast, our model makes many predictions, many of which are justified.

In Fig. 3.4, we show a proposed order of addition for amino acids into the genetic code. We posit that the genetic code evolved from a 1-amino acid code (Gly) to a 4-amino acid code (Val, Ala, Asp, Gly) to an 8-amino acid code (Val, Leu, Ala, Pro, Asp, Glu, Arg, Gly). The selection to add amino acids to the code was to generate novel polypeptide products. Sectoring to a 4-amino acid code required single-base recognition of the 2nd anticodon position. Evolution to an 8-amino acid code required recognition of two anticodon positions, but in the absence of the EF-Tu latch (Loveland et al., 2017; Rozov, Demeshkina, et al., 2016; Rozov et al., 2018; Rozov, Westhof, et al., 2016), recognition of the 1st wobble base and the 3rd anticodon position were both wobbly, and only pyrimidine versus purine recognition was initially achieved. Therefore, to advance beyond an 8-amino acid code required evolution of the EF-Tu latch to confer single-base recognition at the 2nd and 3rd anticodon positions. At the 8-amino acid stage, we posit that genetic code columns 1, 2, and 4 sectored by one mechanism, and column 3 sectored by a slightly different mechanism. Columns 1, 2, and 4 sectored according to the 3rd anticodon position (2nd + 3rd anticodon positions). Column 3 sectored according to the 1st (wobble) position (2nd + 1st anticodon positions). Essentially, in order to expand the genetic code, the primitive ribosome and translation system were "learning" (teaching themselves) to read just two out of three anticodon bases. The evolutionary history of column 3, sectoring on the 1st anticodon wobble base rather than the 3rd anticodon position, caused column 3 to become the most innovated column in the code.

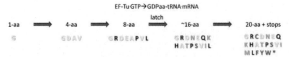

Figure 3.4 A proposed order of addition for amino acids (aa) to the genetic code. Amino acids appear to invade by genetic code rows. Yellow (light gray in print version): Row 4 amino acids; Red (black in print version): Row 2 amino acids; Green (dark gray in print version): Row 3 amino acids; Cyan (gray in print version) Row 1 amino acids and stop codons (asterisk). Leu, Ser and Arg are only scored with a single color. For simplicity, only a final, primary position in the code is scored by color.

For the code to continue to sector requires evolution of the EF-Tu GTPase anticodon-codon latch (Fig. 3.4). From the 8-amino acid code, we posit that the code sectored to approximately a 16-amino acid code (i.e., ∼ Gly, Arg, Asp, Asn, Glu, Gln, Lys, His, Ala, Thr, Pro, Ser, Val, Ile, and Leu). This transition requires evolution of the EF-Tu latch because the 2nd and 3rd anticodon positions must be read with single-base resolution to achieve an ∼ 16-amino acid code, and in the absence of the latch, the 3rd anticodon position is read essentially as a wobble position, with more difficulty than the 2nd position. At the ∼ 16 amino acid stage of evolution, additional amino acids can be added, but as the system evolves enhanced translational accuracy, fidelity restrictions create an ever-more stringent barrier to encoding additional amino acids. The reason for this limitation is that amino acids are added to the code through mischarging of tRNAs followed by selection for a more complex and innovated code. Therefore, addition of amino acids to the code is positively selected, because a richer code makes encoded proteins more complex, but negatively selected because additions increase tRNA-charging and translation errors, forcing additional challenges to the system. Potentially, a 32-assignment code could have evolved (i.e., 29 amino acids + 3 stop codons). We posit that some sectors of the code are difficult to split, however, from 4-codon sectors to 2-codon sectors because these divisions cause error catastrophe, particularly for the left half of the genetic code. Hydrophobic and neutral amino acids encoded by the left half of the code have little amino acid side chain character for recognition by aaRS enzymes, so splitting a 4-codon sector into two 2-codon sectors might cause too frequent tRNA charging errors (Kim et al., 2019; Pak, Kim, & Burton, 2018).

Evolution of the genetic code within columns

Because genetic code columns appear to dominate evolution of the code (Figs. 3.2 and 3.3), how can this be explained? We posit that evolution of the genetic code is dominated by the tRNA anticodon. The second anticodon position, which is most important for translational accuracy, is represented by genetic code columns (Fig. 3.3B). Therefore, the genetic code evolved primarily within columns, rather than within rows, or according to a random distribution. Based on this reasoning, a simple, step-by-step model for genetic code evolution is proposed in Figs. 3.5, 3.6, 3.7, 3.8, and 3.9. Fig. 3.5 shows the initial evolution of the code to encode polyglycine. Fig. 3.6 shows the 4–amino acid code to encode Gly, Asp, Ala, and Val (Chatterjee & Yadav, 2019; Kim et al., 2019; Koonin & Novozhilov, 2009, 2017). Fig. 3.7 shows wobbly sectoring to an 8–amino acid code. Fig. 3.8 shows evolution to a ∼16–amino acid code, after evolution

row		1		2		3		4			column
		U		C		A		G		2nd position	
1A	U	GLY-RBZ	A/GAA	GLY-RBZ	A/GGA	GLY-RBZ	A/GUA	GLY-RBZ	A/GCA	U/C	
1B		GLY-RBZ	U/CAA	GLY-RBZ	U/CGA	GLY-RBZ	U/CUA	GLY-RBZ	U/CCA	A/G	
2A	C	GLY-RBZ	A/GAG	GLY-RBZ	A/GGG	GLY-RBZ	A/GUG	GLY-RBZ	A/GCG	U/C	
2B		GLY-RBZ	U/CAG	GLY-RBZ	U/CGG	GLY-RBZ	U/CUG	GLY-RBZ	U/CCG	A/G	
3A	A	GLY-RBZ	A/GAU	GLY-RBZ	A/GGU	GLY-RBZ	A/GUU	GLY-RBZ	A/GCU	U/C	
3B		GLY-RBZ	U/CAU	GLY-RBZ	U/CGU	GLY-RBZ	U/CUU	GLY-RBZ	U/CCU	A/G	
4A	G	GLY-RBZ	A/GAC	GLY-RBZ	A/GGC	GLY-RBZ	A/GUC	GLY-RBZ	A/GCC	U/C	
4B		GLY-RBZ	U/CAC	GLY-RBZ	U/CGC	GLY-RBZ	U/CUC	GLY-RBZ	U/CCC	A/G	
	1st position	aa-aaRS	Ac	aa-aaRS	Ac	aa-aaRS	Ac	aa-aaRS	Ac	3rd position	
										Wobble	

Figure 3.5 The 1–amino acid code. All tRNAs are tRNA^{Gly}. aaRS designations are indicated, although, through multiple steps, ribozyme (RBZ) aaRS enzymes were initially probably responsible for aminoacylating tRNAs. *Ac*, anticodon.

row		1		2		3		4			column
		U		C		A		G		2nd position	
1A	U	VAL-RBZ	A/GAA	ALA-RBZ	A/GGA	ASP-RBZ	A/GUA	GLY-RBZ	A/GCA	U/C	
1B		VAL-RBZ	U/CAA	ALA-RBZ	U/CGA	ASP-RBZ	U/CUA	GLY-RBZ	U/CCA	A/G	
2A	C	VAL-RBZ	A/GAG	ALA-RBZ	A/GGG	ASP-RBZ	A/GUG	GLY-RBZ	A/GCG	U/C	
2B		VAL-RBZ	U/CAG	ALA-RBZ	U/CGG	ASP-RBZ	U/CUG	GLY-RBZ	U/CCG	A/G	
3A	A	VAL-RBZ	A/GAU	ALA-RBZ	A/GGU	ASP-RBZ	A/GUU	GLY-RBZ	A/GCU	U/C	
3B		VAL-RBZ	U/CAU	ALA-RBZ	U/CGU	ASP-RBZ	U/CUU	GLY-RBZ	U/CCU	A/G	
4A	G	VAL-RBZ	A/GAC	ALA-RBZ	A/GGC	ASP-RBZ	A/GUC	GLY-RBZ	A/GCC	U/C	
4B		VAL-RBZ	U/CAC	ALA-RBZ	U/CGC	ASP-RBZ	U/CUC	GLY-RBZ	U/CCC	A/G	
	1st position	aa-aaRS	Ac	aa-aaRS	Ac	aa-aaRS	Ac	aa-aaRS	Ac	3rd position	
										Wobble	

Figure 3.6 The 4–amino acid code. Colors follow amino acids through code sectoring. Bases indicated in red (dark gray in print version) type are disallowed (wobble A is very rare in Archaea).

row		1		2		3		4			column
		U		C		A		G		2nd position	
1A	U	LEU-IA	A/GAA		A/GGA	ASP-IIA	A/GUA	ARG-IA	A/GCA	U/C	
1B		LEU-IA	U/CAA		U/CGA	GLU-IA	U/CUA	ARG-IA	U/CCA	A/G	
2A	C	LEU-IA	A/GAG		A/GGG	ASP-IIA	A/GUG	ARG-IA	A/GCG	U/C	
2B		LEU-IA	U/CAG		U/CGG	GLU-IA	U/CUG	ARG-IA	U/CCG	A/G	
3A	A	VAL-IA	A/GAU	ALA-IIA	A/GGU	ASP-IIA	A/GUU	GLY-IIA	A/GCU	U/C	
3B		VAL-IA	U/CAU	ALA-IIA	U/CGU	GLU-IA	U/CUU	GLY-IIA	U/CCU	A/G	
4A	G	VAL-IA	A/GAC	ALA-IIA	A/GGC	ASP-IIA	A/GUC	GLY-IIA	A/GCC	U/C	
4B		VAL-IA	U/CAC	ALA-IIA	U/CGC	GLU-IA	U/CUC	GLY-IIA	U/CCC	A/G	
	1st position	aa-aaRS	Ac	aa-aaRS	Ac	aa-aaRS	Ac	aa-aaRS	Ac	3rd position	
										Wobble	

Figure 3.7 The 8–amino acid code. Columns 1, 2, and 4 sector on the 2nd and 3rd anticodon positions. Column 3 sectors on the 2nd and 1st (wobble) anticodon positions, leading to column 3 complexity.

row		1		2		3		4		column
		U		C		A		G		2nd position
1A	U	LEU-IA	A/GAA		A/GGA	ASP-IIB	A/GUA	ARG-ID	A/GCA	U/C
1B		LEU-IA	U/CAA		U/CGA	GLU-IB	U/CUA	ARG-ID	U/CCA	A/G
2A	c	LEU-IA	A/GAG		A/GGG	HIS-IIA	A/GUG	ARG-ID	A/GCG	U/C
2B		LEU-IA	U/CAG		U/CGG	GLN-IB	U/CUG	ARG-ID	U/CCG	A/G
3A	A	ILE-IA	A/GAU	THR-IIA	A/GGU	ASN-IIB	A/GUU	ARG-ID	A/GCU	U/C
3B		ILE-IA	U/CAU	THR-IIA	U/CGU	LYS-IE	U/CUU	ARG-ID	U/CCU	A/G
4A	G	VAL-IA	A/GAC	ALA-IIA	A/GGC	ASP-IIB	A/GUC	GLY-IIA	A/GCC	U/C
4B		VAL-IA	U/CAC	ALA-IID	U/CGC	GLU-IB	U/CUC	GLY-IIA	U/CCC	A/G
	1st position	aa-aaRS	Ac	aa-aaRS	Ac	aa-aaRS	Ac	aa-aaRS	Ac	3rd position Wobble

Figure 3.8 The ~16-amino acid code after evolution of the EF-Tu latch. At this stage, proteins may take on sufficient complexity to replace ribozyme aaRS enzymes.

row		1		2		3		4		column
		U		C		A		G		2nd position
1A	U	PHE-IIC	A/GAA		A/GGA	TYR-IC	A/GUA	CYS-IB	A/GCA	U/C
1B		LEU-IA	U/CAA		U/CGA	STOP	U/CUA	TRP-IC	U/CCA	A/G
2A	c	LEU-IA	A/GAG		A/GGG	HIS-IIA	A/GUG	ARG-ID	A/GCG	U/C
2B		LEU-IA	U/CAG		U/CGG	GLN-IB	U/CUG	ARG-ID	U/CCG	A/G
3A	A	ILE-IA	A/CAU	THR-IIA	A/GGU	ASN-IIB	A/GUU		A/CUU	U/C
3B		MET-IA	U/CAU	THR-IIA	U/CGU	LYS-IE	U/CUU	ARG-ID	U/CCU	A/G
4A	G	VAL-IA	A/GAC	ALA-IIA	A/GGC	ASP-IIB	A/GUC	GLY-IIA	A/GCC	U/C
4B		VAL-IA	U/CAC	ALA-IID	U/CGC	GLU-IB	U/CUC	GLY-IIA	U/CCC	A/G
	1st position	aa-aaRS	Ac	aa-aaRS	Ac	aa-aaRS	Ac	aa-aaRS	Ac	3rd position Wobble

Figure 3.9 The standard 21-assignment genetic code (20-amino acids + stops) in Archaea. Amino acids and stop codons shaded in charcoal were late additions to the code (row 1). Column 1, row 3B (anticodon CAU) is co-occupied with Ile-IA and Met-IA.

of the EF-Tu latch (Loveland et al., 2017; Rozov, Demeshkina, et al., 2016; Rozov et al., 2018; Rozov, Westhof, et al., 2016). Fig. 3.9 shows evolution to the standard genetic code (21 assignments: 20-amino acids + stops).

Column 2

Column 2 of the genetic code sectored evenly into four 4-codon sectors encoding Ala, Thr, Pro, and Ser (Figs. 3.8 and 3.9). Because of early wobbly recognition of the 3rd anticodon position, however, we posit that column 2 may have initially sectored into two 8-codon sectors encoding Ala and Pro (Fig. 3.7). Essentially, the idea is that, as the ribosome evolves to read the 3rd anticodon position, initially only pyrimidine versus purine discrimination was possible. Only after evolution of the EF-Tu anticodon-codon latch, is single base recognition at the 3rd anticodon position achieved (Fig. 3.8). We posit that Ala, which we posit entered the code before Pro, protected the more favored sectors, according to the 2nd and 3rd position anticodon rules C > G > U >> A. Because the 4th row of the code is favored (3rd position C), Ala protected favored sectors from invasion by Pro. After evolution of the EF-Tu latch, the entire anticodon could be read on the ribosome. Thus, Thr could invade the Ala sector with Ala retaining the most favored rows (row 4; 3rd anticodon position C). Ser could invade the Pro sector. Amino acids that entered the code first, therefore, retained the most favored rows in the code. Because column 2 has no 2-codon sectors, column 2 has no sectoring to discriminate amino acids according to the wobble position. In part, we posit that column 2 amino acids (Ala, Thr, Pro, and Ser) have too little character to be more accurately distinguished within their aaRS active sites, limiting further innovation in column 2 from 4-codon sectors to 2-codon sectors (Kim et al., 2019; Pak, Kim, & Burton, 2018). Interestingly, ThrRS-IIA, ProRS-IIA, and SerRS-IIA are closely related aaRS enzymes by structure and sequence (Fig. 3.3). The similarity of these enzymes indicates evolution within genetic code column 2.

Columns 1 and 4

Columns 1 and 4 are posited to be more chaotic versions of sectoring, similar to the sectoring of column 2. Column 1 is posited to have sectored first between Val (8-codon sector) and Leu (8-codon sector) (Fig. 3.7). Evolution of the EF-Tu anticodon-codon latch was necessary to sector further. Val was invaded by Ile, with Val retaining the most favored sectors (row 4; 3rd anticodon position C). We posit that, prior to Met invasion, there were three tRNA^{Ile} anticodons (GAU, UAU, and CAU) (in Archaea, AAU is rarely or never used). Ile (CAU) was invaded by Met (CAU). To retain Met in the code, Ile (UAU) was selected against, because tRNA^{Ile} (UAU) potentially reads both Ile (AUA) and Met (AUG) codons. Different modification enzymes evolved to alter the wobble base of tRNA^{Ile} (CAU) and tRNA^{Met} (CAU) so these tRNAs can be accurately discriminated, specifically reading Ile (AUA) and Met (AUG) codons. At an early time in evolution (i.e., before LUCA), therefore, two tRNA^{Met} (CAU), 1 elongator tRNA^{Met} (CAU) and 1 initiator tRNA^{Met} (CAU), and two tRNA^{Ile} (GAU and CAU) were utilized (Satpati et al., 2014; Voorhees et al., 2013). There are minor alterations to this common situation in some organisms. In eukaryotes, more tRNA^{Ile} anticodons are utilized (UAU and IAU; I for inosine (adenine→inosine)), and additional rules apply (Pak, Kim, & Burton, 2018). Phe was a late addition to the code invading a Leu sector. As a late invader, Phe is relegated to disfavored row 1 (3rd anticodon position A). In previous publications, our laboratory has somewhat confused discussion of Ile and Met sectoring. We apologize. Ile and Met sectoring is a fundamental but simple story in structuring of the code. Essentially, Met invaded a 4-codon Ile sector, but the invasion was never fully resolved and now can never be.

Interestingly, in column 1, Val, Ile, and Leu are hydrophobic amino acids. Furthermore, ValRS-IA, IleRS-IA, MetRS-IA, and LeuRS-IA are closely related enzymes by sequence and structure (Fig. 3.3). We posit that amino acid hydrophobicity and conserved aaRS enzymes demonstrate evolution of the genetic code within column 1. We posit, further, that evolution of the genetic code within columns relates to the central importance of tRNA anticodon position 2. Met is in some ways a special case because Met is a late addition to the code and Met utilizes two tRNA^{Met} (CAU) at the base of code evolution (1 elongator and 1 initiator tRNA^{Met}). Phe is also a late entry into the code. Phe is an aromatic and hydrophobic amino acid. Interestingly, PheRS-IIC helps to discriminate charging of tRNA^{Phe} (GAU) from charging by the class IA enzymes for other column 1 amino acids. All of the aaRS enzymes in column 1 have an editing active site to enhance the accuracy of tRNA charging (Perona & Gruic-Sovulj, 2014). Editing is posited to be important for these aaRS enzymes because hydrophobic amino acids have few discriminating features within the aaRS active site for accurate charging to tRNA. To protect against charging errors, column 1 has 4-codon sectors for Val and Leu. Ile was reduced to a 3-codon sector because of invasion by Met. Met (CAU) and Ile (CAU) share a 1 codon sector, but Met is required for translation starts, and Met was a late addition to the code. Although Met invading the Ile sector was never fully resolved during code establishment, Met invading the Ile sector suggests a mechanism for the previous successful invasions of other sectors of the code by other amino acids.

Column 4 may have sectored as follows. Initially, column 4 split between Gly and Arg (8-codon sectors), with Gly occupying the favored rows 3 and 4 (Fig. 3.7). With evolution of the anticodon-codon EF-Tu latch, Arg invaded row 3, displacing Gly. We posit that Ser invaded column 4 to attain a better anticodon with 2nd anticodon position C (column 4; anticodon GCU). Ser was able to invade column 4 because tRNA^{Ser} is a type II tRNA with an expanded variable loop, which is recognized as a determinant by SerRS-IIA for accurate amino acid placement (Perona & Gruic-Sovulj, 2014). Other tRNAs generally depend strongly on recognition of their anticodon, for amino acid placement by their aaRS, making such a jump within the genetic code impossible, without inducing error catastrophe. The Arg sector could be invaded by Ser because ArgRS-ID cannot charge type II tRNA^{Ser} (GCU) with Arg. Ser is the only amino acid that is split between columns (in this case columns 2 and 4) within the genetic code table. We posit that Ser is the only amino acid to have jumped successfully in evolution of the code, and that only SerRS-IIA and type II tRNA^{Ser} could readily support such a jump. Because ArgRS-ID is a class I aaRS, invasion of the Arg sector by SerRS-IIA may have been facilitated. In column 4, Trp, Cys and a stop codon are late entries into the genetic code that are relegated to disfavored row 1 (3rd anticodon position A). Stop codons are recognized by proteins, not tRNA (Burroughs & Aravind, 2019), so stop codons do not follow tRNA anticodon selection rules. Aromatic amino acids Phe (column 1), Tyr (column 3), and Trp (column 4) are late entries into the genetic code, all found in disfavored row 1 (3rd anticodon position A).

Column 3

Column 3 is the most innovated column in the code. We posit that the reason for such high innovation is because column 3 sectored initially on the 2nd and 1st (wobble) anticodon positions (Fig. 3.7). Columns 1, 2, and 4, by contrast, sectored initially on the 2nd and 3rd anticodon positions, as described above. We posit that column 3 initially sectored between Asp and Glu, with Asp utilizing 1st wobble anticodon position G, and Glu utilizing 1st wobble anticodon position U/C, creating the striped pattern. Because we posit that Asp entered the code before Glu, we posit that a preference is indicated for selection of wobble anticodon position G over U/C. A possible reason for this preference is that only one tRNA (GUN) is necessary to recognize two mRNA codons, rather than two tRNAs (UUN, CUN). In the wobble position, G pairs with C as a Watson—Crick pair, and G forms a wobble pair with U. Full establishment of the three-nucleotide code required the EF-Tu latch. Once the latch evolved, other amino acids could invade column 3. Initially, Asp is posited to have occupied all sectors in column 3 (Fig. 3.6). After invasion by Glu, Asp may have occupied sectors 4A, 3A, 2A, and 1A (Fig. 3.7). Ultimately, Asp retained the most favored row 4A, abandoning all other sectors in column 3 (Fig. 3.9). Glu retained favored row 4B and abandoned less favored rows 3B, 2B, and 1B to invading amino acids. Asn invaded row 3A. Lys invaded row 3B. Gln invaded row 2B. His invaded row 2A. Tyr and stop codons were late entries into the code occupying disfavored row 1 (3rd anticodon position A).

As noted above, we posit that Archaea are most similar to LUCA for translation functions (Kim et al., 2019; Pak et al., 2017; Battistuzzi et al., 2004; Kim et al., 2018). For instance, archaeal tRNAs are more ancient than bacterial tRNAs. Therefore, to understand the initial evolution of the code requires a focus on archaeal systems. Notably, in Archaea, AspRS-IIB, AsnRS-IIB, and HisRS-IIA enzymes are reasonably closely related in structure and sequence (Fig. 3.3A). Furthermore, in Archaea, GluRS-IB, LysRS-IC, and GlnRS-IB are closely related enzymes in structure and sequence (Fig. 3.3A). Evolution of aaRS enzymes, therefore, strongly supports our proposed model for column 3 evolution. Interestingly, in Bacteria, LysRS-IIB is closely related to AspRS-IIB and AsnRS-IIB in structure and sequence. Therefore, we posit that LysRS-IIB in Bacteria was evolved from AspRS-IIB, within column 3, reinforcing the importance of evolution within columns. The switch from LysRS-IC (Archaea) to LysRS-IIB (Bacteria) likely occurred close to the root of the archaeal→bacterial divergence. Clearly, evolution of the genetic code within column 3 (tRNA anticodon position 2 U) is demonstrated.

Evolution within genetic code columns

So, why did the genetic code evolve so clearly along code columns? The obvious answer is that the code initially evolved around the tRNA anticodon 2nd position, which is most important for translational accuracy. Assuming this idea is correct, what is the mechanism? One mechanism is indicated by translation systems in Archaea (and some Bacteria). We identify ancient Archaea, those most closely related to LUCA, as species with the least radiated tRNAomes from $tRNA^{Pri}$ (Kim et al., 2019; Pak, Du, et al., 2018; Pak, Kim, & Burton, 2018). From analysis of tRNAomes, examples of ancient archaeal families include *Pyrococcus*, *Pyrobaculum*, *Staphylothermus*, *Aeropyrum,* and *Sulfolobus*. Ancient Archaea lack GlnRS-IIB. In place of GlnRS-IIB, these species charge $tRNA^{Gln}$ with Glu using GluRS-IIB. Then a Glu-$tRNA^{Gln}$ amidotransferase converts the complex to Gln-$tRNA^{Gln}$ (Nureki et al., 2010; Rampias et al., 2010). As described above, Met also invaded an Ile sector leading to the sharing of the CAU anticodon. Similar mechanisms appear to be common in sectoring the code within columns. A similar mechanism could potentially explain transitions leading to the occupation of the following code sectors: (1) Val→Leu; (2) Val→Ile; (3) Ile→Met; (4) Ala→Pro; (5) Ala→Thr; (6) Pro→Ser; (7) Asp→Asn; (8) Asp→His; and (9) Glu→Gln. In some cases, mischarging of tRNAs may simply occur by invasion of an amino acid from outside the code, as with Ile → Met, rather than metabolic modification of a charged tRNA, as with Glu → Gln and, we assume, Asp → Asn. Some of these sectoring events require additional steps, that is, exchanging AlaRS-IIA with AlaRS-IID, a step that most likely occurred about the time of full sectoring of genetic code column 2. These posited events are best indicated from evolution of aaRS enzymes, which are often closely related within columns (Fig. 3.3). To a lesser extent, these events may be indicated by analysis of tRNAomes, particularly in ancient Archaea (Kim et al., 2019; Pak, Du, et al., 2018). tRNAs that evolve within genetic code columns tend to mutate and/or accumulate modifications rapidly to establish their new identity and to prevent mischarging. In columns 1 and 2, aaRS editing also suppresses mischarging of tRNAs.

An alternate model for sectoring of column 2

Because there are some limits to our clairvoyance describing events from ~4 billion years ago, we posit an alternate model for sectoring of column 2 and for Ser jumping from column 2 to column 4. Column 2 may have initially split between Ala (8-codon sector; rows 3 and 4) and Ser (8-codon sector; rows 1 and 2). After evolution of the EF-Tu latch, Ser may then have invaded the Ala sector, resulting in Ser occupying column 2, rows 1, 2, and 3, and Ala retaining favored row 4. Ser (GGU) may then have jumped to Ser (GCU) (column 4) via a single base change in tRNA^{Ser} (anticodon position 2). Subsequently, Pro invaded row 2, displacing Ser, and Thr invaded row 3, displacing Ser. Such a model has potential advantages over that described above. Notably, Ser jumping to favored column 4 is more easily described. Also, because of amino acid similarity, Ser → Thr appears to be a reasonable transition in row 3. Pro would displace Ser in row 2, leaving Ser in disfavored row 1, but this might not pose a difficulty if Ser had previously occupied favored column 4 row 3A. We give this example of an alternate model as a guide for understanding evolution of the genetic code. The model we advance is a working model. The detail in the model we find to be an advantage. Alternate pathways make slightly different predictions that potentially can be challenged computationally and/or by experiment. We know of no other model for code evolution that could potentially make such detailed or informative predictions.

An alternate model for evolution of stop codons

A appears to be strongly disfavored in the 3rd anticodon position. We, therefore, posit that row 1 of the genetic code may not have initially been populated with amino acids. Instead, row 1 may have represented stop codons, and evolution of the EF-Tu latch may have been required to encode amino acids in row 1. In such a modified model, row 1 would occupy with amino acids only at the ~16 amino acid stage. This suggestion is a minor possible adjustment to the model described above.

aaRS editing

aaRS editing is a feature of some aaRS enzymes [56]. Editing allows a mischarged amino acid to be hydrolyzed from the tRNA 3′-end, using a separate active site from the aminoacylating active site and limiting amino acid misincorporation on the ribosome. Remarkably, in Archaea, aaRS enzymes that edit locate to columns 1 and 2 of the genetic code (Figs. 3.2 and 3.3). A partial exception is SerRS-IIA, which resides in column 4 as well as in column 2. As described above, it appears that Ser jumped from column 2 to column 4. In Archaea, ProRS-IIA does not edit, but, in Bacteria, ProRS-IIA does edit. The amino acids specified in columns 1 and 2 are hydrophobic (column 1) and neutral (column 2). Neutral amino acids in column 2 also have limited hydrogen-bonding capacity (i.e., Ser and Thr side chains are mostly limited to a single H-bond to an aaRS side chain). Editing is more essential for aaRS enzymes that have a decreased capacity to discriminate the amino acid substrate within their active sites. Editing, therefore, is most important for amino acids that are hydrophobic, neutral, and/or have reduced H-bonding capacity (columns 1 and 2). The pattern of aaRS editing demonstrates additional unexpected structure in the genetic code.

Late additions to the genetic code

Phe, Tyr, Trp, Cys, Met, and His are posited to be among the last additions to the genetic code (Brooks et al., 2002; Fournier & Alm, 2015; Mukai et al., 2017). Stop codons, which are read by protein release factors rather than tRNAs, are also posited to be late additions to the code. Phe, Tyr, Trp, and Cys are all in disfavored row 1 of the code (3rd anticodon position A). Stop codons, which do not follow tRNA anticodon rules (Burroughs & Aravind, 2019), are in disfavored row 1 of the code. Met appears to have invaded the Ile sector as a late event. For tRNA^{Met} (CAU) and tRNA^{Ile}

(CAU), wobble C covalent modifications, particularly to tRNA^{Ile} (agmatidine in Archaea; lysidine in Bacteria), are very important to discriminate tRNA^{Met} and tRNA^{Ile}. tRNA^{Met} acceptor stems (As) are important to discriminate elongator tRNA^{Met} (i.e., 5′-As GCCCGGG) from initiator tRNA^{Met} (i.e., 5′-As AGCGGGA). In Archaea, tRNA^{Ile} (CAU) (C → agmatidine) is commonly used to recognize AUA Ile codons without recognition of AUG Met codons. The tRNA^{Ile} (CAU) C ~ A wobble pair is recognized by modification of the C to agmatidine to achieve this discrimination (Mandal et al., 2010; Satpati et al., 2014; Voorhees et al., 2013). Only in eukaryotes is tRNA^{Ile} (UAU) commonly utilized. In eukaryotes, tRNA^{Ile} (IAU; I for inosine) can recognize the three Ile codons (AUC, AUU, AUA).

Alternate genetic code models

Focusing on tight coevolution of metabolic systems to describe patterns of genetic code evolution is likely a mistake. We posit that apparent parallels between genetic code evolution and amino acid metabolism evolution (i.e., evolution of the genetic code within columns) occur primarily because of evolution of the genetic code according to the tRNA anticodon, as described above. If the tRNA anticodon is the central feature determining the history of evolution of the genetic code, the entire scenario makes sense according to Darwinian principles. Otherwise, the sectoring of the code makes little to no sense. That stated, we do accept that coevolution of metabolism and the genetic code occurred. In some cases, modification of amino acids on tRNAs is a mechanism for code sectoring (i.e., Asp → Asn and Glu → Gln). In other cases, amino acids appear to invade from outside the code (i.e., Ile → Met). We argue that, with invasions from outside the code, coevolution of metabolism and the code provided the amino acids that were available to invade the code rather than metabolism driving the sectoring of the code.

Another potential error that we perceive in genetic code models is to build up the code a sector at a time. We do not believe such models are reasonable. Initially, we made the assumption that the entire code (all anticodons and all codons) was populated with tRNA^{Gly} (Fig. 3.5). To our surprise, this simple assumption allowed a rich and highly detailed model for genetic code evolution with clear selection strategies to emerge. We were surprised because we commenced this effort thinking that, after ~4 billion years of evolution, development of such a detailed model was not possible.

In support of our assumption, however, the tRNA anticodon is successfully mutated more rapidly than other tRNA sequences, which are required to support RNA stems and to stabilize tRNA structure. This is because the anticodon stem and loop sticks out from the tRNA. Other anticodon loop bases (7-nt loop positions 1, 2, 6, and 7) are highly constrained in sequence to maintain U-turn geometry, which is necessary to present a 3-nt anticodon. We imagined that tRNA initially was essential genetic evolutionary property (Fig. 3.1). That is to say that tRNA was of value to synthesize polyglycine to stabilize protocells. We imagined that, being of value, tRNA was replicated rapidly in an RNA world, providing many copies for modification and genetic alteration. The other advantage to our model was that the model brought mRNA in line with tRNA, because, initially, all codons encoded glycine. In this way, tRNA anticodons and mRNA codons could more easily coevolve to encode incoming amino acids. In other scenarios, anticodons and codons must converge to encode products, which is more difficult to imagine. As anticodons evolved within the code, mRNA codons coevolved to encode ever more complex products. By contrast, a GC→GCA→GCAU genetic code (based on codons rather than anticodons) (Hartman & Smith, 2019) evolutionary model does not adequately address this difficulty. Such a model begins within a sector and evolves to occupy other sectors. Such models have the critical flaw described above.

Another scheme for evolution of the genetic code is the GNC→SNS→standard code model (N = any base; S = G or C) (Chatterjee & Yadav, 2019). We find some agreement and some disagreement with this model. One objection that we would raise is that the scheme is presented as a codon-centric model rather than an anticodon-centric model. Working out from tRNA is a big advantage, as we hope we have shown. Also, why begin with codon GNC (also anticodon GNC) rather than, for instance, CNG? Anticodon GNC gives Gly, Asp, Ala, and Val as the first four amino acids, which agrees with our assessment (Figs. 3.4 and 3.6). GNC also supports 3rd anticodon position C, which we deem favorable because of the C > G > U >> A anticodon rule. However, initial selection of GNC rather than CNG appears arbitrary. SNS makes some sense because C/G should facilitate 1st and 3rd position anticodon reading before evolution of the EF-Tu anticodon-codon latch. The SNS phase brings Gly, Asp, Ala, Val, Leu, Pro, Glu, Gln, His, and Arg into the code, which is very similar to the recruitment order we favor. Possible limitations of the alternate GNC model include (1) apparent favoring of codons over anticodons; (2) beginning the code arbitrarily with GNC, rather than SNS; (3) no consideration of the EF-Tu latch in anticodon-codon reading; (4) building up the code via sectors rather than filling in and invading; (5) awkward coevolution of anticodons and codons; and (6) little consideration of why 4- and 2-codon sectors are favored over 1-codon sectors. A recent paper on this subject (Chatterjee & Yadav, 2019) utilizes an incorrect tRNA evolution model and an incorrect aaRS evolution model. The model does not fully recognize the extent to which aaRS enzyme evolution parallels the establishment of the code (Fig. 3.3).

Alternate representations of the genetic code

We strongly advocate for the representation of the genetic code shown in Figs. 3.2 and 3.3. Many other representations of the genetic code are possible. Circular representations of the code emphasize selection for $G = C$ pairs in anticodon—codon interactions (Grosjean & Westhof, 2016). As we argue here, however, the Darwinian selection for the tRNA anti-codon positions 2 and 3 appears to be $C > G > U >> A$, rather than for both C and G to be strongly favored in the anticodon. We reject codon tables lacking anticodons because the genetic code evolved around the tRNA anticodon. Recognizing the central importance of tRNA and the tRNA anticodon, as we do here, provides remarkable insight into genetic code evolution.

Life on another planet

If life were to evolve independently on a planet or moon away from Earth, how would organisms evolve coding? Would such organisms evolve tRNA or a tRNA-like coding molecule? Would the coding molecule be RNA? It is very difficult to imagine an alternate scheme for chemical coding of comparable complexity to tRNA-based coding. It is also difficult to evolve a more complex code than the standard 3-nt genetic code. Notably, tRNA has a perfectly structured RNA anticodon loop to support a 3-nt code. The anticodon loop is 7-nt. Significantly, a 6- or 8-nt loop cannot support a compact U-turn geometry or as tight a loop. A tight anticodon loop is necessary to support anticodon-codon interactions during translation. So, with RNA as the genetic material, only a 3-nt code appears reasonable. We recognize that 4-nt codes are conceivable and demonstrated by engineering with tRNA, but (to our knowledge) 4-nt codes are not observed in free-living organisms. To increase the coding potential of the standard genetic code would require scrambling the standard code. For instance, 4-codon sectors on the left half of the code might be split into 2-codon sectors, to add amino acids, but to avoid error catastrophe, hydrophobic amino acids would have to share sectors with amino acids with more distinct character (i.e., charge and/or H-bonding capacity). So, coding capacity might be increased by aggressive engineering, but evolving a much more complex code than the standard code under conditions of natural selection appears unlikely. At most, only a few additional amino acids could be added to the naturally evolved code. The 21-assignment standard code is very close to the theoretical 32-assignment limit.

So, tRNA is a highly specialized molecule constructed from ligation of two different types of 31-nt minihelix (1-D loop minihelix + 2-anticodon loop/T loop minihelices) (Burton, 2020; Kim et al., 2019; Pak et al., 2017). Very likely, tRNA could not be constructed of three identical minihelices (i.e., 3-anticodon loop/T loop minihelices). During prelife, such a molecule would probably be processed into 3 separate 31-nt minihelices by then-existing ribozyme ribo-endonucleases. From the tRNA structure, it is likely that neither the anticodon loop nor the T loop minihelix could be substituted with another minihelix. For instance, as the anticodon loop, a D loop minihelix would likely support a 1-nt rather than a 3-nt code. Significantly, the D loop minihelix cannot support a 7-nt U-turn loop. It is very likely that the D loop minihelix component of tRNA could be reengineered with a different sequence to form a tRNA-like molecule that might serve for coding. Such a tRNA variant might support coding and life but would be highly analogous to tRNA on Earth and would support a similar code with similar coding capacity. The authors of this paper cannot yet imagine a chemical tRNA substitute with advantages over the tRNA found on Earth, to support or engineer a richer coding system.

Predictions

The 3-minihelix model we propose for evolution of tRNA makes many predictions, most of which are confirmed (Burton, 2020). We proposed that the anticodon loop and the T loop are homologs and then showed that they are (Burton, 2020; Kim et al., 2019; Pak et al., 2017). We proposed that the D loop minihelix core was initially a UAGCC repeat and then showed that this prediction was confirmed (Pak et al., 2017). We proposed that the last 5 nt of the D loop and the V loop were derived from acceptor stems, and we confirmed this prediction (Burton, 2020; Kim et al., 2018; Pak et al., 2017).

We proposed that acceptor stems were based on a GCG repeat and its CGC complement. We confirmed this prediction (Burton, 2020; Pak et al., 2017). We posited that the expanded V loop in type II tRNA was initially a $3'$-acceptor stem ligated to a $5'$-acceptor stem and then showed that this was the case (Kim et al., 2018). In ancient Archaea (i.e., *Pyrococcus furiosis*), tRNA^{Pri} (the primordial tRNA) is almost identical to tRNA^{Gly}, indicating that Gly was the first encoded amino acid. We have shown that no 2-minihelix model (Di Giulio, 2019; Di Giulio, 2012; Branciamore & Di Giulio, 2011; Di Giulio et al., 2020; Nagaswamy & Fox, 2003; Widmann et al., 2005) is adequate to describe tRNA evolution, because (for instance) such models are inconsistent with obvious anticodon loop and T loop homology (Burton, 2020; Kim et al., 2019; Pak, Kim, & Burton, 2018; Pak et al., 2017). Only a 3-minihelix model is adequate to describe tRNA evolution. No accretion model, in which tRNA grows a stem at a time, is adequate to describe tRNA evolution because accretion models are inconsistent with (for instance) anticodon loop and T loop homology.

Based on tRNA evolution models and our analysis of aaRS evolution, we propose a working model for evolution of the genetic code (Kim et al., 2019). This model also makes multiple predictions, some of which are already tested. Our model predicts that tRNAomes in ancient organisms are more highly ordered than tRNAomes in more derived organisms, as we have shown (Kim et al., 2018, 2019; Pak, Du, et al., 2018). We predict that a telomerase-like ribozyme can be generated to form RNA repeats (i.e., GCG, CGC, and UAGCC) (Fig. 3.1). We predict many ribozymes that have previously been generated in vitro (Chumachenko et al., 2009; Horning & Joyce, 2016; Illangasekare & Yarus, 2012; McGinness & Joyce, 2003; Paul & Joyce, 2002; Turk et al., 2010; Yarus, 2011; Zhang & Cech, 1998). We predict a peptidyl transferase ribozyme, which has been generated (Zhang & Cech, 1998). We predict that tRNAomes evolve both by divergence and convergence within a limited sequence space. This idea has been partly tested. We predict that Archaea is more similar to LUCA than Bacteria for transcription and translation functions (Battistuzzi et al., 2004; Burton & Burton, 2014; Burton, 2014; Burton et al., 2016; Kim et al., 2018, 2019; Pak, Du, et al., 2018; Pak et al., 2017). This model is strongly supported in multiple ways. We predict the standard code (20 amino acids + stops) is approximately the most complex code that could easily evolve and that the maximum code complexity (32-assignments) can be approached, but not achieved, because of fidelity mechanisms (Kim et al., 2018, 2019; Pak, Du, et al., 2018; Pak, Kim, & Burton, 2018). We predicted evolution of the genetic code within columns (2nd anticodon position) and we demonstrate this (Figs. 3.2 and 3.3). We predicted that the genetic code has order based on its evolutionary history, which we demonstrate.

The frozen accident

The standard genetic code evolved quickly (within ~ 300 million years; Fig. 3.1), and, once formed, did not change very much thereafter. Evidence for this conclusion is the obvious conservation of genetic code evolution patterns reflected in patterns of aaRS evolution (i.e., Fig. 3.3). Innovation in forming the code was driven by tRNA charging modifications and errors (Kim et al., 2018, 2019; Pak, Du, et al., 2018; Pak, Kim, & Burton, 2018). Alteration of Glu-tRNA^{Gln} to Gln-tRNA^{Gln} by a Glu-tRNA^{Gln} amidotransferase is an example of an intermediate in sectoring the genetic code to encode both Glu and Gln (Nureki et al., 2010; O'Donoghue et al., 2011; Rampias et al., 2010). Subsequently, a GlnRS can evolve to supplant the Glu-tRNA^{Gln} amidotransferase. Met invading an Ile 4-codon sector reflects another mechanism to enrich the code (Mandal et al., 2010; Satpati et al., 2014; Voorhees et al., 2013). In this case, the sectoring was never completed. In the case of Met invading Ile, Met appears to begin incorporation by mischarging tRNA^{Ile} to Met-tRNA^{Ile}. Then a duplicated tRNA^{Ile} (CAU) mutates to tRNA^{Met} (CAU), and IleRS duplicates, and one copy evolves to MetRS. So, tRNA charging errors and modifications of amino acids bound to tRNAs drove innovation of the code. Evolving translational fidelity mechanisms, therefore, drives the code toward closure by inhibiting further amino acid additions. Fidelity mechanisms include aaRS editing, evolution of the aaRS synthetic site (to enhance accuracy of amino acid addition), the EF-Tu latch, tRNA modifications, tRNA evolution, and amino acid character (i.e., size, H-bonding, charge, hydrophobicity). It appears, for instance, that aaRS editing in columns 1 and 2 of the code helped to protect 4-codon sectors from division to two 2-codon sectors, limiting the complexity of the code (Kim et al., 2019; Pak, Kim, & Burton, 2018).

Author contributions

L.L. and Z.F.B. involved in conceptualization, methodology, data curation, writing—original draft preparation, writing—review and editing, supervision, project administration. All authors have read and agreed to the published version of the manuscript.

References

Barbieri, M. (2019a). A general model on the origin of biological codes. *BioSystems, 181,* 11–19. https://doi.org/10.1016/j.biosystems.2019.04.010

Barbieri, Marcello (2019b). Evolution of the genetic code: The ambiguity-reduction theory. *Biosystems, 185,* 104024. https://doi.org/10.1016/j.biosystems.2019.104024

Battistuzzi, F. U., Feijao, A., & Hedges, S. B. (2004). A genomic timescale of prokaryote evolution: Insights into the origin of methanogenesis, phototrophy, and the colonization of land. *BMC Evolutionary Biology, 4.* https://doi.org/10.1186/1471-2148-4-44

Bernhardt, H. S. (2016). Clues to tRNA evolution from the distribution of class II tRNAs and serine codons in the genetic code. *Life, 6*(1). https://doi.org/10.3390/life6010010

Bernhardt, H. S., & Patrick, W. M. (2014). Genetic code evolution started with the incorporation of glycine, followed by other small hydrophilic amino acids. *Journal of Molecular Evolution, 78*(6), 307–309. https://doi.org/10.1007/s00239-014-9627-y

Bernhardt, H. S., & Tate, W. P. (2008). Evidence from glycine transfer RNA of a frozen accident at the dawn of the genetic code. *Biology Direct, 3.* https://doi.org/10.1186/1745-6150-3-53

Błażej, P., Wnetrzak, M., Mackiewicz, D., Gagat, P., & Mackiewicz, P. (2019). Many alternative and theoretical genetic codes are more robust to amino acid replacements than the standard genetic code. *Journal of Theoretical Biology, 464,* 21–32. https://doi.org/10.1016/j.jtbi.2018.12.030

Błażej, Paweł, Wnetrzak, M., Mackiewicz, D., Mackiewicz, P., & de Brevern, A. G. (2018). Optimization of the standard genetic code according to three codon positions using an evolutionary algorithm. *PLoS One, 13*(8), e0201715. https://doi.org/10.1371/journal.pone.0201715

Branciamore, S., & Di Giulio, M. (2011). The presence in tRNA molecule sequences of the double hairpin, an evolutionary stage through which the origin of this molecule is thought to have passed. *Journal of Molecular Evolution, 72*(4), 352–363. https://doi.org/10.1007/s00239-011-9440-9

Brooks, D. J., Fresco, J. R., Lesk, A. M., & Singh, M. (2002). Evolution of amino acid frequencies in proteins over deep time: Inferred order of introduction of amino acids into the genetic code. *Molecular Biology and Evolution, 19*(10), 1645–1655. https://doi.org/10.1093/oxfordjournals.molbev.a003988

Burroughs, A. M., & Aravind, L. (2019). The origin and evolution of release factors: Implications for translation termination, ribosome rescue, and quality control pathways. *International Journal of Molecular Sciences, 20*(8), 1981. https://doi.org/10.3390/ijms20081981

Burton, Z. F. (2014). The old and new testaments of gene regulation: Evolution of multisubunit RNA polymerases and co-evolution of eukaryote complexity with the RNAP II CTD. *Transcription, 5.* https://doi.org/10.4161/trns.28674

Burton, Z. F. (2020). The 3-minihelix tRNA evolution theorem. *Journal of Molecular Evolution, 88*(3), 234–242. https://doi.org/10.1007/s00239-020-09928-2

Burton, S. P., & Burton, Z. F. (2014). The σ enigma: Bacterial σ factors, archaeal TFB and eukaryotic TFIIB are homologs. *Transcription, 5*(4). https://doi.org/10.4161/21541264.2014.967599

Burton, Z. F., Opron, K., Wei, G., & Geiger, J. H. (2016). A model for genesis of transcription systems. *Transcription, 7*(1), 1–13. https://doi.org/10.1080/21541264.2015.1128518

Caetano-Anollés, D., & Caetano-Anollés, G. (2016). Piecemeal buildup of the genetic code, ribosomes, and genomes from primordial tRNA building blocks. *Life, 6*(4), 43. https://doi.org/10.3390/life6040043

Chatterjee, S., & Yadav, S. (2019). The origin of prebiotic information system in the peptide/RNA world: A simulation model of the evolution of translation and the genetic code. *Life, 9*(1), 25. https://doi.org/10.3390/life9010025

Chumachenko, N. V., Novikov, Y., & Yarus, M. (2009). Rapid and simple ribozymic aminoacylation using three conserved nucleotides. *Journal of the American Chemical Society, 131*(14), 5257—5263. https://doi.org/10.1021/ja809419f

Demongeot, J., Glade, N., Moreira, A., & Vial, L. (2009). RNA relics and origin of life. *International Journal of Molecular Sciences, 10*(8), 3420—3441. https://doi.org/10.3390/ijms10083420

Demongeot, J., & Moreira, A. (2007). A possible circular RNA at the origin of life. *Journal of Theoretical Biology, 249*(2), 314—324. https://doi.org/10.1016/j.jtbi.2007.07.010

Demongeot, J., & Seligmann, H. (2019). The Uroboros theory of life's origin: 22-nucleotide theoretical minimal RNA rings reflect evolution of genetic code and tRNA-rRNA translation machineries. *Acta Biotheoretica, 67*(4), 273—297. https://doi.org/10.1007/s10441-019-09356-w

Demongeot, J., & Seligmann, H. (2020). RNA rings strengthen hairpin accretion hypotheses for tRNA evolution: A reply to commentaries by Z.F. Burton and M. Di Giulio. *Journal of Molecular Evolution, 88*(3), 243—252. https://doi.org/10.1007/s00239-020-09929-1

Di Giulio, M. (2012). The origin of the tRNA molecule: Independent data favor a specific model of its evolution. *Biochimie, 94*(7), 1464—1466. https://doi.org/10.1016/j.biochi.2012.01.014

Di Giulio, M. (2019). A comparison between two models for understanding the origin of the tRNA molecule. *Journal of Theoretical Biology, 480*, 99—103. https://doi.org/10.1016/j.jtbi.2019.07.020

Di Giulio, M. (2020). An RNA ring was not the progenitor of the tRNA molecule. *Journal of Molecular Evolution, 88*(3), 228—233. https://doi.org/10.1007/s00239-020-09927-3

Di Giulio, I., McFadyen, B. J., Blanchet, S., Reeves, N. D., Baltzopoulos, V., & Maganaris, C. N. (2020). Mobile phone use impairs stair gait: A pilot study on young adults. *Applied Ergonomics, 84*. https://doi.org/10.1016/j.apergo.2019.103009

Doig, A. J. (2017). Frozen, but no accident — why the 20 standard amino acids were selected. *FEBS Journal, 284*(9), 1296—1305. https://doi.org/10.1111/febs.13982

Fournier, G. P., & Alm, E. J. (2015). Ancestral reconstruction of a pre-LUCA aminoacyl-tRNA synthetase ancestor supports the late addition of Trp to the genetic code. *Journal of Molecular Evolution, 80*(3—4), 171—185. https://doi.org/10.1007/s00239-015-9672-1

Grosjean, H., & Westhof, E. (2016). An integrated, structure- and energy-based view of the genetic code. *Nucleic Acids Research, 44*(17), 8020—8040. https://doi.org/10.1093/nar/gkw608

Hartman, H., & Smith, T. F. (2019). Origin of the genetic code is found at the transition between a thioester world of peptides and the phosphoester world of polynucleotides. *Life, 9*(3). https://doi.org/10.3390/life9030069

Horning, D. P., & Joyce, G. F. (2016). Amplification of RNA by an RNA polymerase ribozyme. *Proceedings of the National Academy of Sciences of the United States of America, 113*(35), 9786—9791. https://doi.org/10.1073/pnas.1610103113

Illangasekare, M., & Yarus, M. (2012). Small aminoacyl transfer centers at GU within a larger RNA. *RNA Biology, 9*(1), 59—66. https://doi.org/10.4161/rna.9.1.18039

Jaeger, L., Wright, M. C., & Joyce, G. F. (1999). A complex ligase ribozyme evolved in vitro from a group I ribozyme domain. *Proceedings of the National Academy of Sciences of the United States of America, 96*(26), 14712—14717. https://doi.org/10.1073/pnas.96.26.14712

Kim, D. E., & Joyce, G. F. (2004). Cross-catalytic replication of an RNA ligase ribozyme. *Chemistry and Biology, 11*(11), 1505–1512. https://doi.org/10.1016/j.chembiol.2004.08.021

Kim, Yunsoo, Kowiatek, B., Opron, K., & Burton, Z. (2018). Type-II tRNAs and evolution of translation systems and the genetic code. *International Journal of Molecular Sciences, 19*(10), 3275. https://doi.org/10.3390/ijms19103275

Kim, Y., Opron, K., & Burton, Z. F. (2019). A tRNA- and anticodon-centric view of the evolution of aminoacyl-tRNA synthetases, tRNAomes, and the genetic code. *Life, 9*(2). https://doi.org/10.3390/life9020037

Koonin, E. V. (2017). Frozen accident pushing 50: Stereochemistry, expansion, and chance in the evolution of the genetic code. *Life, 7*(2). https://doi.org/10.3390/life7020022

Koonin, E. V., & Novozhilov, A. S. (2009). Origin and evolution of the genetic code: The universal enigma. *IUBMB Life, 61*(2), 99–111. https://doi.org/10.1002/iub.146

Koonin, E. V., & Novozhilov, A. S. (2017). Origin and evolution of the universal genetic code. *Annual Review of Genetics, 51*, 45–62. https://doi.org/10.1146/annurev-genet-120116-024713

Kunnev, D., & Gospodinov, A. (2018). Possible emergence of sequence specific RNA aminoacylation via peptide intermediary to initiate darwinian evolution and code through origin of life. *Life, 8*(4), 44. https://doi.org/10.3390/life8040044

Kun, Á., & Radványi, Á. (2018). The evolution of the genetic code: Impasses and challenges. *BioSystems, 164*, 217–225. https://doi.org/10.1016/j.biosystems.2017.10.006

Loveland, A. B., Demo, G., Grigorieff, N., & Korostelev, A. A. (2017). Ensemble cryo-EM elucidates the mechanism of translation fidelity. *Nature, 546*(7656), 113–117. https://doi.org/10.1038/nature22397

Ma, W. (2017). What does "the RNA world" mean to "the origin of life"? *Life, 7*(4), 49. https://doi.org/10.3390/life7040049

Mandal, D., Kohrer, C., Su, D., Russell, S. P., Krivos, K., Castleberry, C. M., Blum, P., Limbach, P. A., Soll, D., & RajBhandary, U. L. (2010). Agmatidine, a modified cytidine in the anticodon of archaeal tRNAIle, base pairs with adenosine but not with guanosine. *Proceedings of the National Academy of Sciences, 107*(7), 2872–2877. https://doi.org/10.1073/pnas.0914869107

Mariscal, C., Barahona, A., Aubert-Kato, N., Aydinoglu, A. U., Bartlett, S., Cárdenas, M. L., Chandru, K., Cleland, C., Cocanougher, B. T., Comfort, N., Cornish-Bowden, A., Deacon, T., Froese, T., Giovannelli, D., Hernlund, J., Hut, P., Kimura, J., Maurel, M. C., Merino, N., … James Cleaves, H. (2019). Hidden concepts in the history and philosophy of origins-of-life studies: A workshop report. *Origins of Life and Evolution of Biospheres, 49*(3), 111–145. https://doi.org/10.1007/s11084-019-09580-x

McGinness, K. E., & Joyce, G. F. (2003). In search of an RNA replicase ribozyme. *Chemistry and Biology, 10*(1), 5–14. https://doi.org/10.1016/S1074-5521(03)00003-6

Mukai, T., Reynolds, N. M., Crnković, A., & Söll, D. (2017). Bioinformatic analysis reveals archaeal tRNATyr and tRNAtrp identities in bacteria. *Life, 7*(1). https://doi.org/10.3390/life7010008

Nagaswamy, U., & Fox, G. E. (2003). RNA ligation and the origin of tRNA. *Origins of Life and Evolution of the Biosphere, 33*(2), 199–209. https://doi.org/10.1023/A:1024658727570

Nureki, O., O'Donoghue, P., Watanabe, N., Ohmori, A., Oshikane, H., Araiso, Y., Sheppard, K., Söll, D., & Ishitani, R. (2010). Structure of an archaeal non-discriminating glutamyl-tRNA synthetase: A missing link in the evolution of Gln-tRNAGln formation. *Nucleic Acids Research, 38*(20), 7286–7297. https://doi.org/10.1093/nar/gkq605

O'Donoghue, P., Sheppard, K., Nureki, O., & Söll, D. (2011). Rational design of an evolutionary precursor of glutaminyl-tRNA synthetase. *Proceedings of the National Academy of Sciences of the United States of America, 108*(51), 20485–20490. https://doi.org/10.1073/pnas.1117294108

Opron, K., & Burton, Z. F. (2019). Ribosome structure, function, and early evolution. *International Journal of Molecular Sciences, 20*(1). https://doi.org/10.3390/ijms20010040

Pak, D., Root-Bernstein, R., & Burton, Z. F. (2017). tRNA structure and evolution and standardization to the three nucleotide genetic code. *Transcription, 8*(4), 205–219. https://doi.org/10.1080/21541264.2017.1318811

Pak, D., Du, N., Kim, Y., Sun, Y., & Burton, Z. F. (2018). Rooted tRNAomes and evolution of the genetic code. *Transcription, 9*(3), 137–151. https://doi.org/10.1080/21541264.2018.1429837

Pak, D., Kim, Y., & Burton, Z. F. (2018). Aminoacyl-tRNA synthetase evolution and sectoring of the genetic code. *Transcription, 9*(4), 205–224. https://doi.org/10.1080/21541264.2018.1467718

Paul, N., & Joyce, G. F. (2002). A self-replicating ligase ribozyme. *Proceedings of the National Academy of Sciences of the United States of America, 99*(20), 12733–12740. https://doi.org/10.1073/pnas.202471099

Perona, J. J., & Gruic-Sovulj, I. (2014). Synthetic and editing mechanisms of aminoacyl-tRNA synthetases. *Topics in Current Chemistry, 344*, 1–41. https://doi.org/10.1007/128_2013_456

Pinho, M. G., Kjos, M., & Veening, J. W. (2013). How to get (a)round: Mechanisms controlling growth and division of coccoid bacteria. *Nature Reviews Microbiology, 11*(9), 601–614. https://doi.org/10.1038/nrmicro3088

Pressman, A. D., Liu, Z., Janzen, E., Blanco, C., Müller, U. F., Joyce, G. F., Pascal, R., & Chen, I. A. (2019). Mapping a systematic ribozyme fitness landscape reveals a frustrated evolutionary network for self-aminoacylating RNA. *Journal of the American Chemical Society, 141*(15), 6213–6223. https://doi.org/10.1021/jacs.8b13298

Rampias, T., Sheppard, K., & Söll, D. (2010). The archaeal transamidosome for RNA-dependent glutamine biosynthesis. *Nucleic Acids Research, 38*(17), 5774–5783. https://doi.org/10.1093/nar/gkq336

Ribas de Pouplana, L., Torres, A., & Rafels-Ybern, Á. (2017). What froze the genetic code? *Life, 7*(2), 14. https://doi.org/10.3390/life7020014

Rogers, S. O. (2019). Evolution of the genetic code based on conservative changes of codons, amino acids, and aminoacyl tRNA synthetases. *Journal of Theoretical Biology, 466*, 1–10. https://doi.org/10.1016/j.jtbi.2019.01.022

Rogers, J., & Joyce, G. F. (2001). The effect of cytidine on the structure and function of an RNA ligase ribozyme. *RNA, 7*(3), 395–404. https://doi.org/10.1017/S135583820100228X

Rozov, A., Demeshkina, N., Westhof, E., Yusupov, M., & Yusupova, G. (2016). New structural insights into translational miscoding. *Trends in Biochemical Sciences, 41*(9), 798–814. https://doi.org/10.1016/j.tibs.2016.06.001

Rozov, A., Westhof, E., Yusupov, M., & Yusupova, G. (2016). The ribosome prohibits the G•U wobble geometry at the first position of the codon-anticodon helix. *Nucleic Acids Research, 44*(13), 6434–6441. https://doi.org/10.1093/nar/gkw431

Rozov, A., Wolff, P., Grosjean, H., Yusupov, M., Yusupova, G., & Westhof, E. (2018). Tautomeric G•U pairs within the molecular ribosomal grip and fidelity of decoding in bacteria. *Nucleic Acids Research, 46*(14), 7425–7435. https://doi.org/10.1093/nar/gky547

Saint-Léger, A., Bello, C., Dans, P. D., Torres, A. G., Novoa, E. M., Camacho, N., Orozco, M., Kondrashov, F. A., & De Pouplana, L. R. (2016). Saturation of recognition elements blocks evolution of new tRNA identities. *Science Advances, 2*(4). https://doi.org/10.1126/sciadv.1501860

Satpati, P., Bauer, P., & Åqvist, J. (2014). Energetic tuning by tRNA modifications ensures correct decoding of isoleucine and methionine on the ribosome. *Chemistry - A European Journal, 20*(33), 10271–10275. https://doi.org/10.1002/chem.201404016

Scheffers, D. J., & Pinho, M. G. (2005). Bacterial cell wall synthesis: New insights from localization studies. *Microbiology and Molecular Biology Reviews, 69*(4), 585–607. https://doi.org/10.1128/MMBR.69.4.585-607.2005

Turk, R. M., Chumachenko, N. V., & Yarus, M. (2010). Multiple translational products from a five-nucleotide ribozyme. *Proceedings of the National Academy of Sciences of the United States of America, 107*(10), 4585–4589. https://doi.org/10.1073/pnas.0912895107

Turk, R. M., Illangasekare, M., & Yarus, M. (2011). Catalyzed and spontaneous reactions on ribozyme ribose. *Journal of the American Chemical Society, 133*(15), 6044–6050. https://doi.org/10.1021/ja200275h

Voorhees, R. M., Mandal, D., Neubauer, C., Köhrer, C., Rajbhandary, U. L., & Ramakrishnan, V. (2013). The structural basis for specific decoding of AUA by isoleucine tRNA on the ribosome. *Nature Structural and Molecular Biology, 20*(5), 641–643. https://doi.org/10.1038/nsmb.2545

Widmann, J., Di Giulio, M., Yarus, M., & Knight, R. (2005). tRNA creation by hairpin duplication. *Journal of Molecular Evolution, 61*(4), 524–530. https://doi.org/10.1007/s00239-004-0315-1

Yarus, M. (2011). The meaning of a minuscule ribozyme. *Philosophical Transactions of the Royal Society B: Biological Sciences, 366*(1580), 2902–2909. https://doi.org/10.1098/rstb.2011.0139

Zapun, A., Vernet, T., & Pinho, M. G. (2008). The different shapes of cocci. *FEMS Microbiology Reviews, 32*(2), 345–360. https://doi.org/10.1111/j.1574-6976.2007.00098.x

Zhang, B., & Cech, T. R. (1998). Peptidyl-transferase ribozymes: Trans reactions, structural characterization and ribosomal RNA-like features. *Chemistry and Biology, 5*(10), 539–553. https://doi.org/10.1016/S1074-5521(98)90113-2

CHAPTER 4

Type II tRNAs and evolution of translation systems and the genetic code

Yunsoo Kim[1], Bruce K. Kowiatek[2], Kristopher Opron[3] and Zachary F. Burton[4]

[1]School of Information, University of Michigan, Ann Arbor, Michigan, MI, United States; [2]Allied Health Sciences, Blue Ridge Community and Technical College, Martinsburg, WV, United States; [3]Bioinformatics, Michigan State University, Ann Arbor, MI, United States; [4]Department of Biochemistry and Molecular Biology, Michigan State University, East Lansing, MI, United States

Abbreviations

aaRS	Aminoacyl-tRNA synthetase (i.e., LeuRS)
Ac loop	Anticodon loop
As	Acceptor stems
As*	Acceptor stem remnants
LUCA	Last universal common (cellular) ancestor
T loop	T loop or TΨC loop
V loop	Variable loop

Introduction

Ribosomes, mRNA, translation systems, genetic coding, and aminoacyl tRNA synthetases (aaRS enzymes; i.e., SerRS) evolved around cloverleaf tRNA. The evolution of tRNA, therefore, is the central problem in understanding the evolution of life on earth. A model was determined for the evolution of type-I tRNAs, lacking a V loop expansion (Fig. 4.1) (Pak et al., 2017; Root-Bernstein et al., 2016). The model is based on the ligation of three 31-nt minihelices followed by two symmetrical 9-nt deletions within ligated 3′- and 5′-acceptor stems. The model posits that type-I and type-II V loops are homologous to acceptor stems.

Cloverleaf tRNA evolved from short, defined genetic segments. A 31-nt minihelix is a 17-nt microhelix flanked 5′- and 3′- by 7-nt acceptor stems. Acceptor stems are based on a GCG repeat and its CGC complement, so the primordial tRNA acceptor stems are 5′-GCGGCGG-3′ (5′-As; As for acceptor

The Makings of a Clinical Protocol
ISBN 978-0-323-95749-6
https://doi.org/10.1016/B978-0-323-95749-6.00005-3

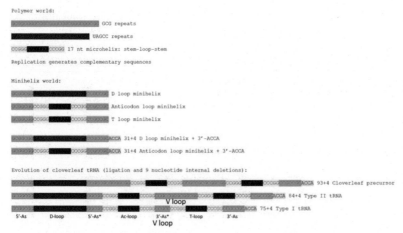

Polymer world:

GCG repeats

UAGCC repeats

17 nt microhelix: stem-loop-stem

Replication generates complementary sequences

Minihelix world:

D loop minihelix

Anticodon loop minihelix

T loop minihelix

31+4 D loop minihelix + 3'-ACCA

31+4 Anticodon loop minihelix + 3'-ACCA

Evolution of cloverleaf tRNA (ligation and 9 nucleotide internal deletions):

93+4 Cloverleaf precursor

84+4 Type II tRNA

75+4 Type I tRNA

5'-As   D-loop   5'-As*   Ac-loop   3'-As*   T-loop   3'-As

V loop

V loop

Figure 4.1 Models for the evolution of type-I and type-II tRNAs. 5′ and 3′ acceptor stems are *shaded green*. The D loop 17-nt microhelix is *shaded magenta*. U-turn stem-loop-stems are *shaded yellow* (stems) and *red* (7-nt U-turn loop). (For interpretation of the references to colour in this figure legend, the reader is referred to the web version of this article.)

stem) and 5′-CCGCCGC-3′ (3′-As). Ligation of a primordial 3′-As and 5′-As, therefore, gives the 14-nt sequence 5′-CCGCCGCGCGGCGG-3′. In generating type-I tRNAs, symmetrical 9-nt deletions leave the sequences 5′-GGCGG-3′ (5′-As*; As* for acceptor stem remnant) and 5′-CCGCC-3′ (3′-As*). The 5′-As* sequence is the last 5-nt of what others describe as the D loop, but which we identify as an acceptor stem remnant (Pak et al., 2017; Root-Bernstein et al., 2016). The 3′-As* sequence represents the primordial 5-nt V loop sequence for type-I tRNAs.

The 17-nt D-loop microhelix is based on a UAGCC repeat, 5′-UAGCCUAGCCUGGCCUA-3′. The G for A substitution in the third UAGCC repeat allows intercalation of D loop G19 between T loop A60 and A61 in cloverleaf tRNA (Pak et al., 2017). The numbering of tRNAs in this paper follows our adjusted numbering system, based on a D loop that lacks deletions from tRNA^Pri (the primordial cloverleaf tRNA). Our numbering, therefore, may vary from what some readers might expect by +3 nt after the D loop (i.e., the anticodon wobble position is listed here as 37 rather than 34) (Fig. 4.2).

The structure of the anticodon loop and strong interactions of the D loop, T loop, and V loop make tRNA a relatively stiff and efficient adapter for translation. The anticodon (Ac loop) and T loop microhelices derive

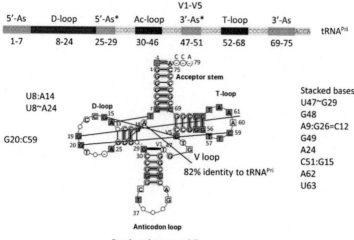

Figure 4.2 A typical *Pyrobaculum aerophilum* (archaea) tRNA has 82% identity with tRNA^Pri. *Red lines* indicate some interactions within the D loop, T loop, and V loop. The typical tRNA has almost two perfect UAGCC repeats (8—17) and identical Ac loop and T loop stems (CCGGG and CCCGG), demonstrating Ac loop and T loop homology. (For interpretation of the references to colour in this figure legend, the reader is referred to the web version of this article.)

from a stem-loop-stem sequence very similar to 5′-CCGGGUU-CAAAACCCGG-3′. The CCGGG and CCCGG complementary stems are strongly supported by sequence analysis (Pak et al., 2017; Root-Bernstein et al., 2016). For tRNA^Pri, there is slight sequence ambiguity within the 7-nt loops, which, significantly, form a U-turn after the second U (between loop positions 2 and 3). The U-turn within the 7-nt Ac loop is necessary to present a 3-nt anticodon to support a 3-nt genetic code (Pak et al., 2017). Without the 7-nt U-turn loop, a 3-nt genetic code would not be possible. In the anticodon loop, loop bases 3—7 stack within the loop as if in a helix, making the 7-nt U-turn Ac loop a compact loop to support a relatively stiff adapter. The T loop has the same 7-nt U-turn loop as the Ac loop, but the intercalation of D loop G19 between T loop A60 and A61 lifts A61 to fill the loop, flipping A62 and U63 out of the T loop (Pak et al., 2017). Interestingly, A62 and U63 participate in a stack of nucleotide bases that are part of the D loop-V loop-T loop interaction (extending to the "elbow").

The 3-minihelix model is supported by inspection of archaeal tRNAs from ancient species such as *Pyrococcus furiosis*, *Staphylothermus marinus*, and

Pyrobaculum aerophilum (Juhling et al., 2009; Pak et al., 2017). There is some controversy in the literature about whether archaea or bacteria are closer relatives to the last universal common ancestor (LUCA), but, in terms of translation systems and tRNA, ancient archaea are clearly most similar to LUCA (Pak, Du, et al., 2018). This can easily be shown by observing typical tRNA diagrams (Juhling et al., 2009). Very clearly, the CCGGG and CCCGG Ac loop and T loop stems are conserved, demonstrating that the Ac loop and the T loop are homologs. Because the Ac loop and T loop are homologs, no model based on only two minihelices can account for tRNA evolution (Pak et al., 2017). Models based on two minihelices require splitting the Ac stem-loop-stem in two to compare tRNA halves, which is inconsistent with Ac loop and T loop homology, and which is evident from inspection (Pak et al., 2017).

The GCG and CGC repeats make up acceptor stems and acceptor stem remnants. The UAGCC repeats in D loop sequences are also apparent. These patterns begin to degrade in bacterial tRNAs with evolution. Significantly, ancient archaeal tRNAs were generated from highly-ordered sequences, repeats, and inverted repeats (i.e., to form stem-loop-stems). Cloverleaf tRNA, therefore, evolved from an ordered and repetitive sequence, identified in some ancient archaea, to a more chaotic sequence in more derived archaea and bacteria.

Results

A model for evolution of Type-II tRNAs

Fig. 4.1 shows a model for the evolution of type-I and type-II tRNAs. The model for type-II tRNAs posits that the primordial length of the V loop expansion is 14 nt (7 nt (3′-As) + 7 nt (5′-As)). The model further posits the homology of V loops with acceptor stems and acceptor stem remnants. Because archaeal tRNAs are more similar to LUCA tRNAs than are bacterial tRNAs, initially, archaeal tRNAs were collected and compared. In archaea, with rare exceptions, only tRNA^{Leu} and tRNA^{Ser} are type-II tRNAs. We find that expanded and 5-nt V loops are misaligned in tRNAdb and gtRNA databases (Chan & Lowe, 2009, 2016; Juhling et al., 2009). In those databases, V loops were aligned to optimize sequence similarities, introducing inappropriate gaps, rather than, as we align them here, by evolutionary comparisons and secondary structures.

Archaeal tRNAs with expanded V loops

Because V loops are variable in length, they are numbered V1 to VN, in which N = length of the V loop. For archaeal $tRNA^{Leu}$, $N = 14$, typically, as expected from the model (Fig. 4.1). For archaeal $tRNA^{Ser}$, $N = 16$, typically. Analysis of tRNAomes (all of the tRNAs for an organism displayed as an evolutionary tree and rooted to $tRNA^{Pri}$) indicates that $tRNA^{Leu}$ ($N = 14$) evolves to $tRNA^{Ser}$ ($N = 16$), indicating that V loop expansions are derived from $N = 14$ (Fig. 4.1) (Pak, Du, et al., 2018). We posit that the initial length of an expanded V loop was $N = 14$ and that longer and shorter V loop expansions are generated by the insertion or deletion of bases most often located approximately to the middle of the V loop.

V loops in cloverleaf tRNA are under different selection pressures than acceptor stems. In Fig. 4.3, some of these interactions are highlighted. Fig. 4.3A shows a set of stacked bases stabilizing interactions of the D loop, V loop, and T loop. Fig. 4.3B—D shows some details of interactions. In archaeal $tRNA^{Leu}$ and $tRNA^{Ser}$, U V1 is selected to form a G29~U V1 wobble base pair, and C VN is selected to form a reverse Watson—Crick base pair (G15:C VN), termed the "Levitt" base pair (Fig. 4.3C) (Chawla et al., 2014; Oliva et al., 2007). In archaea, G15 is often modified to

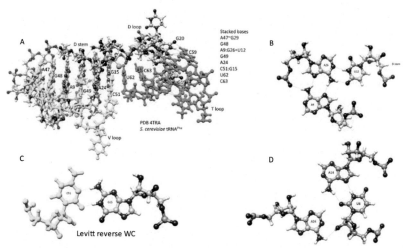

Figure 4.3 D loop-V loop-T loop interactions (the tRNA "elbow"). (A) Stacked bases. (B) Interaction of A9-U12-A26. (C) The Levitt base pair (G15:C51). (D) Interaction of U8-A14-A24. *Blue lines* indicate hydrogen bonds. The image is from PDB 4TRA. (For interpretation of the references to colour in this figure legend, the reader is referred to the web version of this article.)

archaeosine, which stabilizes the G15 (archaeosine):C VN interaction, particularly in the presence of Mg^{2+}. Typical secondary structures of expanded V loops are selected to be different for $tRNA^{Leu}$ and $tRNA^{Ser}$, so that aaRS enzymes make few errors charging $tRNA^{Leu}$, $tRNA^{Ser}$, and other tRNAs. Similarly, in type-I archaeal tRNAs, a G29~U V1 wobble pair and a G15:C V5 reverse Watson-Crick Levitt base pair are selected strongly.

Fig. 4.4 shows V loop expansions in archaea. As noted above, *Pyrococcus* is an ancient archaeal family with significant similarity to LUCA tRNAs (Pak, Du, et al., 2018). Three *Pyrococcus* species are compared for $tRNA^{Leu}$ (Fig. 4.4A) and $tRNA^{Ser}$ (Fig. 4.4B). Using typical tRNA diagrams, for $tRNA^{Leu}$, $N = 14$, typically. For $tRNA^{Ser}$, $N = 15$, typically. The G15:C VN Levitt base pair and the G29~U V1 wobble pair are evident. Secondary structures are sufficiently different for LeuRS and SerRS to discriminate $tRNA^{Leu}$ from $tRNA^{Ser}$.

For all archaea, results are very similar (Fig. 4.4C−F). For $tRNA^{Leu}$ (Fig. 4.4C), $N = 14$, typically. For $tRNA^{Ser}$ (Fig. 4.4D), $N = 16$, typically, indicating further V loop expansion through the archaeal domain compared to the most ancient archaea such as *Pyrococcus* (Fig. 4.4B). Histograms of V loop lengths for archaea are shown in Fig. 4.4E and F. The G15:C VN reverse Levitt base pair and the G29~U V1 wobble pair are evident (Fig. 4.4C and D). Secondary structures are distinct for $tRNA^{Leu}$ and $tRNA^{Ser}$ V loops, so LeuRS, SerRS and other aaRS enzymes can discriminate $tRNA^{Leu}$ and $tRNA^{Ser}$. V loop secondary structures for all archaea are very similar to those observed for *Pyrococcus* tRNAs (Fig. 4.4A and B). As predicted and expected, analysis of archaeal tRNAs with V loop expansions presents a very simple story of evolution that fits to the same model for the evolution of type-I tRNAs (Fig. 4.1).

Evolution of bacterial tRNAs with expanded V loops

Bacteria are expected to be more derived than archaea for tRNA evolution, and bacteria have additional type-II tRNAs that are absent in archaea (Fig. 4.5). In Fig. 4.5A and B, bacterial $tRNA^{Leu}$ and $tRNA^{Ser}$ are compared as typical tRNA diagrams (Juhling et al., 2009). For bacterial $tRNA^{Leu}$ (Fig. 4.5A), $N = 15$, typically, and for $tRNA^{Ser}$ (Fig. 4.5B), $N = 19$, typically, indicating that bacterial tRNAs are more derived from LUCA than archaeal tRNAs. In contrast to archaeal $tRNA^{Leu}$, in bacterial $tRNA^{Leu}$, an atypical A15: U VN Levitt base pair may be

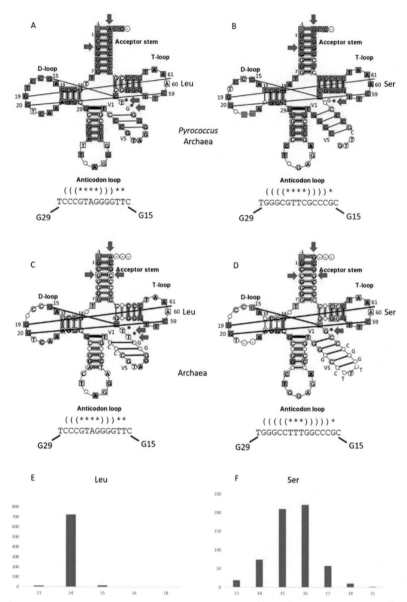

Figure 4.4 Typical type-II tRNAs in archaea. (A) tRNA^Leu in *Pyrococcus*. (B) tRNA^Ser in *Pyrococcus*. (C) tRNA^Leu in archaea. (D) tRNA^Ser in archaea. Some interactions within the D loop, V loop, and T loop are indicated with *red lines*. *Blue arrows* indicate determinants (or antideterminants) for discrimination of tRNAs by aaRS enzymes. *Red asterisks* indicate V loop bases, not in the V loop stem, that may allow discrimination of different V loops (i.e., by LeuRS, SerRS, and other aaRS enzymes). Note that *Pyrococcus* tRNA^Ser has two perfect UAGCC repeats in the D loop (8-UAGCCUAGCC-17). (E) Histogram of N for tRNA^Leu in archaea. (F) Histogram of N for tRNA^Ser in archaea. (For interpretation of the references to colour in this figure legend, the reader is referred to the web version of this article.)

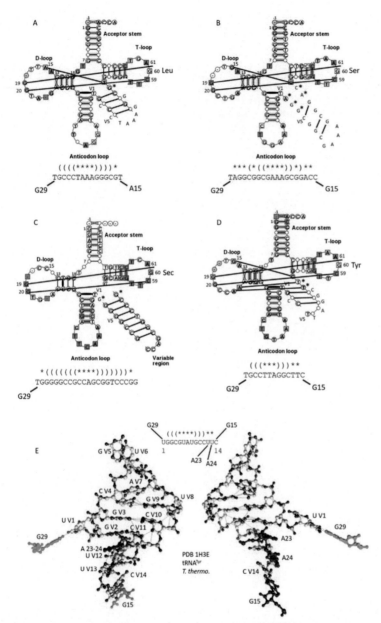

Figure 4.5 Typical type-II tRNAs in bacteria. (A) Bacterial tRNA^Leu. (B) Bacterial tRNA^Ser. (C) tRNA^Sec. (D) tRNA^Tyr. Some interactions within the D loop, V loop, and T loop are indicated with *red lines*. *Red asterisks* indicate V loop bases that are not part of the V loop stem and may allow discrimination of different V loops (i.e., by LeuRS, SerRS, and other aaRS enzymes). (E) The tRNA^Tyr V loop from *Thermus thermophilus* (two views). (For interpretation of the references to colour in this figure legend, the reader is referred to the web version of this article.)

indicated. In bacterial tRNA^{Ser}, a conserved G15:C VN Levitt base pair is typical. G29~U V1 wobble pairs are typical for both tRNA^{Leu} and tRNA^{Ser} in bacteria. V loop secondary structures are distinct for archaeal and bacterial tRNA^{Leu} and tRNA^{Ser}. The bacterial tRNA^{Ser} V loop is particularly floppy, with fewer stabilizing V loop stem base pairs than are present in archaeal tRNA^{Ser} (Fig. 4.4B and D). The tRNA^{Ser} V loop is a major positive determinant for recognition by SerRS (Perona & Gruic-Sovulj, 2014). Because archaeal and bacterial V loop expansions are distinct, it is likely that there is archaeal-bacterial speciation that limits tRNA sharing between domains, i.e., via horizontal gene transfer. Different modifications of tRNAs in the two domains must also suppress tRNA exchanges.

Some bacteria and a few species of archaea utilize tRNA^{Sec} (Sec for selenocysteine) (Fig. 4.5C). To charge tRNA^{Sec}, serine is first attached to tRNA^{Sec}, and is then modified to selenocysteine (Turanov et al., 2011). For tRNA^{Sec}, $N = 22$, typically, and V loop base pairing is extensive, in contrast to bacterial tRNA^{Ser}. The floppiness of the tRNA^{Ser} V loop, therefore, may, in part, help to discriminate tRNA^{Ser} and tRNA^{Sec} to support the tRNA^{Sec} modification pathway. tRNA^{Sec} appears to be derived from tRNA^{Ser}. Because tRNA^{Sec} can also be a type-I tRNA, it appears that the ability to process the V loop from type-II to type-I may have persisted in evolution (Chan & Lowe, 2009, 2016; Juhling et al., 2009). Human tRNA^{Sec} (PDB 3A3A) shows a slightly modified cloverleaf fold lacking a Levitt base pair (Itoh et al., 2009).

In bacteria, but not in archaea, tRNA^{Tyr} is a type-II tRNA (Fig. 4.5D). Bacterial tRNA^{Tyr}, therefore, may have been reassigned relative to tRNA^{Tyr} in archaea. From the analysis of tRNAomes (Pak, Du, et al., 2018), it appears that bacterial tRNA^{Tyr} may have arisen from a tRNA^{Ser}. For bacterial tRNA^{Tyr}, $N = 13$, typically. The G15:C VN reverse Watson-Crick Levitt base pair is typically present, along with the G29~U V1 wobble base pair. Bacterial tRNA^{Tyr} is discriminated from other bacterial type-II tRNAs by the shorter length of its V loop and the distributions of its V loop stem pairs. In Fig. 4.5E, the structure of the tRNA^{Tyr} 14-nt V loop from *Thermus thermophilus* is shown along with its connections from a cocrystal with TyrRS (PDB 1H3E) (Yaremchuk et al., 2002). The G15~C V14 and G29~U V1 connections and expected V loop secondary structures are evident. TyrRS reads V loop bases directly to charge tRNA^{Tyr}.

Statistical analyses

Expanded V loops are derived from acceptor stems

Our model for the evolution of type-II tRNA V loops posits that they arose from a sequence very close to CCGCCGCGCGGCGG, which is a primordial 3′-As ligated to a 5′-As Fig. 4.1. The prediction is tested by comparing the 3′-As to the first 7-nt of V loops and the 5′-As to the last 7-nt of V loops. This approach works for sequences of $N > 13$, so V loops with $N < 14$ were not considered. The first and the last nt were dropped from the comparisons because the first V loop nt (V1) interacts with position 29 (generally G29 ~ U V1), and the last V loop nt (VN) interacts with position 15 (generally G15 ~ C VN; the Levitt base pair). We are assuming that V loops longer than $N = 14$ have insertions near the center of the sequence.

We have used a random permutation test (Pak et al., 2017) to indicate the similarities of two long alignments (n sequences; n is large) of potentially homologous short and aligned sequences. In this comparison, collected alignments of length l ($l1 = l2$) and a number of sequences n ($n1$, $n2$) are aligned. The code utilized requires a comparison of $n1 = n2$ for the two aligned sequences, so, in cases in which $n2 > n1$, 50 random selections of $n2 = n1$ are selected, and the comparison is repeated. In each comparison, 1000 random permutations of $l2$, $n2 = n1$ are compared to $l1$, $n1$. P-values of 50 repetitions are averaged. If the P-value for the comparison is small (i.e., <0.05), this indicates similarity and possible homology. If the P-value for the comparison is larger (i.e., >0.05), this indicates that sequences are more dissimilar and possibly not homologous. The results are shown in Table 4.1.

Table 4.1 Homology of type-II V loops to acceptor stems (arch: archaea; bact: bacteria).

	P-values against archaeal acceptor stems					
V loop	Arch LEU	Arch SER	Bact LEU	Bact SER	Bact TYR	Bact SEC
Average	.001	.001	.020	.999	1.000	.001
	P-values against bacterial acceptor stems					
V loop	Arch LEU	Arch SER	Bact LEU	Bact SER	Bact TYR	Bact SEC
Average	.001	.013	1.000	.277	.020	.860

We conclude that expanded type-II V loops are homologous to acceptor stems (3′-As ligated to 5′-As) (Table 4.1) as predicted by the model for tRNA evolution (Fig. 4.1). Archaeal tRNA^{Leu} and tRNA^{Ser} V loops test as potentially homologous to both archaeal (P-values = .001 and .001) and bacterial (P-values = .001 and .013) acceptor stems. Because 1000 random permutations are compared, the lowest P-value that is possible is .001. This result appears to demonstrate the model shown in Fig. 4.1 for V loop expansions. The bacterial tRNA^{Leu} V loop also tests as potentially homologous to archaeal acceptor stems (P-value = .020). Bacterial tRNA^{Sec} (Sec for selenocysteine) tests as being likely homologous to archaeal acceptor stems (P-value = .001). Interestingly, bacterial tRNA^{Tyr} with an expanded V loop test as potentially homologous to bacterial acceptor stems (P-value = .020) but not to archaeal acceptor stems (P-value = 1). The bacterial tRNA^{Tyr} P-value = .020 result may be attributable to the convergent evolution of C-G rich sequences because bacterial tRNA^{Tyr} is probably derived from sequences similar to bacterial tRNA^{Leu} or bacterial tRNA^{Ser}. Also, similarly to bacterial tRNA^{Tyr}, bacterial tRNA^{Ser} scores as more similar to bacterial acceptor stems (P-value = .277) than archaeal acceptor stems (P-value = .999). As we have shown in other ways, archaeal tRNAs are significantly less radiated from a primordial tRNA than bacterial tRNAs (Pak et al., 2017; Pak, Du, et al., 2018), and this observation is documented from comparisons of type-II tRNA V loop expansions.

Kinship of expanded V loops

The relatedness of expanded V loops is described in Fig. 4.6 using comparisons of P-values. Once again, a P-value < .05 indicates the probable homology of compared alignments. Archaeal tRNA^{Leu} and tRNA^{Ser} are closely related in sequence (Pak, Du, et al., 2018), and all bacterial V loop expansions appear to relate closely to archaeal tRNA^{Leu} and tRNA^{Ser}. These results indicate that archaeal tRNAs are closer to LUCA tRNAs than bacterial tRNAs. In bacteria, tRNA^{Leu} and tRNA^{Ser} have diverged from one another, presumably to support aaRS discrimination in tRNA charging (Perona & Gruic-Sovulj, 2014). We posit that type-II bacterial tRNA^{Tyr} was derived from tRNA^{Ser} through a process of tRNA reassignment, which we have previously described for another tRNA (Pak, Du, et al., 2018). Essentially, a type-I tRNA^{Tyr} (as in archaea) was eliminated and replaced by a type-II tRNA^{Tyr} derived from a tRNA^{Ser}.

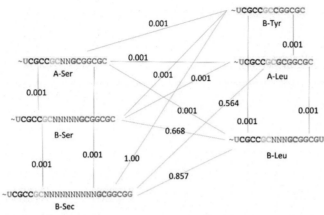

Figure 4.6 Relatedness of V loop expansions. *P*-values are shown. *Blue lines* indicate detected homology. *Orange lines* indicate V loop divergence (i.e., for aaRS discrimination). "A" indicates archaea. "B" indicates bacteria. (For interpretation of the references to colour in this figure legend, the reader is referred to the web version of this article.)

Discussion

The mechanism for type-II tRNA evolution is the same as the mechanism for type-I tRNA evolution but lacks a processing step (Fig. 4.1). The primordial length of the type-II V loop was 14-nt, as observed for archaeal tRNA^{Leu} (Fig. 4.4). We posit that tRNA^{Leu} evolved to tRNA^{Ser}, which has a longer V loop because of insertions near the middle of the loop (Pak, Du, et al., 2018). Both in length and sequence, archaeal tRNA^{Leu} and tRNA^{Ser} appear closer to LUCA tRNAs than bacterial tRNA^{Leu} and tRNA^{Ser}. Type-II bacterial tRNA^{Tyr} and tRNA^{Sec} appear to be derived from tRNA^{Ser}. Bacterial type-II tRNA^{Tyr} appears to have evolved by tRNA reassignment: elimination of a type-I tRNA^{Tyr} and reassignment of a tRNA^{Ser} to tRNA^{Tyr}.

The model for the evolution of type-I tRNAs was developed by inspection, similarly to the solving of a puzzle (Pak et al., 2017; Root-Bernstein et al., 2016). From inspection of typical tRNA diagrams, using ancient archaea, homology of the Ac and T stem-loop-stems was evident (Fig. 4.2). This accounts for 34-nt of tRNA, and the type-I tRNA core (lacking 3′-ACCA) is initially 75 nt. Considering acceptor stems, 48 nt of tRNA are described. This is more than half the tRNA. Eventually, the D loop microhelix was solved as a truncated UAGCC repeat of 17 nt (Fig. 4.2). Because three 17-nt microhelix sequences were present (D loop,

Ac loop and T loop), and the D loop and T loop were flanked on one side by a 7-nt acceptor stem, this indicated that the molecule was derived from three 31-nt minihelices. Therefore, the evolution of type-I tRNAs could be solved by two symmetrical 9-nt deletions, and the two remaining 5-nt sequences were the last 5 nt of the D loop (5'-As*) and the 5-nt V loop (3'-As*) (Fig. 4.1). The length of the primordial cloverleaf tRNA lacking 3'-ACCA was 75 nt (Figs. 4.1 and 4.2). This model for type-I tRNA evolution was strongly supported using a battery of statistical sequence comparisons (Pak et al., 2017). In the current paper, the model for type-I tRNA evolution was extended to describe the evolution of type-II tRNAs with V-loop expansions (Fig. 4.1). The model for type-II tRNAs is identical, except there is no processing of the 3' 14-nt ligated 3'- and 5'-acceptor stems (7-nt + 7-nt), which becomes the expanded V loop. Slight variations of the same model, therefore, completely and cleanly account for the evolution of type-I and type-II tRNAs.

Type-II tRNAs have been processed in evolution to type-I tRNAs, consistent with our model for the evolution of type-I tRNAs (Fig. 4.1). Notably, tRNA^{Ser} and tRNA^{Sec}, which primarily are type-II tRNAs, include type-I tRNAs (Juhling et al., 2009). The model in Fig. 4.1 indicates that the first cloverleaf tRNAs may have included 93, 84 and 75-nt core sequences. Most tRNAs are of the type-I variety, but tRNA^{Leu}, tRNA^{Ser}, tRNA^{Sec,} and bacterial tRNA^{Tyr} remained mostly type II. Because of the apparent appearance of type-I tRNAs from type-II tRNAs in modern lineages, it is possible that type-II tRNAs can still be processed under some conditions to type-I tRNAs. Perhaps, these 93, 84, and 75-nt species of tRNA cloverleafs coexisted, until the selection of primarily type-I tRNAs and a smaller collection of type-II tRNAs. It appears that V loop lengths and sequences were mostly selected to optimize discrimination by ∼20 aaRS enzymes (Perona & Gruic-Sovulj, 2014).

Comparison of tRNA evolution models

Competing models for tRNA evolution have been advanced (Branciamore & Di Giulio, 2011; Bernhardt, 2016; Di Giulio, 2009; Nagaswamy & Fox, 2003; Tamura, 2015; Widmann et al., 2005). Some of these models indicate that two minihelices might be ligated to form a primordial tRNA. Our model, by contrast, requires the ligation of three 31-nt minihelices representing two different 17-nt microhelix core sequences (1:2; the D loop and the homologous Ac and T stem-loop-stems) (Fig. 4.1). We show clearly that the Ac loop and T loop are homologs (Fig. 4.2) (Pak et al., 2017). In a two minihelix model, however, the Ac and the T stem-loop-stem cannot be homologous, because the Ac loop must be bisected to make the comparison, spoiling the alignment. Rather, a two minihelix model predicts that the D loop and the T loop should be similar in sequence, which they clearly are not. As we show here, and as we have shown previously, the D-loop microhelix is based on a UAGCC repeat, which cannot be similar in sequence to a CCGGGUU-CAAAUCCCGG T stem-loop-stem (Pak et al., 2017). In Fig. 4.4B, we show two perfect UAGCC repeats in the D loop, indicating the UAGCC repeat. Another criticism of the two minihelix models is that they appear to require unlikely sequence and structural convergence of the 7-nt U-turn Ac and T loops. If the homology of the Ac and T stem-loop-stems is accepted (Fig. 4.2), only the three minihelix model makes sense. One proposed two minihelix model is based too heavily on the analysis of tRNA introns in the Ac loop of one archaeal species (Di Giulio, 2009). Introns are found in many sites of archaeal tRNAs, not just in the Ac loop (Sugahara et al., 2009). Our three minihelix model is strongly supported by the identification of internal D loop (5′-As*) and V loop (type I: 3′-As*; type II: 3′-As ligated to a 5′-As) homologies to acceptor stems. Our model for type-II tRNAs strongly supports the model we previously proposed for processing a ligated 3′-As and 5′-As (14 nt) by deletion of 9 nt to yield a 3′-As* type-I V loop (Pak et al., 2017), because we identify the previously predicted intermediate in processing to a type-I tRNA as existing in type-II tRNA. Put more simply, type-II tRNA is the predicted intermediate in the processing of type-I tRNA (Fig. 4.1) (Pak et al., 2017). Our model makes strong sequence predictions, which are all justified by statistical tests (Table 4.1) (Pak et al., 2017). So far as we can judge, two minihelix models do not make strong sequence predictions that can be justified by any analysis we can apply.

Evolution of the genetic code

Because tRNA evolution is such a simple story, the evolution of the genetic code and translation systems becomes simpler to understand (Pak, Du, et al., 2018; Pak, Kim, & Burton, 2018). Significantly, the tRNA-centric view provides a simplified understanding of genetic code evolution. As viewed from the perspective of mRNA, in which all 64 codons are used, $>10^{84}$ genetic codes and up to 63 encoded amino acids might be possible (José et al., 2009). Viewed from the perspective of tRNA, however, the genetic code is half the size: a 32-letter code in tRNA versus a 64-letter code in mRNA (Pak, Du, et al., 2018; Pak, Kim, & Burton, 2018). The reason the code in tRNA is smaller than it is in mRNA is that ambiguity in reading the wobble position of tRNA limits the size of the code. Essentially, because codon-anticodon contacts are not fully proofread for the wobble position base on the ribosome, there is only purine versus pyrimidine discrimination at the wobble position, not a single base (A,G,C,U) recognition. The single exception is tRNA^{Ile} (UAU) versus tRNA^{Met} (CAU), which is supported by extensive modifications to tRNA^{Met} (CAU) (Agris et al., 2007, 2018; Väre et al., 2017). Furthermore, for the most part, tRNA^{Ile} (UAU) is only utilized in eukaryotes and not in prokaryotes. The maximum complexity of the genetic code in tRNA, therefore, is $4 \times 4 \times 2$, instead of $4 \times 4 \times 4$ in mRNA. The standard code, therefore, evolved to encode 20 amino acids rather than a larger number. There are additional dimensions to this story described in other works (Pak, Du, et al., 2018; Pak, Kim, & Burton, 2018).

Evolution of tRNA sequence proceeded from order to chaos

Archaeal tRNAs are better preserved from LUCA than bacterial tRNAs, and archaeal tRNAs are more highly-ordered in their sequence (Pak et al., 2017; Root-Bernstein et al., 2016). Ancient archaea such as *Pyrococcus*, *Pyrobaculum*, and *Staphylothermus* have tRNAomes (i.e., typical tRNA diagrams) that are more similar to a LUCA tRNAome than more derived species (Pak, Du, et al., 2018). These tRNAomes are more ordered in sequence because tRNA evolved from repeating sequences (Fig. 4.1). The ancient world of ~4 billion years ago, therefore, in some cases, evolved biological complexity from the ordered sequence, in the form of repeats and inverted repeats. From the analysis of tRNA evolution, therefore, the assumption that biological complexity was generated only from random polymer sequences is incorrect. Processes such as replication slippage and abortive initiation generated repeats and/or short RNA fragments that could be attached by ligation. Evolution from ordered repeats to chaos can clearly be seen in tRNA evolution, by inspection of typical tRNA diagrams for ancient archaea compared to more derived bacteria. Furthermore, mechanisms probably existed in the ancient world to measure the lengths of sequences, because repeats were clipped into functional units of 5 nt (Ac and T loop stems), 7 nt (i.e., acceptor stems, Ac and T loops), and 17 nt (i.e., D loop, Ac loop, and T loop microhelices). Of course, these length selections may represent selections for evolving biological functions.

The inanimate to animate transition

The central advance in biological intellectual property in the evolution of life on earth was cloverleaf tRNA, the adapter that permits biological coding, and around which coding functions evolved (Pak, Du, et al., 2018; Pak, Kim, & Burton, 2018). From conserved tRNA sequences, the pathway of tRNA evolution has been determined (Fig. 4.1). This is a story of building biological complexity from ordered repeats and snap-back stem-loop-stems. So, life on earth was snapped together (ligated) similarly to the children's game of LEGO (trademark) (Fig. 4.1). The inanimate to animate transition is described as a simple model in Fig. 4.7, tracking the evolution of microhelices → minihelices → cloverleaf tRNA → translation systems → cellular life. Life, therefore, evolved from a primitive inanimate polymer world that includes short sequences (i.e., ACCA; abortive initiation), repeats (i.e., GCG, CGC, and UAGCC repeats; replication slippage), and inverted repeats (Ac loop and T loop microhelices; stem-loop-stems, which can attach to form replication primers). Polymers are generated via dehydration reactions, so cycles of hydration and dehydration may be sufficient to describe the generation of the first biopolymers (Bowman et al., 2015; Rodriguez-Garcia et al., 2015).

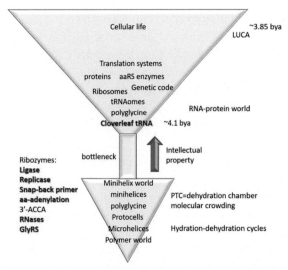

Figure 4.7 Evolution of translation systems. A simple model for the inanimate → animate transition based on the evolution of cloverleaf tRNA from sequence repeats and 31-nt minihelices including stem-loop-stems. See the text for details.

From a strange polymer world that includes 17-nt microhelices (i.e., D loop, Ac loop, and T loop microhelices), a small collection of ribozymes appears necessary and possibly sufficient to generate cloverleaf tRNA and translation systems. These ribozyme activities have been largely reinvented in vitro, and some of these ribozymes can be quite small, indicating that their evolution via simple nonbiotic processes might be possible (Chumachenko et al., 2009; Illangasekare & Yarus, 2012; Kennedy et al., 2010; Turk et al., 2010, 2011; Yarus, 2011, 2015). Hydration-dehydration cycles drive polymerization reactions and concentrate cofactors such as Mg^{2+}.

As we have previously proposed, microhelices, minihelices, and cloverleaf tRNA may have been initially evolved to synthesize polyglycine, which proposed to have been used to stabilize protocells, as in bacterial cell walls (Pak, Du, et al., 2018; Pak, Kim, & Burton, 2018). We note that polyglycine requires a membrane-anchored carbohydrate to form cell wall-like cross-links, so polyglycine by itself is not sufficient for the stabilization of protocells. One evidence for the polyglycine model is that $tRNA^{Pri}$ (the primordial tRNA cloverleaf; Fig. 4.2) is closest in sequence to $tRNA^{Gly}$ in ancient archaea (Pak, Du, et al., 2018). We have described a Darwinian pathway to evolve the 21-letter genetic code (20 amino acids with stops) from a one-letter code synthesizing polyglycine (using any mRNA sequence). Once cloverleaf tRNA evolves, therefore, the evolution of the genetic code appears to be assured. We conclude that the evolution of the tRNA cloverleaf is the major advance in the evolution of biological intellectual property that led to the evolution of the genetic code, translation systems, and cellular life. Once cloverleaf tRNA evolves, Darwinian selection drives the evolution of dependent processes (translation, genetic code, aaRS enzymes), so the evolution of cloverleaf tRNA is the core advance.

As described previously, the evolution of the ribosome requires initially a scaffold on which to mount and move mRNA (a decoding center) and perhaps a mobile peptidyl transferase center (Root-Bernstein et al., 2016). The peptidyl transferase center can be viewed as dehydration and molecular crowding chamber to drive the polymerization of polypeptide chains (Bowman et al., 2015). If the peptidyl transferase center is a "ribozyme", it is not a good one. Every other function of the ribosome is a refinement or add-on: i.e., translocation, proofreading, initiation, and termination functions. Unlike tRNA (Fig. 4.1), the evolutionary source of rRNA is more obscure. Without breakthrough success, our laboratory has attempted to solve this problem in collaboration with Robert Root-Bernstein (MSU). As a cautionary tale, rRNA sequences appear cloverleaf tRNA-like, and ancient archaeal rRNAs appear more tRNA-like than more derived species, indicating that cloverleaf tRNA was one of the building blocks of rRNA sequences. Back translating rRNA sequences (BlastX; NCBI) gives apparent open reading frames, but none of these can clearly be traced to an independent functional gene or close homolog. Generally, in archaeal and bacterial genomes, these long open reading frames are only found in rRNA sequences. Open reading frames with an apparent annotation, i.e., a "cell wall hydrolase" embedded in the peptidyl transferase center of 23S rRNA as a reverse orientation gene, cannot be confirmed to encode a cell wall hydrolase with a known function that exists anywhere as a stand-alone gene or that has notable homology to other cell wall hydrolases or structures. We concluded that this sequence does not encode a cell wall hydrolase, and was wrongly annotated but propagated through the reported annotations of many genomes leading to the potential confusion, including our own. So, although a fairly simple model for ribosome evolution can be proposed, to our knowledge, in contrast to tRNA (Fig. 4.1), the detailed evolution of rRNA sequences and the evolution of the ribosome remain largely unsolved but compelling mysteries.

Materials and methods

Methods and databases have been described previously (Juhling et al., 2009; Pak et al., 2017; Pak, Du, et al., 2018; Pak, Kim, & Burton, 2018; Root-Bernstein et al., 2016). The statistical permutation test is useful for comparing two short, aligned sequences with many examples of each for possible homology (Pak et al., 2017). Most tRNA sequences were taken from gtRNAdb (Chan & Lowe, 2009, 2016). For convenience, typical tRNA diagrams were taken from the older tRNAdb (Juhling et al., 2009). V loop sequences were added to typical tRNA diagrams by hand. Structures were analyzed using UCSF Chimera (Chumachenko et al., 2009; Yang et al., 2012).

Conclusions

We conclude that the evolution of cloverleaf tRNA, as described here (Fig. 4.1), drove the inanimate to animate transition in the evolution of life on earth (Fig. 4.7). The inanimate world is characterized by a strange polymer world with unexpected order that includes sequence repeats (i.e.,

GCG, CGC and UAGCC repeats), short abundant potentially functional sequences (i.e., ACCA), and snap-back primers (stem-loop-stems, inverted repeats; microhelices, minihelices). Ordered polymers led to the evolution of microhelices, minihelices, and cloverleaf tRNA (Fig. 4.1). Once cloverleaf tRNA evolved, the evolution of the genetic code, the ribosome, translation systems, and cellular life were assured. Strangely, very few uncertainties remain in this amazing story recorded and told in genetic sequences that evolved about 4 billion years ago.

Author contributions

Conceptualization, B.K., K.O., and Z.F.B.; Data curation, Y.K., and Z.F.B.; Formal analysis, Y.K., and Z.F.B.; Investigation, Y.K., and Z.F.B.; Methodology, Y.K., and Z.F.B.; Project administration, B.K., K.O. and Z.F.B.; Software, Y.K., and K.O.; Supervision, K.O.; Validation, Y.K.; Visualization, Z.F.B.; Writing—original draft, Z.F.B.; Writing—review and editing, Z.F.B. The auhtors declare no conflict of interest and no additional funding in the writing of this chapter.

References

Agris, P. F., Eruysal, E. R., Narendran, A., Väre, V. Y. P., Vangaveti, S., & Ranganathan, S. V. (2018). Celebrating wobble decoding: Half a century and still much is new. *RNA Biology, 15*(4—5), 537—553. https://doi.org/10.1080/15476286.2017.1356562

Agris, P. F., Vendeix, F. A. P., & Graham, W. D. (2007). tRNA's wobble decoding of the genome: 40 Years of modification. *Journal of Molecular Biology, 366*(1), 1—13. https://doi.org/10.1016/j.jmb.2006.11.046

Bernhardt, H. S. (2016). Clues to tRNA evolution from the distribution of class II tRNAs and serine codons in the genetic code. *Life, 6*(1). https://doi.org/10.3390/life6010010

Bowman, J. C., Hud, N. V., & Williams, L. D. (2015). The ribosome challenge to the RNA world. *Journal of Molecular Evolution, 80*(3—4), 143—161. https://doi.org/10.1007/s00239-015-9669-9

Branciamore, S., & Di Giulio, M. (2011). The presence in tRNA molecule sequences of the double hairpin, an evolutionary stage through which the origin of this molecule is thought to have passed. *Journal of Molecular Evolution, 72*(4), 352—363. https://doi.org/10.1007/s00239-011-9440-9

Chan, P. P., & Lowe, T. M. (2009). GtRNAdb: A database of transfer RNA genes detected in genomic sequence. *Nucleic Acids Research, 37*(1), D93—D97. https://doi.org/10.1093/nar/gkn787

Chan, P. P., & Lowe, T. M. (2016). GtRNAdb 2.0: An expanded database of transfer RNA genes identified in complete and draft genomes. *Nucleic Acids Research, 44*(1), D184—D189. https://doi.org/10.1093/nar/gkv1309

Chawla, M., Abdel-Azeim, S., Oliva, R., & Cavallo, L. (2014). Higher order structural effects stabilizing the reverse watson-crick guanine-cytosine base pair in functional RNAs. *Nucleic Acids Research, 42*(2), 714—726. https://doi.org/10.1093/nar/gkt800

Chumachenko, N. V., Novikov, Y., & Yarus, M. (2009). Rapid and simple ribozymic aminoacylation using three conserved nucleotides. *Journal of the American Chemical Society, 131*(14), 5257—5263. https://doi.org/10.1021/ja809419f

Di Giulio, M. (2009). A comparison among the models proposed to explain the origin of the tRNA molecule: A synthesis. *Journal of Molecular Evolution, 69*(1), 1—9. https://doi.org/10.1007/s00239-009-9248-z

Illangasekare, M., & Yarus, M. (2012). Small aminoacyl transfer centers at GU within a larger RNA. *RNA Biology, 9*(1), 59—66. https://doi.org/10.4161/rna.9.1.18039

Itoh, Y., Chiba, S., Sekine, S. I., & Yokoyama, S. (2009). Crystal structure of human selenocysteine tRNA. *Nucleic Acids Research, 37*(18), 6259—6268. https://doi.org/10.1093/nar/gkp648

José, M. V., Govezensky, T., García, J. A., & Bobadilla, J. R. (2009). On the evolution of the standard genetic code: Vestiges of critical scale invariance from the RNA world in current prokaryote genomes. *PLoS ONE, 4*(2). https://doi.org/10.1371/journal.pone.0004340

Juhling, F., Morl, M., Hartmann, R. K., Sprinzl, M., Stadler, P. F., & Putz, J. (2009). tRNAdb 2009: Compilation of tRNA sequences and tRNA genes. *Nucleic Acids Research, 37*(Database), D159—D162. https://doi.org/10.1093/nar/gkn772

Kennedy, R., Lladser, M. E., Wu, Z., Zhang, C., Yarus, M., De Sterck, H., & Knight, R. (2010). Natural and artificial RNAs occupy the same restricted region of sequence space. *RNA, 16*(2), 280—289. https://doi.org/10.1261/rna.1923210

Nagaswamy, U., & Fox, G. E. (2003). RNA ligation and the origin of tRNA. *Origins of Life and Evolution of the Biosphere, 33*(2), 199—209. https://doi.org/10.1023/A:1024658727570

Oliva, R., Tramontano, A., & Cavallo, L. (2007). Mg2+ binding and archaeosine modification stabilize the G15-C48 Levitt base pair in tRNAs. *RNA, 13*(9), 1427—1436. https://doi.org/10.1261/rna.574407

Pak, D., Du, N., Kim, Y., Sun, Y., & Burton, Z. F. (2018). Rooted tRNAomes and evolution of the genetic code. *Transcription, 9*(3), 137—151. https://doi.org/10.1080/21541264.2018.1429837

Pak, D., Kim, Y., & Burton, Z. F. (2018). Aminoacyl-tRNA synthetase evolution and sectoring of the genetic code. *Transcription, 9*(4), 205—224. https://doi.org/10.1080/21541264.2018.1467718

Pak, D., Root-Bernstein, R., & Burton, Z. F. (2017). tRNA structure and evolution and standardization to the three nucleotide genetic code. *Transcription, 8*(4), 205—219. https://doi.org/10.1080/21541264.2017.1318811

Perona, J. J., & Gruic-Sovulj, I. (2014). Synthetic and editing mechanisms of aminoacyl-tRNA synthetases. *Topics in Current Chemistry, 344*, 1—41. https://doi.org/10.1007/128_2013_456

Rodriguez-Garcia, M., Surman, A. J., Cooper, G. J. T., Suárez-Marina, I., Hosni, Z., Lee, M. P., & Cronin, L. (2015). Formation of oligopeptides in high yield under simple programmable conditions. *Nature Communications, 6*. https://doi.org/10.1038/ncomms9385

Root-Bernstein, R., Kim, Y., Sanjay, A., & Burton, Z. F. (2016). tRNA evolution from the proto-tRNA minihelix world. *Transcription, 7*(5), 153—163. https://doi.org/10.1080/21541264.2016.1235527

Sugahara, J., Fujishima, K., Morita, K., Tomita, M., & Kanai, A. (2009). Disrupted tRNA gene diversity and possible evolutionary scenarios. *Journal of Molecular Evolution, 69*(5), 497—504. https://doi.org/10.1007/s00239-009-9294-6

Tamura, K. (2015). Origins and early evolution of the tRNA molecule. *Life, 5*(4), 1687—1699. https://doi.org/10.3390/life5041687

Turanov, A. A., Xu, X. M., Carlson, B. A., Yoo, M. H., Gladyshev, V. N., & Hatfield, D. L. (2011). Biosynthesis of selenocysteine, the 21st amino acid in the genetic code, and a novel pathway for cysteine biosynthesis. *Advances in Nutrition, 2*(2), 122–128. https://doi.org/10.3945/an.110.000265

Turk, R. M., Chumachenko, N. V., & Yarus, M. (2010). Multiple translational products from a five-nucleotide ribozyme. *Proceedings of the National Academy of Sciences of the United States of America, 107*(10), 4585–4589. https://doi.org/10.1073/pnas.0912895107

Turk, R. M., Illangasekare, M., & Yarus, M. (2011). Catalyzed and spontaneous reactions on ribozyme ribose. *Journal of the American Chemical Society, 133*(15), 6044–6050. https://doi.org/10.1021/ja200275h

Väre, V. Y. P., Eruysal, E. R., Narendran, A., Sarachan, K. L., & Agris, P. F. (2017). Chemical and conformational diversity of modified nucleosides affects tRNA structure and function. *Biomolecules, 7*(1). https://doi.org/10.3390/biom7010029

Widmann, J., Di Giulio, M., Yarus, M., & Knight, R. (2005). tRNA creation by hairpin duplication. *Journal of Molecular Evolution, 61*(4), 524–530. https://doi.org/10.1007/s00239-004-0315-1

Yang, Z., Lasker, K., Schneidman-Duhovny, D., Webb, B., Huang, C. C., Pettersen, E. F., Goddard, T. D., Meng, E. C., Sali, A., & Ferrin, T. E. (2012). UCSF Chimera, MODELLER, and IMP: An integrated modeling system. *Journal of Structural Biology, 179*(3), 269–278. https://doi.org/10.1016/j.jsb.2011.09.006

Yaremchuk, A., Kriklivyi, I., Tukalo, M., & Cusack, S. (2002). Class I tyrosyl-tRNA synthetase has a class II mode of cognate tRNA recognition. *EMBO Journal, 21*(14), 3829–3840. https://doi.org/10.1093/emboj/cdf373

Yarus, M. (2011). The meaning of a minuscule ribozyme. *Philosophical Transactions of the Royal Society B: Biological Sciences, 366*(1580), 2902–2909. https://doi.org/10.1098/rstb.2011.0139

Yarus, M. (2015). Ahead and behind: A small, small RNA world. *RNA, 21*(4), 769–770. https://doi.org/10.1261/rna.051086.115

CHAPTER 5

An additional, complementary mechanism of action for folic acid in the treatment of megaloblastic anemia

Bruce K. Kowiatek
Allied Health Sciences, Blue Ridge Community and Technical College, Martinsburg, WV, United States

Abbreviations

DHFR	Dihydrofolate Reductase;
DNA	deoxyribonucleic acid;
NADH and NAD+	Nicotinamide Adenine Dinucleotide (reduced and oxidized forms, respectively);
THF	Tetrahydrofolate

Introduction

Dietary deficiency of the vitamin enzyme cofactor folic acid (Fig. 5.1) has serious effects on human health, including chromosome breaks, birth defects such as neural tube defects, increased risk of colon cancer, brain dysfunction, and heart disease (Sousa et al., 2007). Its deficiency is also the cause of megaloblastic anemia, or megaloblastosis, an anemic blood disorder characterized by larger-than-normal red blood cells (RBCs), or megaloblasts (National Institutes of Health. MedlinePlus: Megaloblastic Anemia, 2012). Two hallmarks of megaloblasts are, first, elevated levels of the cell membrane phospholipid phosphatidylcholine (PC) (Wallentin et al., 1977) (Fig. 5.2) and, second, increased amounts of the nucleotide deoxyuridine

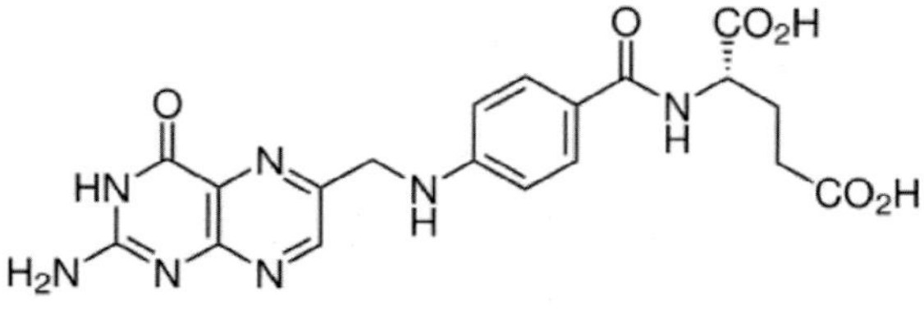

Figure 5.1 Folic acid.

The Makings of a Clinical Protocol
ISBN 978-0-323-95749-6
https://doi.org/10.1016/B978-0-323-95749-6.00006-5

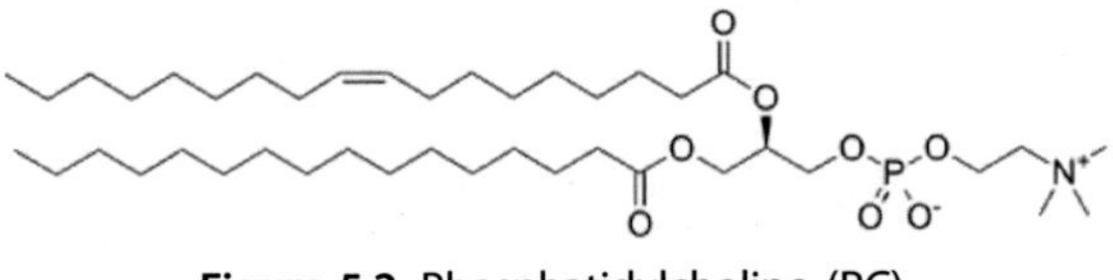

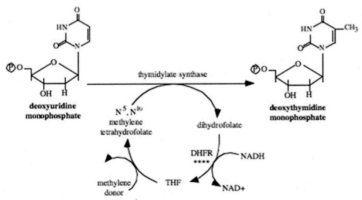

Figure 5.2 Phosphatidylcholine (PC).

5?-monophosphate (dUMP), which is otherwise usually methylated by the folic acid derivatives 5,10-methylenetetrahydrofolate and 7,8-dihydrofolate, via the thymidylate synthase enzyme pathway, which also includes the enzyme dihydrofolate reductase (DHFR), the coenzyme nicotinamide adenine dinucleotide in both its reduced and oxidized forms (NADH and NAD+, respectively), and the folic acid derivative tetrahydrofolate (THF), to deoxythymidine 5?-monophosphate (dTMP), or thymidylate, as part of normal deoxyribonucleic acid (DNA) synthesis (Horton et al., 1996) (Fig. 5.3).

Administration of supplemental folic acid addresses the impaired DNA synthesis causing megaloblastosis; however, despite the possibility of high doses of folic acid–5–15 mg daily for up to four months-the extremely rapid initial onset of action-30 to 60 min-when administered orally (IPCS Inchem. Folic Acid, 1991) is not in keeping with this accepted mechanism, which can take up to nearly 22 h, even under enzymatic control (Black & Abramson, 2003). This would suggest an additional, complementary nonenzymatic mechanism of nucleotide methylation at work. Such nonenzymatic methylation of nucleotides has been observed previously

Figure 5.3 Thymidylate synthase pathway. Abbreviations: *DHFR*, dihydrofolate reductase; *NADH and NAD+*, nicotinamide adenine dinucleotide (reduced and oxidized forms, respectively); *THF*, tetrahydrofolate.

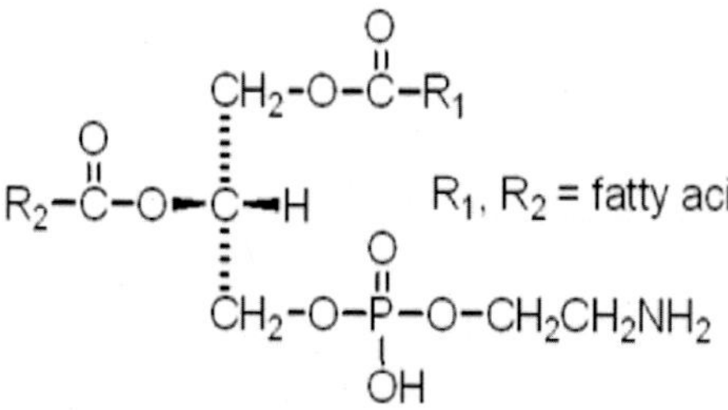

$$CH_2-O-\overset{\overset{\displaystyle O}{\|}}{C}-R_1$$
$$R_2-\overset{\overset{\displaystyle O}{\|}}{C}-O-\overset{|}{C}-H \qquad R_1, R_2 = \text{fatty acid residues}$$
$$CH_2-O-\overset{\overset{\displaystyle O}{\|}}{P}-O-CH_2CH_2NH_2$$
$$\overset{|}{OH}$$

Figure 5.4 Phosphatidylethanolamine (PE).

in vitro with the metabolite cofactor S–adenosylmethionine (SAM–e) (Barrows & Magee, 1982); it is, therefore, proposed here, with in vitro evidence put forth, that rapid, nonenzymatic methylation by folic acid, via 5,10–methylenetetrahydofolate and 7,8–dihydrofolate intermediates, of dUMP to form thymidylate using PC as a methyl donor, leaving the demethylated cell membrane phospholipid phosphatidylethanolamine (PE) (Fig. 5.4), is a viable additional and complementary mechanism in helping reduce megalobastic RBCs to normal size and function.

Experimental

All research was conducted at The Wellness Pharmacy in Winchester, VA, the USA from February 2018 through March 2018. All sterile manipulations were performed using an aseptic technique in a NUAIRE Biological Safety Cabinet Class II Type A/B3 laminar flow hood. All pH measurements were made using a Horiba TwinpH waterproof B-213 Compact pH Meter. All Gas Chromatography/Mass Spectrometry (GC/MS) data were obtained using a Hewlett Packard 5890 Series II Gas Chromatograph. Compounds were identified as the trimethylsilyl (TMS) derivatives by using the agreement of the retention times with those of standards as the criterion for identification and the following ions for determination: m/z 321.0488 for thymidylate (MassBank Record: PR100611. Thymidine-5?-Monophosphate, 2011), and m/z 255.13 for PE (MassBank Record: UT001356. Phosphatidylethanolamine, 2011). All in-laboratory photography was obtained using a Casio EX-Z57 digital camera. All Pyrex glassware was sterilized at 130°C for 1 h (Black & Microbiology, 1993) via autoclave using a Quincy Lab Inc. Model 30 GC Lab Oven. All chemical supplies were purchased from the Professional Compounding Centers of America (PCCA) in Houston, TX, USA.

All measurements of chemicals were standardized to 0.1 Molarity (M) ±5% using an Ohaus Analytical Plus electronic balance accurate to within ±0.0001 g. Final test and control samples were obtained via filtration through a sterile 0.2 μm EPS, Inc. Medi–Dose Group Disposable Disc Filter Unit and corresponding sterile Monoject syringe. Three trials per step were performed and recorded and the data presented here represent the average of that total data. All data collected fell within a statistically acceptable ±5% ($P = .05$) internal margin of variance (Bolton, 1997) with no outliers. As a control, 25 mL (mL) of sterile 0.9% sodium chloride in water (normal saline (NS)), pH 7.4, was heated to 37°C and maintained, with 0.45 g of PC (95% ± 5%) added and allowed to melt uniformly throughout (Fig. 5.5). To this was added 0.015 g of folic acid (Fig. 5.6) and the pH adjusted first downward to two via titration with 0.1 M hydrochloric acid (HCl), then upward to 10 via titration with 0.1 M sodium hydroxide (NaOH) solution. The pH was then adjusted back to 7.4 via HCl.

As a test, to a second, similarly prepared mixture at pH 7.4 and 37°C, 0.45 g of highly solubilized dUMP (95% ± 5%) was added (Fig. 5.7), subsequent pH measured, and its contents compared to the first mixture (Fig. 5.8). Two samples were procured from the second mixture for GC/MS analysis, one to test for the presence of thymidylate and one to test for the presence of PE.

Figure 5.5 Control mixture of 25 mL NS, pH 7.4, at 37°C with 0.45 g PC (95% ± 5%) melted uniformly throughout.

Figure 5.6 Control mixture from Fig. 5.5 with 0.15 g folic acid added.

Figure 5.7 Test mixture, pH 6.2, at 37°C with 0.45 g dUMP (95% ± 5%) added.

Figure 5.8 Comparison of test (*left*) and control (*right*) mixtures.

Results and discussion

At physiologic salinity, pH, and temperature (Seely et al., 2007), no pre-cipitation of any of the components of the control mixture was observed; furthermore, no precipitation was observed in this mixture as a function of pH within the stability range of a therapeutically relevant quantity of folic acid (De Brouwer et al., 2007) and PC (Ho et al., 1987). Precipitation, however, was observed in the test mixture, beginning immediately upon the addition of dUMP and reaching completion within 10 min after its addition. The pH of the test mixture was 6.2. Such precipitation ruled out as a function of pH, indicated an increase in hydrophobicity, as would be the case if hydrophilic dUMP were being converted into more hydrophobic thymidylate (Anan-dagopu et al., 2008). Also, the change in the appearance of the mixture from opaque to translucent indicated the conversion of PC into PE (Ho et al., 1987). GC/MS analysis confirmed the presence of both thymidylate (Fig. 5.9) (95% ± 5%) and PE (Fig. 5.10) (95% ± 5%).

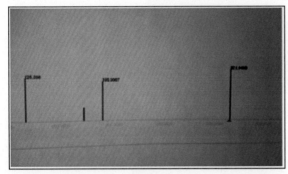

Figure 5.9 GC/MS of TMS-thymidylate (95% ± 5%) was detected in the test mixture. Note peak at m/z 321.0488.

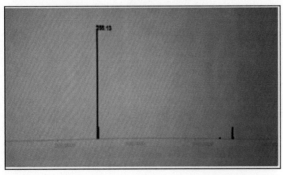

Figure 5.10 GC/MS of TMS-PE (95% ± 5%) was detected in the test mixture. Note peak at m/z 255.13.

The mechanism at work here appears to be extremely rapid and nonenzymatic in nature. It most likely involves successive demethylation of PC by folic acid to form, first, a 5,10-methylenetetrahydrofolate intermediate and PE, with the former then donating its methylene group to methylate dUMP into thymidylate, leaving a second, 7,8-dihydrofolate intermediate, which then quickly oxidizes back to folic acid. Such a supposition is supported by observations of these two intermediates' formation by transient, acidic changes in pH by De Brouwer et al. (2007). This transient nature, however, along with the extremely rapid rate of reaction observed in this experiment, precludes their detection here, although its inferred presence seems to best account for the observed results.

Such a rapid, nonenzymatic mechanism would also account for the quick onset of action of folic acid therapy in the treatment of megaloblastic anemia, acting in an additional and complementary fashion to that in already elucidated enzymatic mechanisms. It also establishes folic acid, alongside SAM-e, as a powerful agent involved in the nonenzymatic methylation of endogenous nucleotides.

Acknowledgments

I thank the staff of The Wellness Pharmacy for the generous use of their facility and equipment. I also especially thank my loving family for their generous gift of time in the performance of these experiments and the writing of this article. This article is dedicated to the memory of Raymond Burnell Knepp (1942–2009), founder of The Wellness Pharmacy.

References

Anandagopu, P., Suhanya, S., Jayaraj, V., & Rajasekaran, E. (2008). Role of thymine in protein coding frames of mRNA sequences. *Bioinformation, 2*(7), 304–307. https://doi.org/10.6026/97320630002304

Barrows, L. R., & Magee, P. N. (1982). Nonenzymatic methylation of DNA by s-adenosylmethionine in vitro. *Carcinogenesis, 3*(3), 349–351. https://doi.org/10.1093/carcin/3.3.349

Black, J. G. (1993). Emergence of special fields of microbiology: immunology. *Microbiology: Principles and Applications, 2*(1), 23–25.

Black, G. E., & Abramson, F. P. (2003). Measuring DNA synthesis rates with stable isotopes. *Analytical Chemistry, 75*(3).

Bolton, S. (1997). *Pharmaceutical statistics* (3rd ed.). Marcel Dekker.

De Brouwer, V., Zhang, G. F., Storozhenko, S., Van Der Straeten, D., & Lambert, W. E. (2007). PH stability of individual folates during critical sample preparation steps in prevision of the analysis of plant folates. *Phytochemical Analysis, 18*(6), 496–508. https://doi.org/10.1002/pca.1006

Horton, H., Moran, L. A., & Ochs. (1996). *Principles of biochemistry* (2nd ed.). Upper Saddle River: Prentice Hall.

Ho, R. J. Y., Schmetz, M., & Deamer, D. W. (1987). Nonenzymatic hydrolysis of phosphatidylcholine prepared as liposomes and mixed micelles. *Lipids, 22*(3), 156–158. https://doi.org/10.1007/BF02537295

IPCS Inchem. (1991). *Folic acid.*

MassBank Record: PR100611. (2011). *Thymidine-5?-monophosphate.*

MassBank Record: UT001356. (2011). *Phosphatidylethanolamine.*

National Institutes of Health. (2012). *MedlinePlus: Megaloblastic anemia.*

Seely, Stephens, T., & Tate, P. (2007). *Anatomy and physiology* (8th ed.). McGraw-Hill.

Sousa, M. M. L., Krokan, H. E., & Slupphaug, G. (2007). DNA-uracil and human pathology. *Molecular Aspects of Medicine, 28*(3–4), 276–306. https://doi.org/10.1016/j.mam.2007.04.006

Wallentin, L., Berlin, R., & Vikrot, O. (1977). Studies on plasma lipid and phospholipid composition in pernicious anemia before and after specific treatment. *Acta Medica Scandinavica, 201*(1–6), 161–165. https://doi.org/10.1111/j.0954-6820.1977.tb15674.x

Non-enzymatic methylation of cytosine in RNA by S-adenosylmethionine and implications for the evolution of translation

Bruce K. Kowiatek
Allied Health Science, Blue Ridge Community and Technical College, Martinsburg, WV, United States

Abbreviations

DNA	deoxyribonucleic acid;
RNA	ribonucleic acid;
SAH	S-adenosyl-homocysteine;
SAM	S-adenosylmethionine;
tRNA	transfer RNA

Introduction

The nonenzymatic methylation of cytosine (C) to form 5-methylcytosine (5-mC) in deoxyribonucleic acid (DNA) by the intracellular methyl group donor S-adenosylmethionine (SAM) (Figs. 6.1 and 6.2) resulting in S-adenosylhomocysteine (SAH) and minor thymine (T) via spontaneous deamination (Fig. 6.3), implicated in certain point mutagenic cancers, has been widely known since the 1980s (Mazin et al., 1985), as has the proposed Watson-Crick mechanism of the adenine moiety (A) (Fig. 6.4) of

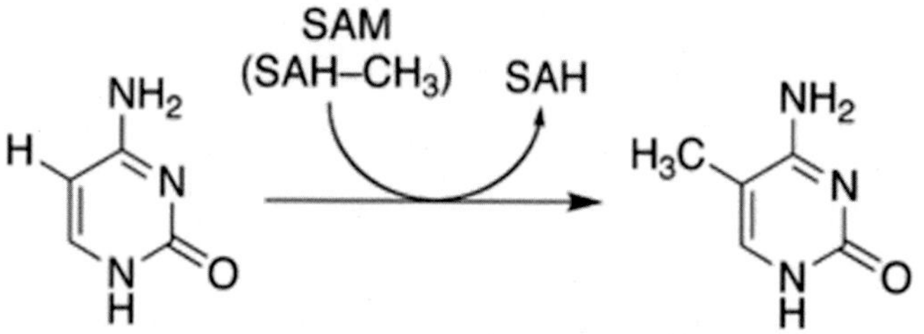

Figure 6.1 Nonenzymatic methylation of C via SAM to form 5-mC and SAH.

The Makings of a Clinical Protocol
ISBN 978-0-323-95749-6
https://doi.org/10.1016/B978-0-323-95749-6.00004-1

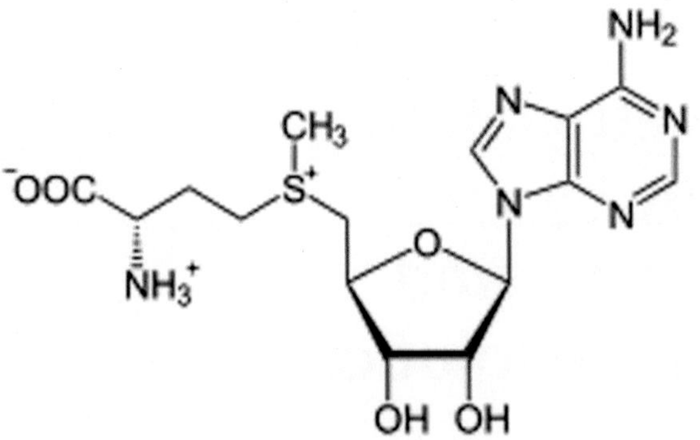

Figure 6.2 4: S-adenosylmethionine (SAM).

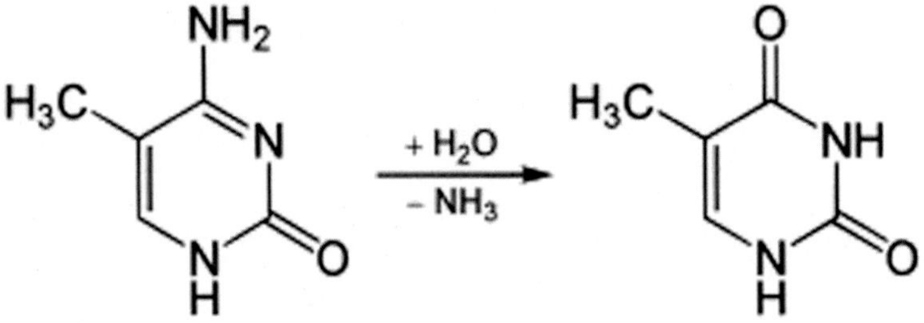

Figure 6.3 Spontaneous deamination of 5-mC to form T.

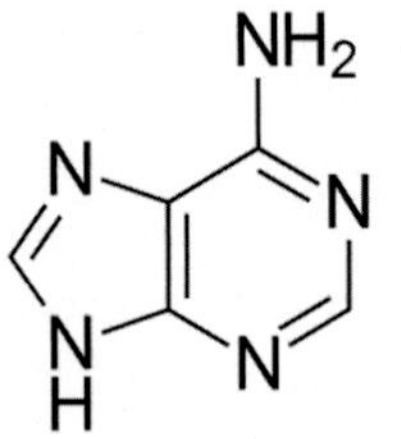

Figure 6.4 Adenine moiety (A) of SAM.

SAM base-pairing with T or uracil (U) (Fig. 6.5) (Hancock, 1984). Such analogous base-pairing and nonenzymatic methylation in ribonucleic acid (RNA), however, has not been as widely addressed, particularly with respect to the origins and evolution of the process of translation initiation in the context of the hypothesized RNA world that preceded the current DNA-protein world (Robertson & Joyce, 2012). It is posited here, with spectrophotometric evidence put forth, that such base-pairing and

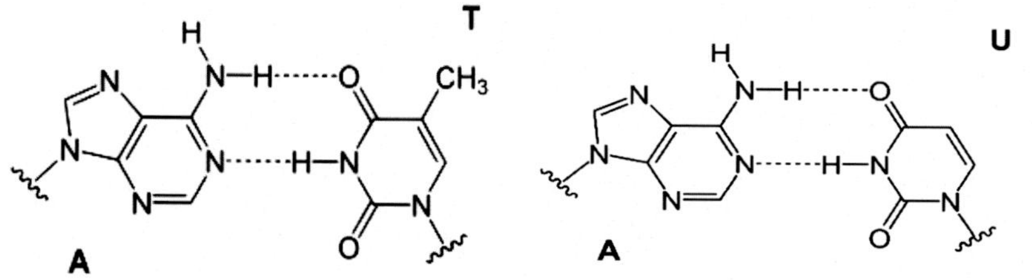

Figure 6.5 Watson-Crick base-pairing mechanism of the adenine moiety (A) of SAM with T and U.

nonenzymatic methylation with subsequent deamination in RNA may constitute a rudimentary form of metabolism and self-replication with implications for the origins and evolution of translation initiation, possibly including the origin and evolution of the transfer RNA (tRNA) molecule (Fig. 6.6).

During translation initiation, the small ribosomal subunit binds to the messenger RNA (mRNA) initiation sequence, whereupon a tRNA

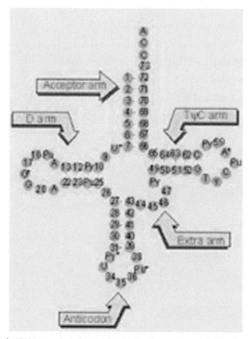

Figure 6.6 Typical tRNA molecule with typically conserved bases. Note conserved TΨC bases in the T-loop section of the TΨC arm. Abbreviations: Ψ, pseudouridine; *Pu*, purine (A or G); and *Py*, pyrimidine (C or U).

$$
\begin{array}{cc}
& \text{O} \\
& \| \\
& \text{H}-\text{C} \qquad \text{H} \\
\text{NH}_2 & \text{N} \\
| & | \\
\text{H}-\text{C}-\text{COOH} & \text{H}-\text{C}-\text{COOH} \\
| & | \\
(\text{CH}_2)_2 & (\text{CH}_2)_2 \\
| & | \\
\text{S} & \text{S} \\
| & | \\
\text{CH}_3 & \text{CH}_3 \\
\text{Methionine} & \textit{N}\text{-Formylmethionine}
\end{array}
$$

Figure 6.7 Methionine (Met, M), the amino acid composition of SAM; N-formylmethionine (fMet) found in bacteria.

molecule, universally carrying the amino acid component of SAM, methionine (Met, M), formylated to N-formylmethionine (fMet) in bacteria (Fig. 6.7), and binds to the mRNA triplet nucleotide (nt) Adenine Uracil Guanine (AUG) initiation codon sequence (Fig. 6.8); while other mRNA initiation codon sequences exist in prokaryotes such as archaea and bacteria, all still contain UG as their second and third respectively, and all still carry Met or fMet, regardless of their position in the genetic code (Belinky et al., 2017).

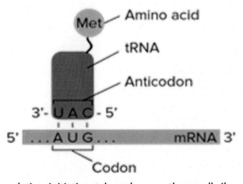

Figure 6.8 Translation initiation takes place on the small ribosomal subunit.

Experimental

All research was conducted at Blue Ridge Community and Technical College in Martinsburg, WV, the USA from January 2019 through May 2019. All pH measurements were made using a Horiba TwinpH waterproof B-213 Compact pH meter. All spectrophotometric data were obtained using a Beckman Coulter DU 530 UV/Vis Spectrophotometer. UV spectra were compared to the standard peak Absorbance (logarithm epsilon) value of 267 nm for T (Koplik, n.d.). All Pyrex glassware was sterilized at 130°C for 1 h (Black & Microbiology, 1993) via autoclave using a Quincy Lab Inc. Model 30 GC Lab Oven. All chemical supplies were purchased from Millipore Sigma, St Louis, MO, USA. All measurements of chemicals were standardized to 0.1 M ±5% using an Ohaus Analytical Plus electronic balance accurate to within ±0.0001 g. Three trials per step were performed and recorded with the data presented here representing the average of that total data. All data collected fell within a statistically acceptable ±5% ($P = .05$) internal margin of variance with no outliers. As a baseline, 0.1 g of T as ribothymidine (Fig. 6.9) was added to 30 mL of sterile deionized water, pH 7.0, heated to 37°C and maintained for 30 min. A sample was then procured, and placed in a crystal quartz cuvette, and its UV/Vis spectrum was obtained (Fig. 6.10).

To this mixture, 0.1 g of C as cytidine (Fig. 6.11) and 0.1 g of SAM were added, with a resultant concentration of 0.3 g/30 mL or 1% and maintained at 37°C for 30 min. Concentration, pH, temperature, and time were all based on the aforementioned nonenzymatic DNA experiments (Rydberg & Lindahl, 1982). A sample was then procured, and placed in a crystal quartz cuvette, and its UV/Vis spectrum was obtained (Fig. 6.12).

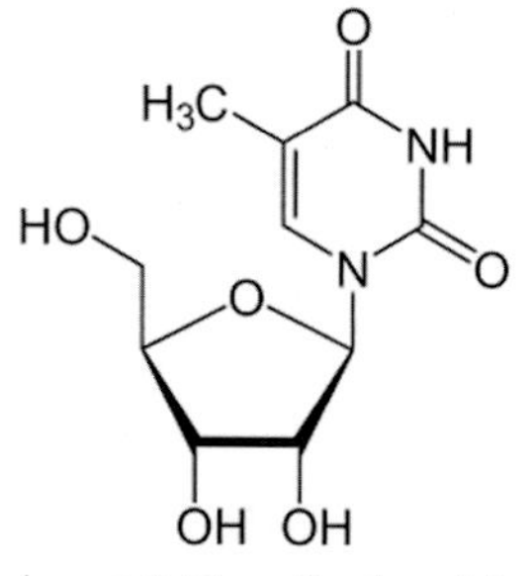

Figure 6.9 T as ribothymidine.

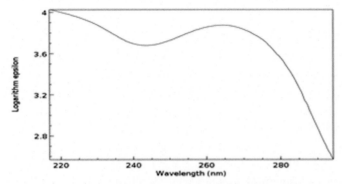

Figure 6.10 Baseline UV/Vis spectrum of T as ribothymidine. Note Absorbance peak of 3.9 at 267 nm.

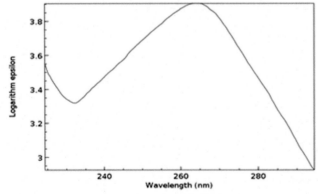

Figure 6.11 C as cytidine.

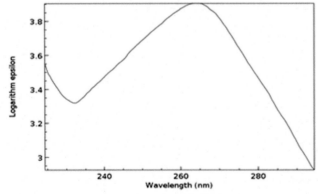

Figure 6.12 UV/Vis spectrum of T as ribothymidine following the addition of C as cytidine and SAM. Note Absorbance peak of 4.1 at 267 nm.

Results and discussion

The Absorbance peak of T as ribothymidine treated with C as cytidine and SAM at 4.1 versus that of the baseline at 3.9 represents a quantitative increase of 20% in the former according to the Beer–Lambert Law (Beer & Law, n.d.), far in excess of the 5% possible by chance. While any proposed base pairing between T as ribothymidine and the adenine moiety (A) of SAM would await future confirmation, Watson–Crick base-pairing rules would presuppose its likelihood and, although somewhat circuitous, upon base-pairing, the pathway from the nonenzymatic methylation of C as cytidine via SAM, first to the formation of 5-mC and then to T as ribothymidine via spontaneous deamination, may represent a rudimentary metabolism as the first instance of a prebiotically plausible molecule, T (Choughuley et al., 1977), utilizing other prebiotically plausible molecules, SAM and C (Biscans, 2018), to replicate itself, with SAH possibly being remethylated to regenerate SAM via nonenzymatic methylation by membrane phosphatidylcholine (PC), similar to the nonenzymatic methylation of folic acid by PC (Kowiatek, 2018), with PC itself potentially being prebiotic, as PC-like molecules have been detected in at least one meteorite (Wirick et al., 2005). Indeed, a primitive form of gene expression and/or silencing might be at work here as well, with both a phenotype (pyrimidine) and genotype (T), which may or may not be expressed, depending on whether 5-mC base pairs with G, halting deamination and thus T formation. Furthermore, the ideal RNA template sequence to facilitate such replication with fidelity would be the mRNA initiation sequence AUG, with UG ensuring it minimally (Fig. 6.13). Such triplet nucleotide sequences have also been shown experimentally to act as templates for RNA-catalyzed RNA ribozymes when nucleotide triphosphates were used as substrates (Attwater et al., 2018).

5' T SAM C 3'

| | |

3' A U G 5'

Figure 6.13 Ideal RNA template sequence to facilitate T self-replication with fidelity, with minimal sequence highlighted in *red*.

Additionally, base-pairing to the AUG template sequence may account for the conserved TΨC sequence in the T-loop section of the TΨC arm of tRNA, with Ψ representing pseudouridine (Fig. 6.14), a modified base found in RNA, which base pairs with any other base (Kierzek et al., 2014), possibly accounting for the origin of this conserved sequence (Fig. 6.15).

The TΨC arm and T-loop are hypothesized to be the most ancient part of tRNA, which is itself the most modified and methylated nucleic acid, postulated to possibly be the most ancient surviving artifact of the RNA world, evolved from either two homologous halves or, more likely, three homologous minihelices, and while the T-loop in all but one archaea, the genus *Pyrococcus,* with archaea believed to be the most ancient of organisms (Kierzek et al., 2014), possesses the modified base N1 methyl pseudouridine (Fig. 6.16) instead of T (Kim et al., 2018), pseudouridine is itself a pre-biotically implausible molecule and may presuppose a role in some capacity for the more prebiotically plausible T (Wurm et al., 2012), although a scenario could be envisaged in which N1 methyl pseudouridine feasibly supplants the role of T in the above-described settings. Considering these

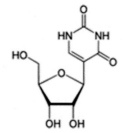

Figure 6.14 Pseudouridine.

5' T Ψ   C 3'

| | |

3' A U G 5'

Figure 6.15 AUG RNA template for the conserved TΨC sequence of the T-loop section of the TΨC arm of tRNA.

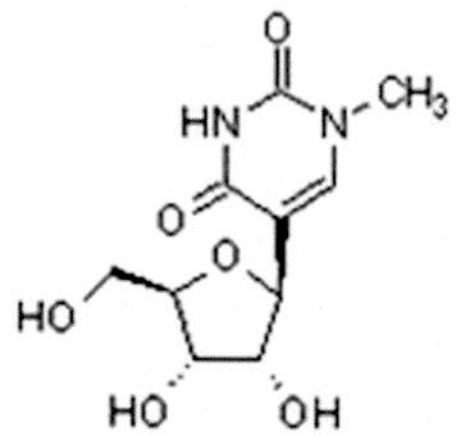

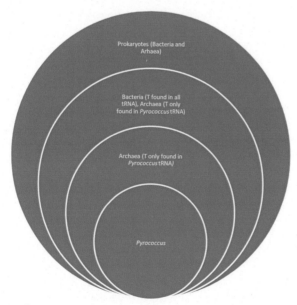

Figure 6.16 N1 methyl pseudouridine.

factors, despite the SAM-dependent presence of T in its tRNA being attributed to ancient horizontal gene transfer from bacteria (Dworkin, 1997), the archaeal genus *Pyrococcus* could still be considered the likeliest candidate for the Last Universal Common Ancestor (LUCA) (Fig. 6.17).

The experimentally verified nonenzymatic methylation of C by SAM to form 5-mC and its subsequent spontaneous deamination to T may very well represent, therefore, the first, or very close to the first, possibly by way of N1methyl pseudouridine, a self-replicating molecule. Additionally, it may be the process responsible for the selection of UG and AUG as the sequence for the mRNA translation initiation codons, the conserved TΨC sequence in the T-loop of the TΨC arm of tRNA, and methionine as the universally carried

Prokaryotes (Bacteria and Arhaea)

Bacteria (T found in all tRNA), Archaea (T only found in *Pyrococcus* tRNA)

Archaea (T only found in *Pyrococcus* tRNA)

Pyrococcus

Figure 6.17 Stacked Venn diagram representing the presence of T in the tRNA of Prokaryotes.

starting amino acid in proteins. Finally, the presence of T in an RNA world in which U was otherwise employed may have been part of the bridge to the present DNA-protein world in which T is favored over U.

Prospectus

Although the nonenzymatic methylation of C by SAM to 5-mC and its subsequent spontaneous deamination to T in DNA at physiologic pH and the temperature has been implicated in some C to T point mutagenic cancers, cancers in general display a global hypomethylation of their DNA (Ehrlich, 2009); SAM, therefore, sold as the over-the-counter (OTC) supplement SAM-e, may still possibly play a role as a potential rescue adjunct therapy in certain cancers, particularly those treated with alkylating chemotherapeutic agents, much in the same way that folic acid is used as a rescue adjunct therapy when using antifolate chemotherapeutic agents (Hagner & Joerger, 2010). Clinical trials in support of this prospectus are, therefore, justifiable.

Acknowledgments

I thank my fellow faculty and staff at Blue Ridge Community and Technical College for the generous use of their facilities and equipment. I also thank Dr. Zachary Burton, Professor Emeritus at Michigan State University (MSU) for invaluable insight and discussion, without whom this paper would not have been possible, and I especially thank my loving family for their generous gift of time in the performance of these experiments and the writing of this paper.

References

Attwater, J., Raguram, A., Morgunov, A. S., Gianni, E., & Holliger, P. (2018). Ribozyme-catalysed RNA synthesis using triplet building blocks. *eLife, 7.* https://doi.org/10.7554/eLife.35255

Beer, L., & Law. (n.d.).

Belinky, F., Rogozin, I. B., & Koonin, E. V. (2017). Selection on start codons in prokaryotes and potential compensatory nucleotide substitutions. *Scientific Reports, 7*(1). https://doi.org/10.1038/s41598-017-12619-6

Biscans, A. (2018). Exploring the emergence of RNA nucleosides and nucleotides on the early earth. *Life, 8*(4), 57. https://doi.org/10.3390/life8040057

Black, Jacqueline (1993). Microbiology. *Journal of Applied & Environmental Microbiology, 2*(6), 334—334. https://doi.org/10.12691/jaem-2-6-6.

Choughuley, A. S. U., Subbaraman, A. S., Kazi, Z. A., & Chadha, M. S. (1977). A possible prebiotic synthesis of thymine: Uracil-formaldehyde-formic acid reaction. *BioSystems, 9*(2—3), 73—80. https://doi.org/10.1016/0303-2647(77)90014-4

Dworkin, J. P. (1997). Attempted prebiotic synthesis of pseudouridine. *Origins of Life and Evolution of the Biosphere, 27*(4), 345–355. https://doi.org/10.1023/A:1006579819734

Ehrlich, M. (2009). DNA hypomethylation in cancer cells. *Epigenomics, 1*(2), 239–259. https://doi.org/10.2217/epi.09.33

Hagner, N., & Joerger, M. (2010). Cancer chemotherapy: targeting folic acid synthesis. *Cancer Management and Research, 19*(2), 293–301. https://doi.org/10.2147/CMR.S10043

Hancock, R. L. (1984). Theoretical mechanisms for synthesis of carcinogen-induced embryonic proteins: XII mutational and non-mutational mechanism as subsets of a more general mechanism. Part A—ethionine. *Medical Hypotheses, 15*(3), 323–331. https://doi.org/10.1016/0306-9877(84)90022-7

Kierzek, E., Malgowska, M., Lisowiec, J., Turner, D. H., Gdaniec, Z., & Kierzek, R. (2014). The contribution of pseudouridine to stabilities and structure of RNAs. *Nucleic Acids Research, 42*(5), 3492–3501. https://doi.org/10.1093/nar/gkt1330

Kim, Y., Kowiatek, B., Opron, K., & Burton, Z. F. (2018). Type-II tRNAs and evolution of translation systems and the genetic code. *International Journal of Molecular Sciences, 19*(10). https://doi.org/10.3390/ijms19103275

Koplik, R. (n.d.). Ultraviolet and visible spectrometry.

Kowiatek, B. K. (2018). An additional, complementary mechanism of action for folic acid in the treatment of megaloblastic anemia. *Biomedical Journal of Scientific & Technical Research, 11*(2). https://doi.org/10.26717/bjstr.2018.11.002071

Mazin, A. L., Gimadutdinov, O. A., Turkin, S. I., Burtseva, N. N., & Vaniushin, B. F. (1985). Neénzimaticheskoe metilirovanie DNK pod deĭstviem S-adenozilmetionina s obrazovaniem iz tsitozina ostatkov minornogo timina i 5-metiltsitozina. *Molekulyarnaya Biologiya, 19*(4), 903–914.

Robertson, M. P., & Joyce, G. F. (2012). The origins of the RNA World. *Cold Spring Harbor Perspectives in Biology, 4*(5), 1. https://doi.org/10.1101/cshperspect.a003608

Rydberg, B., & Lindahl, T. (1982). Nonenzymatic methylation of DNA by the intracellular methyl group donor S-adenosyl-L-methionine is a potentially mutagenic reaction. *The EMBO Journal, 1*(2), 211–216. https://doi.org/10.1002/j.1460-2075.1982.tb01149.x

Wirick, S., Cody, G., & Flynn, G. (2005). Detection of a water soluble component of the Tagish Lake meteorite. *Lunar and Planetary Science, XXXVI.*

Wurm, J. P., Griese, M., Bahr, U., Held, M., Heckel, A., Karas, M., Soppa, J., & Wöhnert, J. (2012). Identification of the enzyme responsible for N1-methylation of pseudouridine 54 in archaeal tRNAs. *RNA, 18*(3), 412–420. https://doi.org/10.1261/rna.028498.111

CHAPTER 7

Early evolution of transcription systems and divergence of Archaea and Bacteria

Lei Lei[1] and Zachary F. Burton[2]

[1]School of Biological Sciences, University of New England, Biddeford, ME, United States; [2]Department of Biochemistry and Molecular Biology, Michigan State University, East Lansing, MI, United States

Abbreviations

BH	bridge helix
BRE	TFB-recognition element
CLR	cyclin-like repeat (TFB HTH domains)
DNAP	DNA polymerase
DPBB	double-Ψ-β-barrel
HTH	helix-turn-helix
InR	initiator element
LUCA	last universal (cellular) common ancestor
Pfu	*Pyrococcus furiosis*
Pol	DNA polymerases (i.e., PolA, PolB, PolC, and PolD)
PPE	promoter-proximal element
RNAP	RNA polymerase
RRM	RNA-recognition motif
SBHM	sandwich barrel hybrid motif
Sso	*Sulfolobus* solfataricus
TBP	TATA-box binding protein
TFB	transcription factor B
TFE	transcription factor E
TIM	triose phosphate isomerase
TL	trigger loop.

Introduction

The purpose of this review is to provide a conceptual overview of transcription systems in the early phase of their evolution, in order to explain how RNA polymerases (RNAPs), general transcription factors, and promoters may have evolved. The review also touches on the divergence of Archaea and Bacteria that appears to have partly been driven by the divergence of transcription systems. The proper way to view structures is

The Makings of a Clinical Protocol
ISBN 978-0-323-95749-6
https://doi.org/10.1016/B978-0-323-95749-6.00010-7

137

using molecular graphics such as UCSF ChimeraX (Goddard et al., 2018; Pettersen et al., 2021). Viewing structures in two dimensions are challenging to the human eyes and mind. We recommend downloading ChimeraX, running tutorials and using it to follow along with this manuscript. For instance, some figures in this paper are difficult to fully appreciate without a more 3-dimensional representation.

Our opinion is that analyzing the structure-function dynamics of any protein requires a combination of approaches: that is, (1) structure analysis; (2) evolution; (3) functional studies; and (4) dynamics. To appreciate structural analysis and dynamics, Cryo-electron microscopy becomes an ever more powerful tool. Cryo-EM provides ensembles of structures often indicating a dynamic progression through a reaction mechanism. Evolutionary studies have the potential to dissect a protein into its component parts to better appreciate how the protein came to have its eventual form and function. In some cases, structural studies have not been combined fully with evolutionary studies, and the historic naming of protein domains can be confusing. Also, very large structures are difficult to analyze unless they can be broken into component parts. We see two potential problems. Without an evolutionary view, structures may be difficult to understand and analyze. Also, the evolution literature can be complex and challenging to read unless one is reasonably expert or determined. In this paper, we attempt to apply a combination of structural and evolutionary principles to the analysis and description of multisubunit RNAPs, general transcription factors, and promoters.

Evolution of 2-DPBB RNAPs and DNAPs

2-Double-Ψ-β-barrel type RNAPs

Near the dawn of the evolution of life on Earth, RNAPs of the 2-DPBB type evolved (Aravind et al., 2005; Fouqueau et al., 2017; Iyer et al., 2003; Lane & Darst, 2010a, 2010b; Madru et al., 2020; Sauguet, 2019; Werner & Grohmann, 2011; Zatopek et al., 2021). These enzymes are found in all domains of life and some viruses. 2-DPBB RNAPs can be either RNA template-dependent or DNA template-dependent, indicating that this important class of enzyme may have arisen in an RNA world before DNA genomes became prominent. The DPBB is a particular fold of cradle-loop barrel (Fig. 7.1) (Alva et al., 2008; Coles et al., 2005, 2006). The crossing chains make a Ψ pattern, hence the barrel name. 2−DPBB type RNAPs have 2-DPBBs at their active sites (Fig. 7.2). Loops from the barrels hold

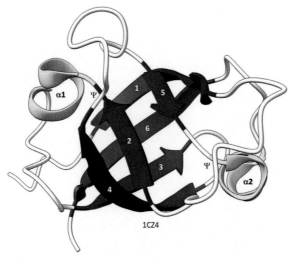

Figure 7.1 Bacterial VAT (VCP—like ATPase) includes a simple DPBB. ChimeraX was used for molecular graphics. The structure is PDB 1CZ4. B-sheets are *red*; α-helices are *yellow*. Ψ indicates the Ψ pattern of crossing peptide chains. (For interpretation of the references to colour in this figure legend, the reader is referred to the web version of this article.)

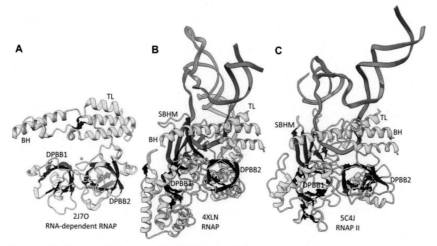

Figure 7.2 The catalytic core of 2-DPBB type RNAPs. (A) An RNA template-dependent RNAP from *Neurospora crassa* (PDB 2J7O). (B) A bacterial multisubunit RNAP (PDB 4XLN). (C) A human multisubunit RNAP (PDB 5C4J). α-helices are *yellow*; β-sheets are *red*; Mg is *green*; RNA is *magenta*; template DNA is *green*; nontemplate DNA is *blue*. BH indicates the *bridge helix*. TL indicates the *trigger loop*. The active site is identified by the Mg (Mg1) and the 3'-end of the RNA (B and C). (For interpretation of the references to colour in this figure legend, the reader is referred to the web version of this article.)

the 2 Mg^{2+} that retain the phosphates of the NTP substrate and activate the RNA 3′-O to catalyze NMP addition. In addition to the 2-DPBBs, both RNA and DNA template-dependent RNAPs have a bridge helix and trigger loop, indicating that these elements are ancient (Iyer & Aravind, 2012; Qian et al., 2016; Salgado et al., 2006). In DNA template-dependent RNAPs, the β-subunit DPBB1 has a sandwich-barrel hybrid motif (SBHM) inserted into one of the barrel loops (Fouqueau et al., 2017; Iyer & Aravind, 2012; Lane & Darst, 2010a, 2010b). The SBHM loop extension forms the historically named "flap" or "wall" motif in multisubunit RNAPs.

Barrels are frequent motifs in ancient evolution. In the earliest evolution, barrels were selected to form compact, structured units with reasonable solubility and structural closure (Burton et al., 2016). For instance, 8−β-sheet barrels [$(β−α)_8$; that is, TIM barrels (TIM for triose phosphate isomerase)] are found in most glycolytic enzymes. Rossmann folds appear to be sheets that are rearranged from $(β − α)_8$ barrels. Most of the citric acid cycle is made up of Rossmann fold proteins. So, much of core metabolism was generated from barrels and, also, from refolded barrels rendered into more linear sheets. Cradle-loop barrels are a similar ancient evolution story (Alva et al., 2008). If early evolution was partly a race to form stable and soluble scaffolds, the formation of barrels helped to build these and, among other possible advantages, helped to avoid the generation of β−sheet amyloids and liquid-liquid phase-separated compartments that resisted ordered protein folding. Clearly, barrels were a successful evolutionary innovation that, once formed, persisted throughout evolution. From this point of view, an important evolutionary event can be viewed as the race to form stable and soluble protein structures with a degree of structural closure. Barrels were typically formed in evolution by repeated motif duplications, so barrels often won races to higher-order structure, solubility, and closure. After the generation of barrels, primitive catalytic sites could be modified to generate many new, more efficient, and more specific enzyme functions. So, for instance, in metabolism, an enzyme with broad specificity built around an 8−β-sheet barrel was duplicated genetically many times and then refined, generating specialist enzymes that formed a more sophisticated and integrated pathway (i.e., glycolysis).

Similarly, the DPBB evolved by duplication of a $β − β − α − β$ unit followed by refolding into a barrel (Alva et al., 2008; Burton et al., 2016). In Fig. 7.1, a $β − β − α − β − − β − β − α − β$ DPBB enzyme domain is shown in which the basic DPBB form is preserved without much modification (Coles et al., 1999). The β—sheets are numbered 1—6, so that the chain can be traced. The α—helices are numbered 1 and 2. The Ψ patterns of the crossing chains are indicated. The ability to identify a DPBB helps with understanding the 2—DPBB enzyme patterns when analyzing more complex structures. Because of modifications of the pattern during evolution or disorder in structures, DPBBs can be a challenge to identify and, in a complex structure, can be potentially difficult to locate.

2—DPBB type enzymes include RNA template—dependent RNAPs (found in some Eukaryotes), multi—subunit RNAPs (found in all domains and some viruses), and DNA template—dependent DNAPs (PolD in most Archaea) (Figs. 7.2 and 7.3) (Fouqueau et al., 2017; Iyer & Aravind, 2012; Iyer et al., 2003; Koonin et al., 2020; Lane & Darst, 2010a, 2010b; Werner & Grohmann, 2011). In 2—DPBB type enzymes, the basic $\beta - \beta - \alpha - \beta - - \beta - \beta - \alpha - \beta$ form can be modified by insertions into barrel loops. In RNA template-dependent 2—DPBB RNAPs, neither DPBB1 (corresponding to the β—subunit DPBB1 in 2—DPBB bacterial RNAPs) nor DPBB2 (corresponding to the β′-subunit DPBB2 in 2—DPBB bacterial RNAPs) includes very large inserts or modifications in the basic DPBB pattern (Iyer & Aravind, 2012; Qian et al., 2016; Salgado et al., 2006).

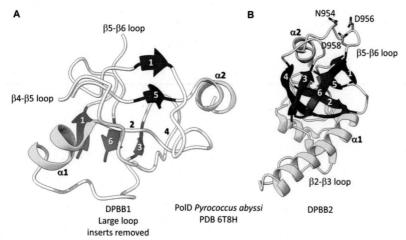

Figure 7.3 The two DPBBs of a DNA template-dependent DNAP (archaeal PolD) (PDB 6T8H). (A) DPBB1 is somewhat disordered in the structure, so not all β-sheets were scored as such by ChimeraX. In (B) DPBB2, N954, D956 and D958 may hold the active site Mg (missing in the structure).

In DNA template-dependent 2—DPBB type RNAPs, by contrast, there are large identifying inserts (Iyer & Aravind, 2012). Significantly, the β—subunit (referring to bacterial RNAPs) DPBB1, includes a sandwich-barrel hybrid motif (SBHM) inserted between β2 and β3 after α1. The SBHM can be recognized because it includes long β—sheets. The SBHM forms the "flap" or "wall" domain of the RNAP that contacts σ (Bacteria) and TFB (Archaea) general transcription factors. The SBHM also contacts the general elongation factors NusG (Bacteria) and Spt5/Spt4 (Archaea). Because the SBHM is missing in RNA template—dependent RNAPs of the 2—DPBB type, the SBHM is considered to be a feature for the transcription of DNA templates (Iyer & Aravind, 2012). Because the SBHM interacts with initiation factors, the SBHM is considered to be evolved to facilitate initiation from DNA templates. A large mostly α—helical insert is found between DPBB1 β5 and β6, after α2. This insert is only partially homologous in archaeal and bacterial RNAPs and appears to make domain-specific contacts to RNAP rather than contacts to transcription factors. In some structures, DPBB1 is somewhat disordered in 2—DPBB DNA template-dependent RNAPs, making some of the β—sheets difficult to discern. The β′-subunit DPBB2 (referring to bacterial RNAPs) has a largely α—helical insert between β2 and β3 (distinct from the SBHM that includes long β—sheets). In Archaea, the insert between DPBB2 β2 and β3 is referred to as a RAGNYA domain that includes β—sheets and α—helices (Balaji & Aravind, 2007; Iyer & Aravind, 2012). The archaeal and bacterial DPBB2 β2—β3 inserts are very different in sequence and make domain-specific contacts to TFB and σ for initiation.

Found in many Archaea, PolD is DNA template-dependent DNAPs of the 2—DPBB form engaged in genomic replication (Koonin et al., 2020; Madru et al., 2020; Raia et al., 2019; Sauguet, 2019). In these enzymes, DPBB1 includes two large inserts, one between β4 and β5 and one between β5 and β6. In available structures, PolD DPBB1 appears to be somewhat disordered, similarly to DPBB1 (β—subunit of bacterial RNAPs) in some structures of DNA template-dependent RNAPs. The significance of this possible similarity in some structures is not known to us. One idea is that DPBB1 is somewhat more dynamic because it accommodates the presence and absence of substrate to a larger extent than DPBB2, which holds active site Mg1 more tightly than DPBB1 holds Mg2. We would be interested to know whether dNTP binding tightens the PolD DPBB1 and whether similar changes might occur in multisubunit RNAPs with NTP binding. In PolD, DPBB2 includes an insert between β1 and β2. The inserts in the DNA template-dependent DNAPs (PolD) discriminate PolD enzymes from multisubunit RNAPs and RNA template-dependent RNAPs and indicate how these more complex enzymes diverged from RNA template-dependent RNAPs of the 2-DPBB form (Koonin et al., 2020).

The story of the evolution of these ancient 2-DPBB-type enzymes cannot now be told with certainty, but we construct a possible narrative. We posit that RNA template-dependent RNAPs may have evolved in an RNA-dominated world prior to LUCA (Iyer & Aravind, 2012; Koonin et al., 2020). These enzymes include no large inserts in their DPBBs, indicating that RNA template-dependent RNAPs probably comprise the most ancient 2-DPBB enzyme form. DNA template-dependent RNAPs (multisubunit RNAPs) and DNAPs (PolD) appear to have radiated mostly independently from the primitive form, although, multisubunit RNAPs and PolD may share one or two Zn motifs that are missing from 2-DPBB RNA template-dependent RNAPs (see below). multisubunit RNAPs and Pol D, however, have distinct DPBB loop inserts. To our knowledge, comparative sequence analyses of these enzymes provide limited insight into details of their divergence, because sequences among enzyme classes are only weakly conserved (Madru et al., 2020; Sauguet, 2019; Zatopek et al., 2021). Because PolD is ancient, this 2-DPBB type enzyme may be the initial evolved DNA template-dependent DNAP for genomic replication (i.e., at LUCA), and other DNAPs, that is, PolA, PolB, and PolC, may have evolved later (Koonin et al., 2020).

RNA template-dependent RNAPs and multisubunit RNAPs have a recognizable bridge helix and trigger loop (Fig. 7.2), and these features are altered and rearranged in DNA template-dependent DNAPs (PolD) of the 2-DPBB type (see below) (Madru et al., 2020). It appears that 2-DPBB multisubunit RNAPs from Archaea and Eukaryotes and PolD from Archaea may share a Zn-finger motif that is missing from RNA template-dependent RNAPs and bacterial multisubunit RNAPs. We posit that Archaea are older than Bacteria and closer to LUCA (Battistuzzi et al., 2004; Lei & Burton, 2020; Long et al., 2020) but also see (Castelle & Banfield, 2018; Da Cunha et al., 2017, 2018; Eme et al., 2018; Forterre, 2015). We, therefore, posit that this Zn-finger was lost in bacterial multisubunit RNAPs, which appear to be a simplified form compared to archaeal multisubunit RNAPs. We posit that bacterial RNAPs were driven to diverge from archaeal RNAPs primarily because bacterial RNAPs coevolved with bacterial σ factors.

RNAP catalytic subunits (a guided tour)

Our view is that Archaea are older than Bacteria, and, therefore, Archaea are closer to LUCA (Battistuzzi et al., 2004; Lei & Burton, 2020; Long et al., 2020; Marin et al., 2017). For other views, see (Castelle & Banfield, 2018; Da Cunha et al., 2017, 2018; Eme et al., 2018; Forterre, 2015). Because of horizontal gene transfer, some phylogenetic analyses may be misleading in determining the deep branching of prokaryotic domains. We believe Bacteria were derived from Archaea. Our opinions are based on ancient evolution studies of transcription systems, tRNA, aminoacyl-tRNA synthetases, ribosomes, and the genetic code. In every comparison we have made, Archaea appear to be the more ancient lineage, and Bacteria appear to be more innovated and more derived evolutionarily from root sequences. Therefore, to describe the multisubunit RNAP catalytic subunits, we use an archaeal RNAP as an example. The RNAP we selected is from *Saccharolobus shibatae* (PDB 2WB1) (Korkhin et al., 2009). The catalytic subunits include 2WB1_A and 2WB1_C (_A and _C indicate the chain designation), which correspond to the β′ subunit of bacterial RNAP, a subunit that is split in some Archaea. 2WB1_B corresponds to the β subunit of bacterial RNAPs. We compare similar motifs in DNAP PolD to emphasize the early evolution of RNAPs.

Fig. 7.4 shows the Rpo1N (2WB1_A; A′) and Rpo1C (2WB1_C; A″) chains. We describe some recognizable protein motifs, reading from the N–

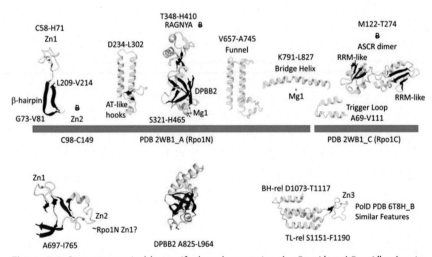

Figure 7.4 Some recognizable motifs that characterize the RpoA′ and RpoA″ subunits of archaeal RNAP, correspond to the β′ subunit of bacterial RNAP.

terminus of the 2WB1_A chain through the 2WB1_C chain. Zn1 is very close to the 2WB1_An N-terminus. Evolutionarily related motifs in PolD are indicated below the blue bar. Zn1 in 2WB1_A may correspond to archaeal DNAP PolD Zn2, based on its position in the structure and its distance from a Zn motif in chain 2WB1_B (Madru et al., 2020). The N-terminal β-sheet of the β-hairpin is next, followed by 2WB1_A Zn2, which is missing in bacterial RNAP. Next is the C-terminal β-hairpin. From D234 to L302 is a helix–loop–helix motif that connects the AT-like hooks (Iyer & Aravind, 2012). The AT-like hook loop contacts single-stranded DNA in the RNAP open complex and elongation complex. Next is the DPBB2 barrel. Between DPBB2 β2 and β3 after α1 is the RAGNYA insert. In Bacteria, a DPBB2 β2–β3 insert after α1 shows no detectable homology and is primarily α-helical (see below). DPBB2 holds Mg1 within the loop between DPBB2 β5 and β6 (NADFDGD). The "funnel" is located in the primary sequence between the DPBB2 and the bridge helix. In the open transcription complex or elongation complex, the DNA template bends by about 90 degrees, and DNA strands separate over the bridge helix. DNA PolD has a similar DPBB2 and, also, modified structures that are probably genetically related to the bridge helix and trigger loop, although these features in PolD appear to be rearranged and repurposed (see below).

The *Saccharolobus shibatae* RNAP is separated into two genes relative to the bacterial RNAP β' subunit, and the subunit separation is between the bridge helix and the trigger loop. The trigger loop is near the archaeal Rpo1C subunit (2WB1_C) N-terminus. The RNAP trigger loop appears to correlate with the PolD "clamp" structure (PDB 6T8H_B; S1151–F1190) (Madru et al., 2020). The *Saccharolobus shibatae* RNAP is separated into two genes relative to the bacterial RNAP β' subunit, and the subunit separation is between the bridge helix and the trigger loop. The trigger loop is near the archaeal Rpo1C subunit (2WB1_C) N-terminus. The RNAP trigger loop appears to correlate with the PolD "clamp" structure (PDB 6T8H_B; S1151–F1190) (Iyer & Aravind, 2012). The ASCR dimer motif is missing in bacterial RNAP and may have been lost by deletion.

In Fig. 7.5, a comparison is shown of bacterial RNAP DPBB2, the bridge helix, and the trigger loop (Fig. 7.5A) and related features in DNAP PolD (Fig. 7.5B). In Fig. 7.5A, an α-helical domain separates DPBB2 β2 and β3. The α-helical loop insert corresponds to and may have replaced the RAGNYA region in archaeal RNAP (Fig. 7.4). The bacterial RNAP β'

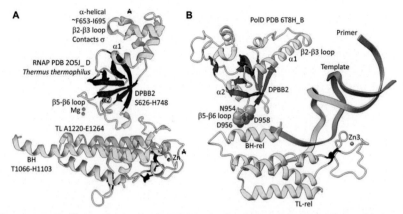

Figure 7.5 Similarities between the DPBB2, bridge helix, and trigger loop of bacterial RNAP and related motifs in DNAP PolD. (A) Bacterial RNAP features. (B) Related PolD features. The similarly placed Zn motifs are not thought to be homologous. "A" with a *double strike through* indicates that a feature of bacterial RNAP is not present in archaeal RNAP.

subunit includes a Zn motif separating the bridge helix and the trigger loop that is missing in Archaea (Fig. 7.5A). PolD also has a Zn motif (Zn3) separating its bridge helix–related and trigger loop–related features, although we do not think these Zn motifs in bacterial RNAP and PolD are related by homology. Rather these Zn motifs may be the result of convergent evolution. In bacterial RNAP, the trigger loop is closer to the active site than the bridge helix and closes over the NTP substrate to expel water from the active site and tighten the substrate for addition to the RNA chain (Vassylyev, Vassylyeva, Zhang, et al., 2007). In the image in Fig. 7.5A, the trigger loop is in the closed and catalytic conformation. In PolD, the trigger loop-related feature is further from the active site than the bridge helix-related feature. In PolD, the bundle of C-terminal α-helices (bridge helix-related and trigger loop-related features) bind DNA and, also, the proofreading PolD subunit (DP1; the 2-DPBBs are part of the DP2 subunit) (Fig. 7.6). The DP1 subunit includes an exonuclease domain. Loops from the bridge helix-related and trigger loop-related PolD features also contact the sliding clamp that maintains PolD processivity (Madru et al., 2020). It appears, therefore, that, although bridge helix- and trigger loop-related features in PolD and RNAPs may be related by evolution, they fulfill different roles.

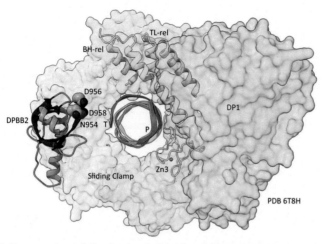

Figure 7.6 Repurposing of the bridge helix-related (BH-rel) and trigger loop-related (TL-rel) motifs in PolD. The DPBB2 (*light red* with *red* β-sheets) and BH-rel, Zn3, and TL-rel region (*yellow*) is shown for the DP2 2-DPBB subunit. (T) template DNA (*blue*); (P) primer DNA (*gold*). The sliding clamp trimer is shown (*green, beige,* and *orange*). The DP1 subunit is *blue*. Active site residues that hold Mg1 are indicated in space-filling representation. (For interpretation of the references to colour in this figure legend, the reader is referred to the web version of this article.)

The archaeal RNAP Rpo2 subunit corresponds to the β–subunit in bacterial RNAP. Features of the Rpo2 RNAP subunit (PDB 2WB1_B; B) are shown in Fig. 7.7. There is a 2-lobed N–terminal domain extending from positions 1—722. The DPBB1 extends from G723 to K995. There are two notable inserts in DPBB1 loops. Between β2 and β3, just after α1, an SBHM is inserted (Iyer & Aravind, 2012). The SBHM is characterized by long β-sheets. In archaeal RNAP, the SBHM is referred to as the "wall" domain, which interacts with the general transcription factor TFB. In bacterial RNAP, the SBHM has been referred to as the "flap" domain, which interacts with the bacterial σ factor. Between β5 and β6, just after α2, an α–helical segment is inserted (∼N914-R985). At the C–terminus of the Rpo2 chain, a Zn finger is located in archaeal RNAPs but missing in bacterial RNAPs. Although the sequences are different, this Zn finger may correspond to Zn1 in archaeal DNAP PolD (Madru et al., 2020). As in PolD, the Rpo2 Zn finger and the Rpo1N Zn1 are close in space in archaeal RNAP, similar to PolD Zn1 and Zn2.

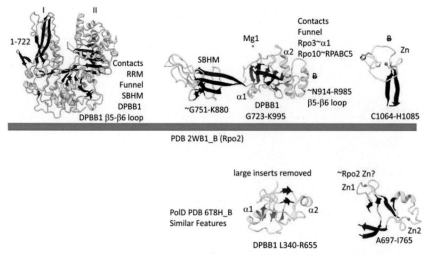

Figure 7.7 Some recognizable motifs in the Rpo2 subunit of archaeal RNAP (corresponding to the β subunit of bacterial RNAP).

The description of the catalytic subunits of multisubunit RNAPs here is incomplete. The intention is to provide some visible and conceptual guide posts for researchers as they begin to probe and familiarize themselves with RNAP structures. Also, we emphasize features that appear most important for interactions between general transcription factors and the RNAP catalytic center (see below). A more detailed description of RNAP evolution and domains is provided by Iyer and Aravind (2012). Reviews of the subunit structures of multisubunit RNAPs are also published elsewhere (Jun et al., 2011; Osman & Cramer, 2020).

2-Mg mechanism of transcription by multisubunit RNAPs

We have described the basic catalytic core of multisubunit RNAPs: 2-DPBBs, a bridge helix, and a trigger loop (Fig. 7.2B and C). These enzymes utilize a 2-Mg mechanism for transcription (Fig. 7.8) (Vassylyev, Vassylyeva, Zhang, et al., 2007). The 2-Mg (Mg1 and Mg2) are held by acidic groups (E and D) on loops of the 2-DPBBs. DPBB1 includes 685-ED-686 (*Thermus thermophilus* RNAP numbering) located on the DPBB1 loop between β4 and β5. D686 appears to interact with Mg2 during phosphodiester bond formation. Mg2 is loosely held in the RNAP structure. DPBB2 includes the highly conserved sequence 737-NADFDGD-743 within the loop between β5 and β6. D739, D741, and D743 strongly

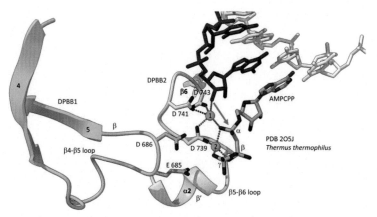

Figure 7.8 The 2 Mg mechanism for transcription by RNAP. The structure (PDB 2O5J) is from *Thermus thermophilus*.

hold Mg1. It is thought that Mg1 remains bound to RNAP, but Mg2 may exchange with each NTP addition. Mg2 normally enters the RNAP bound to the NTP as NTP-Mg. The NADFDGD motif in multisubunit RNAPs corresponds to 954-NCDGDED-961 in archaeal *Pyrococcus abyssi* DNAP PolD (Madru et al., 2020), although, in PolD, the active site Mg1 is held by N954, D956, and D958, so the Mg1-contacting residues are slightly shifted in PolD (Zatopek et al., 2021). In *Neurospora crassa* RNA template-dependent RNAP, Mg1 is held by 1005-GGDYDGD-1011 (Qian et al., 2016; Salgado et al., 2006). Acidic groups retaining Mg1 at the active enzyme site are highly conserved in 2-DPBB type enzymes, although PolD has slightly shifted the set of interacting residues. In the simplest cradle loop barrel enzymes, similar acidic groups can be identified in the same DPBB location (just before β3 and β6), indicating that the initial evolution of DPBBs may have been to chelate Mg (Coles et al., 1999).

Fig. 7.8 shows the 2-Mg mechanism for RNA polymerization. The 3′-O of the RNA chain attacks the α-phosphate of the incoming NTP substrate to add a single NMP unit to the chain and to release pyrophosphate (Vassylyev, Vassylyeva, Perederina, et al., 2007; Vassylyev, Vassylyeva, Zhang, et al., 2007). Mg1 is held tightly by D739, D741, and D743 within the NADFDGD loop between β5 and β6 of the DPBB2 (β′subunit). Mg2 enters with the NTP substrate and probably interacts with D686 of the DPBB1 (β subunit). Mg2 probably leaves with pyrophosphate.

Evolution of archaeal and bacterial GTFs

Because we posit that Archaea are older than Bacteria, we first consider general transcription factors (GTFs) in Archaea (Blombach et al., 2015; Jun et al., 2011). To recognize a core promoter, Archaea utilize TBP (TATA-box binding protein), TFB (transcription factor B) and TFE (transcription factor E). It appears that Bacteria evolved σ factors from TFB and lost TBP and TFE in evolution. Fig. 7.9 shows a promoter-TBP-TFB complex from Archaea (Littlefield et al., 1999). Fig. 7.9A is a detail of the image in Fig. 7.9B to indicate the helix-turn-helix (HTH) motif of the most C-terminal HTH domain. TBP contacts the 8-nt TATA-box. TBP includes a C-terminal repeat sequence that forms a pseudodimer of β-sheet folds to align with pseudodimeric DNA. TBP occupies the minor groove of the DNA. TFB includes two cyclin-like repeats (CLR) formed as 5-α-helix bundles that bind DNA upstream and downstream of TATA (Lagrange et al., 1998; Renfrow et al., 2004). The last 3-helices of each CLR comprise a typical HTH DNA-binding motif (Fig. 7.9A). HTH motifs are comprised of H1-T1-H2-T2-H3 (H for helix; T for turn). Characteristically, H1 braces H2 and H3. H2 is generally a short helix. The N-terminus of H3 penetrates the major groove of DNA and makes most sequence-specific contacts. Fig. 7.9A emphasizes the typical DNA contacts of HTH2 of TFB to the BREup (TFB-recognition element upstream of

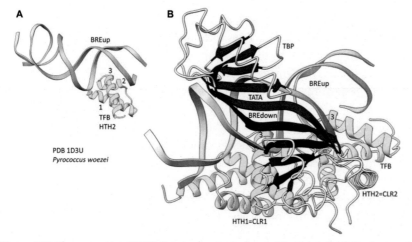

Figure 7.9 The promoter-TBP-TFB complex in Archaea. (A) A detail of the image in panel (B), shows that TFB HTH units are typical and make typical contacts to the major groove of DNA. (B) The promoter-TBP-TFB complex. HTH1 and HTH2 are the last 3 helices of 5-helix cyclin-like repeats (CLR1 and CLR2).

TATA) of the archaeal promoter. Fig. 7.9B is a more complex image that includes TBP and CLR1 and CLR2 of TFB. H3 of CLR1 and CLR2 interacts with the major groove of DNA at BREdown and BREup. TFE is another GTF in Archaea that does not make extensive sequence-specific contacts to DNA (Blombach et al., 2015). In Bacteria, TBP and TFE appear to have been lost in evolution. The TFB C-terminal CLR/HTH repeats appear to have been duplicated and modified in evolution to generate bacterial σ factors.

Bacterial σ factors are homologs of TFB (Burton & Burton, 2014; Burton, 2014; Burton et al., 2016; Iyer & Aravind, 2012) (Fig. 7.10). This idea was first postulated by Aravind and coworkers, based on the similarities of HTH units. Similar to TFB, σ factors were initially strings of HTH units. For instance, σA appears to be derived from 4-HTH units (HTH1-4). We posit that σA was derived from duplication of the TFB C-terminus CLR/ HTH units. σ54, by contrast, might be derived from 6 to 7 (or possibly 8) HTH units. σ54 might have resulted from the early duplication of σA. The more N-terminal HTH units in both σA and σ54 are more degenerate, and, therefore, less recognizable. Here, we consider the four most

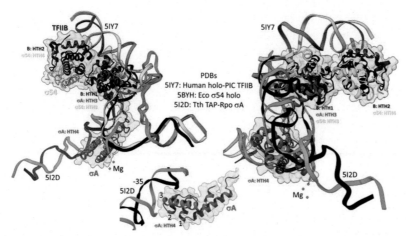

Figure 7.10 Bacterial σ factors and human TFIIB are homologs. Two views and one detail are shown. Two initiation complexes (human and *Thermus thermophilus*) and a σ54 holoenzyme structure (*Escherichia coli*) were overlaid. σA HTH3, σ54 HTH3, and TFIIB (B) HTH1 overlay at the upstream edge of the transcription bubble. σ54 HTH4 and TFIIB HTH2 partly overlay upstream (i.e., BREup). The detail is of σA HTH4 showing characteristic HTH contacts to the promoter −35 region. RNA is cyan. Mg is green. Upstream DNA strands are labeled: 5IY7: (*pink*) nontemplate; (*yellow*) template; and 5I2D: (*green*) nontemplate; (*blue*) template. (For interpretation of the references to colour in this figure legend, the reader is referred to the web version of this article.)

C-terminal HTH units, which are in common comparing σA and σ54, and number them 1 → 4, from the N-terminal end, so HTH4 is the most C-terminal σ HTH unit. TFB, by contrast, includes two HTH units, numbered HTH1 and HTH2, C-terminal to an N-terminal Zn finger domain. So, HTH4 in σA and σ54 corresponds to HTH2 in TFB. HTH3 in σA and σ54 corresponds to HTH1 in TFB. The concept of σ and TFB homology is necessary to consider archaeal and bacterial divergence and the evolution and divergence of promoters.

To further support the homology of σ factors and TFB, we prepared overlays of initiation complexes from bacterial and human systems (Fig. 7.10). Human TFIIB is a close homolog of archaeal TFB. RNAP and other GTFs were removed from the image to attempt simplification. Fig. 7.10 is an overlay of three structures: (1) a human preinitiation complex (PDB 5IY7) (He et al., 2016), (2) a bacterial σA early initiation complex, with a short RNA (PDB 5I2D) (Feng et al., 2016), and (3) a bacterial σ54 holoenzyme (PDB 5BYH) (Yang et al., 2015). Because the image is somewhat busy, two views and a detailed view are shown. TFIIB HTH1, σA HTH3, and σ54 HTH3 colocalize at the upstream end of the transcription bubble. TFIIB HTH2 and σ54 HTH4 partly overlay in the upstream DNA region. By contrast, σA HTH4 follows the diverging trajectory of the upstream DNA to which HTH4 binds at the −35 promoter region (detail image). Notice that σA HTH4 makes typical HTH contacts to the −35 region of the bacterial promoter (Fig. 7.10; detail image), just as TFB makes typical HTH contacts to BREup and BREdown (Fig. 7.9). We conclude from the overlay of these structures that HTH4 and HTH3 of bacterial σ factors correspond to HTH2 and HTH1 of human TFIIB (Burton & Burton, 2014; Burton, 2014; Burton et al., 2016; Iyer & Aravind, 2012).

Promoter-specific regulatory HTH factors

We speculate that GTFs TBP and TFB may have been present at LUCA as part of the earliest mechanisms for opening and managing DNA templates. In Archaea and Bacteria, many promoter-specific transcription factors are dimeric HTH or winged-HTH (HTH factors with β-sheet "wings") factors (Aravind et al., 2005; Iyer & Aravind, 2012). These promoter-specific HTH factors may somehow have been derived by simplification of the CLR domains of TFB (5-α-helix bundles), followed generally by homodimerization. We note that bacterial σ factor HTH units are simplified from the TFB 5-helix CLR formats, from which σ factors appear to be derived (Burton & Burton, 2014; Iyer & Aravind, 2012). The HTH motif was, therefore, a core founding feature in Archaea and Bacteria of the early evolution of both transcriptional GTFs (TFB and σ) and regulatory (HTH and winged-HTH factors) mechanisms.

Evolution of archaeal and bacterial promoters

A model for the divergence of archaeal and bacterial promoters is described (Fig. 7.11). Because of the long passage of time, we are not certain that all aspects of a core promoter model can precisely be stated. The model is presented in order to provide a simple possible narrative that may stimulate more sophisticated bioinformatics approaches to this problem than we were able to do. Also, the model is based partly on our opinion that Archaea is most similar to LUCA, that Bacteria are more derived, and that Bacteria evolved from Archaea (Battistuzzi et al., 2004; Lei & Burton, 2020; Long et al., 2020; Marin et al., 2017). There are reasons to consider this idea. A recent paper indicated that LUCA was most similar to Archaea and that Bacteria were derived from Archaea. tRNAs and tRNAomes (all the tRNAs for an organism) are simpler and more similar to the primordial tRNA sequence in Archaea (Kim et al., 2019; Lei & Burton, 2020; Pak et al., 2018). Also, aminoacyl-tRNA synthetases and the genetic code are simpler to model in Archaea than in Bacteria, indicating that Archaea are more similar to LUCA than Bacteria.

Fig. 7.11 compares a bacterial σA promoter and its GTF contacts and an archaeal promoter and its GTF contacts. The bacterial promoter shows sequences characteristic of a strong promoter with multiple contacts to different regions of σA. Bacteria lack TBP and TFE, which we posit may have been lost during bacterial divergence. Bacteria include RNase HIII,

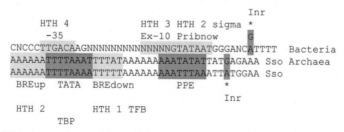

Figure 7.11 Comparison of bacterial σA promoters and archaeal promoters from *Sulfolobus solfataricus* (Sso; an ancient Archaea). See the text for details. Inr for initiator element.

which includes a TBP fold (Brindefalk et al., 2013), however, this possibly indicates that Bacteria had TBP as a transcription factor from Archaea and then lost TBP in evolution, as we propose. According to the structural overlay (Fig. 7.10), however, possibly indicates that Bacteria had TBP as a transcription factor from Archaea and then lost TBP in evolution, as we propose. According to the structural overlay (Burton & Burton, 2014; Burton, 2014; Burton et al., 2016; Iyer & Aravind, 2012). Bacterial σA HTH4 contacts the −35 region of promoters [i.e., (−34)-TTGACA-(-29)]. Archaeal TFB HTH2 contacts the BREup (TFB-recognition element upstream of the TATA-box). TBP binds the 8-nt TATA-box [i.e., (−30)-TTTTAAAA-(-23) in *Sulfolobus solfataricus*] (Ao et al., 2013), but TBP is missing in Bacteria. Bacterial σA HTH3 partly contacts the Extended −10 sequence in double-stranded DNA, found in some promoters, and then resides on double-stranded DNA at the upstream edge of the transcription bubble, as the promoter opens (Fig. 7.10). Archaeal TFB HTH1 contacts the BREdown (TFB-recognition element downstream of the TATA-box) (an A/T-rich sequence downstream from TATA in *Sulfolobus solfataricus*) (Fig. 7.9B). After promoter opening, TFB HTH1 occupies double-stranded DNA just upstream of the transcription bubble (Fig. 7.10).

The Promoter-Proximal Element (PPE) is an A/T-rich sequence in *Sulfolobus solfataricus* promoters upstream of the transcription start [i.e., ~(−11)-AATATTAA-(−4)] (Ao et al., 2013). To us, the PPE resembles a TATA-box and may be derived from one. The PPE appears to be positioned similarly to the bacterial Pribnow box [i.e., (−12)-TATAAT-(−7)] and is similar in sequence. We, therefore, posit that the Pribnow box of bacterial promoters may be derived from an archaeal PPE sequence. Notably, the Pribnow box is recognized by σA HTH2, which is a modified HTH with interesting characteristics. The σA HTH2 opens the bacterial promoter by flipping bases. A(−11) is first flipped out followed by T(−7), leading to promoter opening (Boyaci et al., 2019; Feklistov & Darst, 2011; Feklístov et al., 2014).

Archaeal promoters typically have an initiator sequence surrounding +1, and the transcription start (Ao et al., 2013). Many promoters have (−1)-TATG-(+3). In this case, no 5′-untranslated sequence may be present in the mRNA, which may initiate translation at (+1)-AUG-(+3). (−1)-TGAG-(+3) is also common. In this case, translation generally initiates at a downstream AUG. The initiator element is thought to be recognized directly by RNAP. Bacteria also have an initiator sequence (Cassiano & Silva-Rocha, 2020). Both Archaea and Bacteria utilize ribosome attachment sequences (i.e., AGGA) on some mRNAs with a corresponding interaction sequence near the 3′-end of 16S rRNA (i.e., UCCU).

Interactions of DPBB loops with GTFs

One hypothesis might be that multisubunit RNAP DPBB loops that include inserts contact GTFs in a domain-specific fashion. The idea underlying this hypothesis is that DPBBs form the catalytic center and hold the active site Mg1 and Mg2. The RNAP active site is deeply sequestered within the RNAP core, limiting access to the catalytic center. Inserts in the DPBB loops might allow GTFs to bind closer to the RNAP periphery to communicate with catalytic functions. Because archaeal GTFs and TFB are so different from bacterial σ factors, TFB and σ might be expected to interact with DPBB loops with distinct, domain-specific inserts.

Figs. 7.12 and 7.13 show domain-specific functional contacts of DPBB loops with GTFs. Fig. 7.12 shows a simplified view of a human pre-initiation complex (PDB 5IYD) (He et al., 2016). Most of the factors in the structure have been removed to simplify the image. The human DPBB1 SBHM (β2-β3 insert) contacts TFIIB HTH1/CLR1 located at the

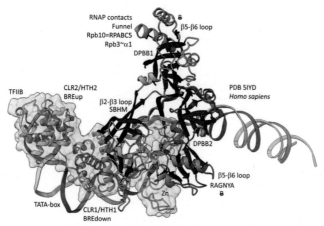

Figure 7.12 Archaea/Eukaryote-specific contacts of TFB/TFIIB with DPBB insert loops. β-sheets are *red*. Other features of Rpb1 are *blue* and Rpb2 are *light red*. TFIIB is orange with transparent space-filling representation. "B" with a *double strike through* indicates a contact specific to Archaea and not found or very different in Bacteria. (For interpretation of the references to colour in this figure legend, the reader is referred to the web version of this article.)

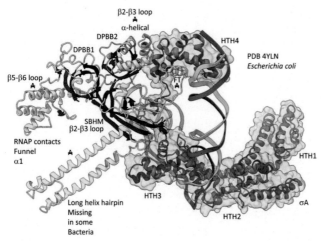

Figure 7.13 Bacteria-specific contacts of σA with DPBB insert loops. β-sheets are *red*. Other β′ features are *beige*, and β features are *yellow*. σA is *green* with transparent space-filling representation. FT for flap tip helix. "A" with a *double strike through* indicates a feature found in Bacteria but very different or not identified in Archaea. (For interpretation of the references to colour in this figure legend, the reader is referred to the web version of this article.)

upstream edge of the transcription bubble. Interestingly, the human DPBB2 RAGNYA β2–β3 insert, specific for Archaea and Eukaryotes, contacts the N-terminal Zn finger of TFIIB. In Fig. 7.13, a detail of the *Escherichia coli* RNAP initiation complex is shown (PDB 4YLN) (Zuo & Steitz, 2015). Bacterial σA HTH3, at the upstream end of the transcription bubble, contacts the SBHM. Thus, homologous GTFs in Archaea (TFB) and Bacteria (σA) make domain-specific contacts to their domain-specific SBHMs. In Bacteria, the flap tip helix is an extension of the SBHM that contacts the σA HTH4, bound to the −35 promoter region. Interestingly, the *Escherichia coli* RNAP SBHM includes a long helix hairpin motif as an insert, missing in Archaea and many Bacteria (i.e., missing in *Thermus thermophilus*, an ancient Bacteria). The long helix hairpin insert contacts σA HTH3 in the initiating complex. The DPBB2 β2–β3 insert in *Escherichia coli* RNAP is an α-helical motif that substitutes for the very different RAG-NYA insert in Archaea, which contacts the N-terminal Zn motif in TFIIB (Fig. 7.12). The corresponding DPBB2 β2–β3 α-helical insert in Bacteria makes domain-specific contacts to αA HTH4, bound at the −35 promoter region (Fig. 7.13).

The DPBB1 β5-β6 insert shows some homology in Archaea and Bacteria but, also, a significant domain-specific character, so we attempted to identify a GTF that might contact this region. We were unsuccessful. So far as we can discern, the β5-β6 DPBB1 inserts in Archaea and Bacteria make domain-specific contacts to other regions of RNAP (Fig. 7.14). In Archaea, the β5-β6 DPBB1 insert contacts: (1) the Rpo1N funnel (A′; homolog of β′ in Bacteria); (2) Rpo10 (N; homolog of RPABC5 in Eukarya); and (3) Rpo3 (C; homolog of α1 in Bacteria). In Bacteria, the β5-β6 DPBB1 insert makes similar domain-specific contacts to RNAP (not shown).

During transcription elongation, TFB and σ factors cycle off RNAP and are replaced by the elongation factor homologs Spt5/Spt4 in Archaea and

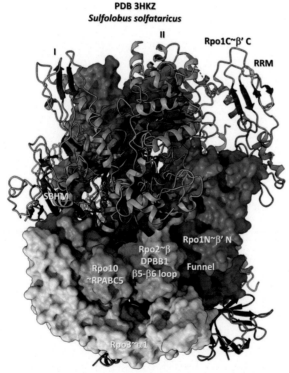

Figure 7.14 The DPBB1 β5-β6 loop (space-filling representation) contacts RNAP. In Archaea, the DPBB1 β5-β6 loop contacts the Rpo1N (homolog of β′ in Bacteria) funnel, the Rpo2 (homolog of β in Bacteria) N-terminal domain (lobe II), Rpo3 (homolog of α1 in Bacteria) and Rpo10 (homolog of RPABC5 in Eukarya). The SBHM contacts lobe I of the N-terminal Rpo2 domain and Rpo3.

NusG in Bacteria (Blombach et al., 2015; Hartzog & Fu, 2013; Tomar & Artsimovitch, 2013; Wang & Artsimovitch, 2021; Werner, 2012; Yakhnin & Babitzke, 2014). These elongation factors occupy approximately the same positions on RNAP as HTH2 and HTH3 of bacterial σA (not shown). These elongation factors, therefore, make domain-specific contacts to the SBHM of their DPBB1 (i.e., see PDB 5TBZ) (Liu & Steitz, 2017). Contacts to GTFs are also specific to the initiation and elongation phases of the transcription cycle. For instance, in Bacteria, the flap tip helix contacts σA during initiation (Fig. 7.13) but does not contact NusG during elongation.

Divergence of Archaea and Bacteria

The evolution of life on Earth appears to be a simple outline with overwhelming detail. According to our view, prelife evolved to LUCA, which we interpret as an ancient Archaea. Archaea diverged to generate Bacteria, which became a more flexible and, in many ways, more successful prokaryotic domain, restricting Archaea somewhat to the margins (i.e., to extremophile environments). Multiple Archaea and Bacteria fused to form Eukaryotes, which have occupied many new niches on Earth (Castelle & Banfield, 2018; Eme et al., 2018; Forterre, 2015). Ancient Archaea, therefore, are very similar to LUCA. Bacteria are more innovated than Archaea and more derived evolutionarily. Because of their mitochondria and complex genomes and development, Eukaryotes have many new capacities lacking in Archaea and Bacteria. We refer to the splitting of the archaeal and bacterial domains as "the great divergence," and we consider this event to be one of the most important advances in the evolution of life as we know it on Earth.

There are several defining differences comparing Archaea and Bacteria: that is, (1) evolution of TFB (Archaea) versus σ factors (Bacteria); (2) utilization of DNAPs PolD and PolB (Archaea) versus PolC (Bacteria) (Koonin et al., 2020), and (3) archaeal versus bacterial membranes (Lane, 2020; Lane & Martin, 2012). Above, we have discussed the divergence of archaeal and bacterial GTFs and promoters in some detail. We consider modifications of bacterial transcription systems to be fundamental and possibly the founding difference in the great divergence of Bacteria from Archaea. For instance, the evolution of bacterial σ factors appears to have driven the simplification and divergence of bacterial RNAPs from archaeal ancestors.

References

Alva, V., Koretke, K. K., Coles, M., & Lupas, A. N. (2008). *Current Opinion in Structural Biology, 18*(3), 358–365. https://doi.org/10.1016/j.sbi.2008.02.006

Ao, X., Li, Y., Wang, F., Feng, M., Lin, Y., Zhao, S., Liang, Y., & Peng, N. (2013). *Journal of Bacteriology, 195*(22), 5216–5222. https://doi.org/10.1128/JB.00768-13

Aravind, L., Anantharaman, V., Balaji, S., Babu, M. M., & Iyer, L. M. (2005). *FEMS Microbiology Reviews, 29*(2), 231–262. https://doi.org/10.1016/j.femsre.2004.12.008

Balaji, S., & Aravind, L. (2007). *Nucleic Acids Research, 35*(17), 5658–5671. https://doi.org/10.1093/nar/gkm558

Battistuzzi, F. U., Feijao, A., & Hedges, S. B. (2004). *BMC Evolutionary Biology, 4.* https://doi.org/10.1186/1471-2148-4-44

Blombach, F., Salvadori, E., Fouqueau, T., Yan, J., Reimann, J., Sheppard, C., & et al. (2015). (Vol. 4).

Boyaci, H., Chen, J., Jansen, R., Darst, S. A., & Campbell, E. A. (2019). *Nature, 565*(7739), 382–385. https://doi.org/10.1038/s41586-018-0840-5

Brindefalk, B., Dessailly, B. H., Yeats, C., Orengo, C., Werner, F., & Poole, A. M. (2013). *Nucleic Acids Research, 41*(5), 2832–2845. https://doi.org/10.1093/nar/gkt045

Burton, Z. F. (2014). *Transcription, 5.* https://doi.org/10.4161/trns.28674

Burton, S. P., & Burton, Z. F. (2014). *Transcription, 5*(4). https://doi.org/10.4161/21541264.2014.967599

Burton, Z. F., Opron, K., Wei, G., & Geiger, J. H. (2016). *Transcription, 7*(1), 1–13. https://doi.org/10.1080/21541264.2015.1128518

Cassiano, M. H. A., & Silva-Rocha, R. (2020). bioRxiv. https://doi.org/10.1101/2020.05.05.079335.

Castelle, C. J., & Banfield, J. F. (2018). *Cell, 172*(6), 1181–1197. https://doi.org/10.1016/j.cell.2018.02.016

Coles, M., Diercks, T., Liermann, J., Gröger, A., Rockel, B., Baumeister, W., Koretke, K. K., Lupas, A., Peters, J., & Kessler, H. (1999). *Current Biology, 9*(20), 1158–1168. https://doi.org/10.1016/S0960-9822(00)80017-2

Coles, M., Djuranovic, S., Söding, J., Frickey, T., Koretke, K., Truffault, V., Martin, J., & Lupas, A. N. (2005). *Structure, 13*(6), 919–928. https://doi.org/10.1016/j.str.2005.03.017

Coles, M., Hulko, M., Djuranovic, S., Truffault, V., Koretke, K., Martin, J., & Lupas, A. N. (2006). *Structure, 14*(10), 1489–1498. https://doi.org/10.1016/j.str.2006.08.005

Da Cunha, V., Gaia, M., Gadelle, D., Nasir, A., Forterre, P., & Rokas, A. (2017). *PLOS Genetics, 13*(6), e1006810. https://doi.org/10.1371/journal.pgen.1006810

Da Cunha, V., Gaia, M., Nasir, A., Forterre, P., & Rokas, A. (2018). *PLOS Genetics, 14*(3), e1007215. https://doi.org/10.1371/journal.pgen.1007215

Eme, L., Spang, A., Lombard, J., Stairs, C. W., & Ettema, T. J. G. (2018). *Nature Reviews Microbiology, 16*(2), 120. https://doi.org/10.1038/nrmicro.2017.154

Feklistov, A., & Darst, S. A. (2011). *Cell, 147*(6), 1257–1269. https://doi.org/10.1016/j.cell.2011.10.041

Feklístov, A., Sharon, B. D., Darst, S. A., & Gross, C. A. (2014). *Annual Review of Microbiology, 68*, 357–376. https://doi.org/10.1146/annurev-micro-092412-155737

Feng, Y., Zhang, Y., & Ebright, R. H. (2016). *Science, 352*(6291), 1330–1333. https://doi.org/10.1126/science.aaf4417

Forterre, P. (2015). *Frontiers in Microbiology, 6.* https://doi.org/10.3389/fmicb.2015.00717

Fouqueau, T., Blombach, F., & Werner, F. (2017). *Annual Review of Microbiology, 71*, 331–348. https://doi.org/10.1146/annurev-micro-091014-104145

Goddard, T. D., Huang, C. C., Meng, E. C., Pettersen, E. F., Couch, G. S., Morris, J. H., & Ferrin, T. E. (2018). *Protein Science, 27*(1), 14–25. https://doi.org/10.1002/pro.3235

Hartzog, G. A., & Fu, J. (2013). *Biochimica et Biophysica Acta—Gene Regulatory Mechanisms, 1829*(1), 105–115. https://doi.org/10.1016/j.bbagrm.2012.08.007

He, Y., Yan, C., Fang, J., Inouye, C., Tjian, R., Ivanov, I., & Nogales, E. (2016). *Nature, 533*, 359–365. https://doi.org/10.1038/nature17970

Iyer, L. M., & Aravind, L. (2012). *Journal of Structural Biology, 179*(3), 299–319. https://doi.org/10.1016/j.jsb.2011.12.013

Iyer, L. M., Koonin, E. V., & Aravind, L. (2003). *BMC Structural Biology, 3*, 1–23. https://doi.org/10.1186/1472-6807-3-1

Jun, S. H., Reichlen, M. J., Tajiri, M., & Murakami, K. S. (2011). *Critical Reviews in Biochemistry and Molecular Biology, 46*(1), 27–40. https://doi.org/10.3109/10409238.2010.538662

Kim, Y., Opron, K., & Burton, Z. F. (2019). *Life, 9*(2). https://doi.org/10.3390/life9020037

Koonin, E. V., Krupovic, M., Ishino, S., & Ishino, Y. (2020). *BMC Biology, 18*(1). https://doi.org/10.1186/s12915-020-00800-9

Korkhin, Y., Unligil, U. M., Littlefield, O., Nelson, P. J., Stuart, D. I., Sigler, P. B., Bell, S. D., & Abrescia, N. G. A. (2009). *PLoS Biology, 7*(5). https://doi.org/10.1371/journal.pbio.1000102

Lagrange, T., Kapanidis, A. N., Tang, H., Reinberg, D., & Ebright, R. H. (1998). *Genes and Development, 12*(1), 34–44. https://doi.org/10.1101/gad.12.1.34

Lane, N. (2020). *Current Biology, 30*(10), R471–R476. https://doi.org/10.1016/j.cub.2020.03.055

Lane, W. J., & Darst, S. A. (2010a). *Journal of Molecular Biology, 395*(4), 671–685. https://doi.org/10.1016/j.jmb.2009.10.062

Lane, W. J., & Darst, S. A. (2010b). *Journal of Molecular Biology, 395*(4), 686–704. https://doi.org/10.1016/j.jmb.2009.10.063

Lane, N., & Martin, W. F. (2012). *Cell, 151*(7), 1406–1416. https://doi.org/10.1016/j.cell.2012.11.050

Lei, L., & Burton, Z. F. (2020). *Life, 10*(3). https://doi.org/10.3390/life10030021

Littlefield, O., Korkhin, Y., & Sigler, P. B. (1999). *Proceedings of the National Academy of Sciences of the United States of America, 96*(24), 13668–13673. https://doi.org/10.1073/pnas.96.24.13668

Liu, B., & Steitz, T. A. (2017). *Nucleic Acids Research, 45*(2), 968–974. https://doi.org/10.1093/nar/gkw1159

Long, X., Xue, H., & Wong, J. T.-F. (2020). *Evolutionary Bioinformatics, 16*. https://doi.org/10.1177/1176934320908267, 117693432090826.

Madru, C., Henneke, G., Raia, P., Hugonneau-Beaufet, I., Pehau-Arnaudet, G., England, P., Lindahl, E., Delarue, M., Carroni, M., & Sauguet, L. (2020). *Nature Communications, 11*(1). https://doi.org/10.1038/s41467-020-15392-9

Marin, J., Battistuzzi, F. U., Brown, A. C., & Hedges, S. B. (2017). *Molecular Biology and Evolution, 34*(2), 437–446. https://doi.org/10.1093/molbev/msw245

Osman, S., & Cramer, P. (2020). *Annual Review of Cell and Developmental Biology, 36*, 1–34. https://doi.org/10.1146/annurev-cellbio-042020-021954

Pak, D., Du, N., Kim, Y., Sun, Y., & Burton, Z. F. (2018). *Transcription, 9*(3), 137–151. https://doi.org/10.1080/21541264.2018.1429837

Pettersen, E. F., Goddard, T. D., Huang, C. C., Meng, E. C., Couch, G. S., Croll, T. I., Morris, J. H., & Ferrin, T. E. (2021). *Protein Science, 30*(1), 70–82. https://doi.org/10.1002/pro.3943

Qian, X., Hamid, F. M., El Sahili, A., Darwis, D. A., Wong, Y. H., Bhushan, S., Makeyev, E. V., & Lescar, J. (2016). *Journal of Biological Chemistry, 291*(17), 9295–9309. https://doi.org/10.1074/jbc.M115.685933

Raia, P., Carroni, M., Henry, E., Pehau-Arnaudet, G., Brûlé, S., Béguin, P., Henneke, G., Lindahl, E., Delarue, M., Sauguet, L., & Stock, A. M. (2019). *PLOS Biology, 17*(1), e3000122. https://doi.org/10.1371/journal.pbio.3000122

Renfrow, M. B., Naryshkin, N., Lewis, L. M., Chen, H. T., Ebright, R. H., & Scott, R. A. (2004). *Journal of Biological Chemistry, 279*(4), 2825–2831. https://doi.org/10.1074/jbc.M311433200

Salgado, P. S., Koivunen, M. R. L., Makeyev, E. V., Bamford, D. H., Stuart, D. I., & Grimes, J. M. (2006). *PLoS Biology, 4*(12), 2274–2281. https://doi.org/10.1371/journal.pbio.0040434

Sauguet, L. (2019). *Journal of Molecular Biology, 431*(20), 4167–4183. https://doi.org/10.1016/j.jmb.2019.05.017

Tomar, S. K., & Artsimovitch, I. (2013). *Chemical Reviews, 113*(11), 8604–8619. https://doi.org/10.1021/cr400064k

Vassylyev, D. G., Vassylyeva, M. N., Perederina, A., Tahirov, T. H., & Artsimovitch, I. (2007). *Nature, 448*(7150), 157–162. https://doi.org/10.1038/nature05932

Vassylyev, D. G., Vassylyeva, M. N., Zhang, J., Palangat, M., Artsimovitch, I., & Landick, R. (2007). *Nature, 448*(7150), 163–168. https://doi.org/10.1038/nature05931

Wang, B., & Artsimovitch, I. (2021). *Frontiers in Microbiology, 11.* https://doi.org/10.3389/fmicb.2020.619618

Werner, F. (2012). *Journal of Molecular Biology, 417*(1–2), 13–27. https://doi.org/10.1016/j.jmb.2012.01.031

Werner, F., & Grohmann, D. (2011). *Nature Reviews Microbiology, 9*(2), 85–98. https://doi.org/10.1038/nrmicro2507

Yakhnin, A. V., & Babitzke, P. (2014). *Current Opinion in Microbiology, 18*(1), 68–71. https://doi.org/10.1016/j.mib.2014.02.005

Yang, Y., Darbari, V. C., Zhang, N., Lu, D., Glyde, R., Wang, Y. P., Winkelman, J. T., Gourse, R. L., Murakami, K. S., Buck, M., & Zhang, X. (2015). *Science, 349*(6250), 882–885. https://doi.org/10.1126/science.aab1478

Zatopek, K. M., Alpaslan, E., Evans, T. C., Sauguet, L., & Gardner, A. F. (2021). *Nucleic Acids Research, 48*(21), 12204–12218. https://doi.org/10.1093/nar/gkaa986

Zuo, Y., & Steitz, T. A. (2015). *Molecular Cell, 58*(3), 534–540. https://doi.org/10.1016/j.molcel.2015.03.010

CHAPTER 8

The 3-minihelix tRNA evolution theorem

Zachary F. Burton

Department of Biochemistry and Molecular Biology, Michigan State University, East Lansing, MI, United States

Models for tRNA evolution

The following models have been advanced to describe evolution of tRNA. First, a 2-minihelix model was posited (Di Giulio, 2012, 2019; Giulio, 2009; Nagaswamy & Fox, 2003). As shown below, no 2-minihelix model can be correct. Second, an accretion model was proposed (Caetano-Anollés & Caetano-Anollés, 2016). In an accretion model, tRNA evolved a stem at a time. Because of the mechanism of sequence growth, any accretion model is a random sequence model, which cannot apply to tRNA, because tRNA has clear internal homologies. As I show, tRNA was generated from ordered sequences: repeats and inverted repeats of known sequence (Kim et al., 2018, 2019; Pak et al., 2017, 2018). Third, a 3-minihelix model was proposed by my laboratory and strongly supported using statistical tests, sequence logos, and data from typical tRNAs (Kim et al., 2018, 2019; Pak et al., 2017, 2018). I consider the 3-minihelix model to be proven correct. Finally, an RNA ring (Uroboros) model has been suggested and supported based on statistical analysis (Demongeot & Seligmann, 2019a, 2019b, 2019c, 2019d, 2019e). I cannot judge the meaning of the statistics in support of this fourth model, but I do demonstrate that this model is also not correct. In addition to being an overly theoretical model, the RNA ring (Uroboros) model is an accretion and random sequence model, which is inconsistent with highly conserved and ordered sequences and internal tRNA homologies.

Because tRNA forms the core of genetic coding, tRNA is the central intellectual property in the evolution of life on Earth. Remarkably, sequence repeats and inverted repeats gave rise to tRNA, so tRNA is the product of surprisingly ordered processes in the ancient prelife world. Many have ventured that the first macromolecules were generated from random polymers, but, at least for tRNA, this is an incorrect assumption (Kim et al., 2018, 2019; Pak et al., 2017, 2018). Remarkably, a history of evolution is

The Makings of a Clinical Protocol
ISBN 978-0-323-95749-6
https://doi.org/10.1016/B978-0-323-95749-6.00001-6

preserved in existing tRNA sequences, making tRNA a living fossil representing sequences predating cellular life from ~4 billion years ago. To discuss the various tRNA evolution models, I will first demonstrate how tRNA evolved. Next, I falsify alternate models.

Numbering of tRNAs

Classic numbering is based on eukaryotic tRNAs with 3-nt deleted within the D loop, causing potential confusion. In this paper, I use both classic numbering and a local system consistent with evolution, described in the figures.

The 3-minihelix model

Fig. 8.1 shows the 3-minihelix model for evolution of tRNA^{Pri} (a primordial tRNA) (Kim et al., 2018, 2019; Pak et al., 2017). The model accounts for evolution of both type I and type II tRNAs. Type II tRNAs have an expanded V loop (V for variable). Both type I and type II tRNAs arose from a 93-nt precursor. A single internal 9-nt deletion generates a type II tRNA (initially 84-nt; ignoring 3′-ACCA). Two internal 9-nt deletions generate a type I tRNA (initially 75-nt; ignoring 3′-ACCA). According to the model, a type II tRNA is an expected intermediate in processing the 93-nt precursor to a type I tRNA. The 93-nt tRNA precursor was generated from ligation of three 31-nt minihelices via a ribozyme RNA ligase. I posit that, in the ancient world, ribozyme RNA ligases prepared RNAs for complementary replication by linking multiple RNAs together and by attaching a snap-back primer (i.e., an anticodon or T stem-loop-stem 17-nt microhelix or 31-nt minihelix) (Fig. 8.1). 31-nt minihelices and tRNAs were processed out by ribozyme endonucleases and exonucleases. Essentially, all of the necessary ribozymes to construct and

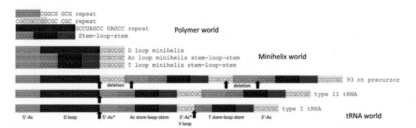

Figure 8.1 The 3-minihelix model for tRNA evolution. Colors indicate internal tRNA homologies.

process the 93-nt tRNA precursor and to utilize tRNAs have been selected or approximated in vitro (Carter & Wills, 2018; Chumachenko et al., 2009; Kim & Joyce, 2004; McGinness et al., 2002; McGinness & Joyce, 2003; Paul & Joyce, 2002; Pressman et al., 2019; Rogers & Joyce, 2001; Turk et al., 2010; Zhang & Cech, 1998).

A D loop minihelix has the sequence 5′-GCGGCGGUAGC-CUAGCCUAGCCUACCGCCGC. The D loop minihelix includes a 7-nt 5′-acceptor stem (5′-GCGGCGG), a D loop 17-nt microhelix (5′-UAGCCUAGCCUAGCCUA), and a 7-nt 3′-acceptor stem (5′-CCGCCGC). The anticodon stem-loop-stem minihelix and the T stem-loop-stem minihelix were identical with sequences ∼5′-GCGGCGGCCGGGUUAAAAACCCGGCCGCCGC. The anticodon and T stem-loop-stem minihelices include 5′- and 3′-acceptor stems (5′-GCGGCGG and 5′-CCGCCGC) and a 17-nt microhelix ∼5′-CCGGGUU/AAAAACCCGG (/ indicates the U-turn). The only sequence ambiguity is within the 7-nt U-turn loops (∼UU/AAAAA). Very clearly, 31-nt minihelices arose from repeats and inverted repeats. Recently, the homology of the anticodon stem-loop-stem and the T stem-loop-stem was questioned (Di Giulio, 2019). The anticodon stem-loop-stem and the T stem-loop-stem, however, have extensive sequence identity. The homology of these 17-nt sequences is obvious and has been confirmed by inspection and by statistical test (Pak et al., 2017).

The tRNA^{Pri} sequence is preserved in existing tRNA sequences with very few systematic changes. Fig. 8.2 shows a typical tRNA from an ancient archaea compared to tRNA^{Pri} (Kim et al., 2018, 2019; Pak et al., 2017, 2018). Fig. 8.2A shows a tRNA structure colored as shown in Figs. 8.1 and 8.2C. A typical tRNA is a composite of many tRNA sequences and is fully described in the tRNA database (Juhling et al., 2009). The match of a typical archaeal tRNA and tRNA^{Pri} is remarkably good (compare Fig. 8.2B and C). Note particularly that, after ∼4 billion years of evolution, the anticodon stem-loop-stem and the T stem-loop-stem are nearly identical sequences and structures. In fact, throughout the sequence, there are very few deviations from the tRNAPri sequence, and all deviations can be explained by the pressures of tRNA folding and the requirements of the anticodon loop. Also, *Pyrobaculum aerophilum* is just one ancient archaea, so further comparisons are useful.

Fig. 8.3 shows three archaeal typical tRNA diagrams (Juhling et al., 2009) for *Pyrococcus* Gly (tRNA^{Gly} rom three *Pyrococcus* species), *Pyrococcus furiosis*, and *Staphylothermus marinus*. The typical *Pyrococcus* tRNA^{Gly} shows

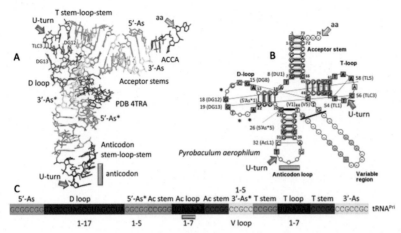

Figure 8.2 Comparison of a typical archaeal tRNA. (A) A tRNA structure (PDB 4TRA). Coloring is as shown in Figs. 8.1 and 8.2C. The positions of the anticodon, amino acid (aa) attachment site and U-turns are indicated. (B) A typical *Pyrobaculum aerophilum* tRNA (DNA sequence is shown). *Thin red lines* indicate interactions discussed in the text. The *bold red line* in the V loop demarcates type I tRNA with a 5-nt V loop from type II tRNAs with V loop expansions (initially 14-nt). *Red asterisks* indicate D loop bases that are deleted in some tRNAs and that disrupt the standard numbering system. *Blue arrows* indicate U-turns in the homologous anticodon and T loops. Standard tRNA numbering is shown. (C) tRNAPri. A local numbering system is indicated. (For interpretation of the references to colour in this figure legend, the reader is referred to the web version of this article.)

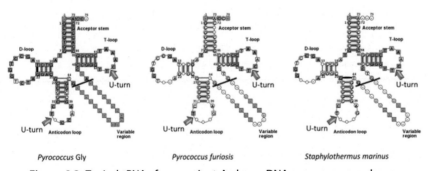

Figure 8.3 Typical tRNAs from ancient Archaea. DNA sequences are shown.

that the D loop 17-nt microhelix is 17-nt, rather than ~14-nt. Classical tRNA numbering can be confusing because numbering was based on eukaryotic tRNAs with a 3-nt deletion in the D loop 17-nt microhelix. The typical tRNA^{Gly} is surprisingly close in sequence to tRNAPri, indicating that the primordial tRNA was a tRNA^{Gly} (Kim et al., 2019). Typical tRNAs from *Pyrococcus furiosis* and *Staphylothermus marinus* are also very close to the tRNA^{Pri} sequence, as is also shown in Fig. 8.2B for *Pyrobaculum aerophilum*. Very clearly, typical archaeal tRNAs are very similar to tRNA^{Pri}. Also, the anticodon stem-loop-stem and the T stem-loop-stem are clearly homologs.

In further support of the 3-minihelix model for tRNA evolution, Fig. 8.4 shows archaeal tRNA (Juhling et al., 2009) sequence logos. For almost every base in tRNA, the top consensus base is the tRNA^{Pri} sequence. Minor exceptions are explained by tRNA folding. The acceptor stems fit the tRNA^{Pri} model very well. In tRNAPri, the 5′–acceptor stem was a GCG repeat (5′-GCGGCGG). The 3′-acceptor stem was a complementary CGC repeat (5′-CCGCCGC). The 5′-As* (As* for acceptor stem fragment) sequence was homologous to the 5′-acceptor stem, as indicated. The 5′-As* sequence was selected to pair in the D stem, selecting some typical sequence changes from tRNA^{Pri}. Similarly, the V loop interacts with the D loop and T loop in the tRNA "elbow" (the bend in the tRNA L shape (Fig. 8.2A)), selecting some sequences. Also, the 5′-As* and 3′-As* were selected not to pair with each other because their prior strong pairing in a 31-nt minihelix was inconsistent with the tRNA fold. Very

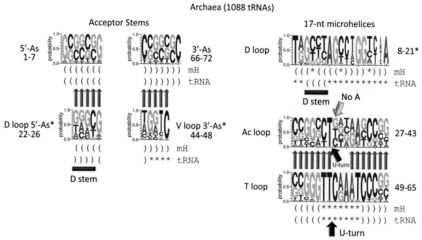

Figure 8.4 Sequence logos demonstrate the 3-minihelix (mH) model. *Blue arrows* indicate homologies. *Red arrows* indicate positions of U-turns. *Red bars* indicate the D stem. The *yellow arrow* indicates that there is no A in the anticodon wobble position. (For interpretation of the references to colour in this figure legend, the reader is referred to the web version of this article.)

clearly, the anticodon loop and the T loop are homologs. The U–turn is in the same position in the 7–nt loops, which have similar loop sequences. The 5–nt anticodon loop and T loop stems (initially 5′-CCGGG and 5′-CCCGG) are very similar in sequence to tRNAPri. As noted below, homology of the anticodon stem-loop-stem and T stem-loop-stem is only consistent with the 3-minihelix tRNA evolution model. Statistical comparison of the anticodon and the T stem-loop-stem 17-nt microhelices gives a P value of $P = .001$, indicating a 1000:1 probability that the anticodon and T stem-loop-stems are homologs (Pak et al., 2017). As the test was run, 0.001 was the highest homology score obtainable.

Analyzing typical tRNA sequences from ancient archaeal species (Fig. 8.5), only a few tRNA[Pri] sequences require explanation. The 5′-acceptor stem was clearly derived from a GCG repeat (GCGGCGG), and the complementary 3′-acceptor stem was derived from a CGC repeat

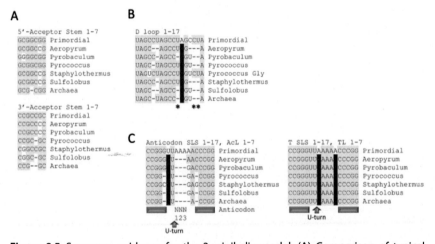

Figure 8.5 Sequence evidence for the 3-minihelix model. (A) Comparison of typical archaeal tRNA sequences and tRNA[Pri] for the 5′- and 3′-acceptor stems. The 5′-acceptor stem was derived from a GCG repeat (GCGGCGG). The 3′-acceptor stem was derived from a CGC repeat (CCGCCGC). (B) Comparison of typical archaeal tRNA sequences and tRNA[Pri] for the D loop. The D loop was derived from a UAGCC repeat (UAGCCUAGCCUAGCCUA). (C) Comparison of typical archaeal tRNA sequences and tRNAPri for the anticodon and the T stem-loop-stems. The anticodon and the T stem-loop-stems were derived from ~CCGGGUU/AAAAACCCGG (/ indicates a U-turn). Sequence ambiguity is only within the 7-nt loops, not in the stems (*green bars*). The positions of U-turns are indicated with *blue arrows*. *Red shading* indicates sequences that may diverge from tRNA[Pri] and that are discussed in the text. (For interpretation of the references to colour in this figure legend, the reader is referred to the web version of this article.)

(CCGCCGC). In the acceptor stems, there are no systematic deviations from the model. Any deviations in individual tRNAs are explained by the requirement to accurately attach amino acids at the 3'-XCCA sequence (X is the discriminator base, which is mostly A in archaea; 3'-A binds the amino acid at the ribose ring) (Fig. 8.2A). The D loop appears to be derived from a UAGCC repeat (initially UAGCCUAGCCUAGCCUA). The first two UAGCC repeats (D1-UAGCCUAGCC-D10) can clearly be identified in typical tRNA sequences. Some D loop sequences can be deleted in tRNAs without disrupting the fold. The only D loop base that requires explanation is G18 (DG12; D loop G12), which was initially A in the 3rd UAGCC repeat. DG12 was altered to G rather than A in order to better intercalate between T loop positions TLA4 and TLA5 (TL = UU/ CAAAU; / = U-turn). I conclude that the D loop was initially a UAGCC repeat (Kim et al., 2019, 2018; Pak et al., 2017).

The 5'-As* (As* indicates a truncated acceptor stem) and 3'-As* sequences were initially paired as a minihelix stem. This pairing was disrupted by mutations to support the tRNA fold, causing slight typical deviations from the tRNA^{Pri} model (Pak et al., 2017). The 5'-As* sequence was also altered to complement the D loop D1-UAGCCUAGCC-D10 sequence to form the 5'-D stem, which was initially D3-GCCU-D6. Because the 5'-As* sequence was initially GGCGG, the initial D 3'-stem was GGCG, with two mismatches to the D 5'-stem GCCU. By altering the 5'-As* sequence from GGCG to GGGC, the D stem was stabilized. I note that GGGC is the most probable archaeal sequence for 5'-As* (Fig. 8.4) as expected. GGGC also reduces interfering pairing interactions between the 5'-As* and the 3'-As* sequence (initially CCGCC).

The 3'-As* became the V loop. The first base of the V loop (typically VU1) was selected to become U to form a wobble pair with 5'-As* G5. The second V loop base (typically VG2) was selected to be primarily G to pack against the D stem G10 = C25 (DG3 = 5'As*C4) base pair, with primary interactions with G10. The third V loop base (typically VG3) remained mostly G to interact with the D stem U13 = G22 (DU6 = 5'As*G1) base pair, with H-bonding to G22. The 4th V loop base (typically VU4) flips out of the tRNA interior. In archaeal tRNAGly, V4 is typically deleted (Fig. 8.3). The 5th base of the V loop (typically VC5) was selected to remain C because it forms the Leavitt reverse Watson-Crick base pair with G15 (DG8) (Kim et al., 2018). Minor deviations in the V loop from the tRNA^{Pri} sequence, therefore, can be explained according to the tRNA fold. Statistical tests strongly support the concept that the 5'-As* and

3′-As* (V loop) sequences were derived from acceptor stem and tRNA^{Pri} sequences, as required by the 3-minihelix tRNA evolution model (Fig. 8.1) (Pak et al., 2017). There are no deviations from the tRNA^{Pri} model in the stems surrounding the anticodon and the T loop, so, the anticodon loop and the T loop are homologs, and their initial stem sequences are known with certainty (Kim et al., 2018, 2019; Pak et al., 2017, 2018). The reader can confirm this by inspection (Figs. 8.2, 8.3, 8.4, and 8.5C). This conclusion is strongly supported by sequence logos (Fig. 8.4) and statistical tests (Pak et al., 2017). An insurmountable problem for tRNA evolution models, other than the 3-minihelix model advocated here, is that none of the alternative models can account for homology of the anticodon stem-loop-stem and the T stem-loop-stem. Alternate models are essentially random sequence models that require adjustment with accretion (random indels) with very weak or no sequence predictions.

The anticodon loop and the T loop sequences require consideration. Because these loops are initially the same sequence, but the loops have adjusted to the tRNA fold, there is some ambiguity in the initial tRNA^{Pri} loop sequence. The tRNA^{Pri} sequence, therefore, is not known with certainty for the anticodon and T loops. I posit that the primordial anticodon loop was close to UU/AAAAA, and diverged typically to CU/NNNAA. NNN is the anticodon. As shown in Fig. 8.4, the anticodon wobble base (34; AcL3) is never (or almost never) A in Archaea, showing that genetic code complexity was limited by tRNA not mRNA (Kim et al., 2019). Other sequences are tolerated in the anticodon loop and have been selected for the functions and specificities of different tRNAs.

The typical T loop in Archaea is almost invariant as UU/CAAAU (Chan & Lowe, 2016; Juhling et al., 2009). T loop position TL4 can be either A or G. T loop position C56 (TLC3) is selected to form a slightly bent Watson-Crick base pair with D loop G19 (DG13) (Fig. 8.2B). Intercalation of G18 (DG12) between loop positions TLA4 and TLA5 elevates loop base 5 to fill the loop, forcing loop bases TLA6 and TLU7 out of the loop, where they form base stacking interactions with the Leavitt G = C (reverse Watson-Crick) base pair in a line of stacked bases (D loop, V loop, and T loop) that extends all the way to and through the D stem. T loop TLU7, which is flipped out of the T loop, is selected to fit a small pocket of bases stabilizing the tRNA fold. T loop base A58 (TLA5) forms a non-Watson-Crick base pair to T loop U54 (TLU1), explaining the conserved base identities. A Watson-Crick TLU1 = TLA5 pair would disrupt the U-turn geometry of the T loop.

The 3-minihelix model, therefore, fully accounts for 75/75 bases in tRNA and almost all of their specific base identities. No other tRNA evolution model comes close to this level of prediction or validation. The 3-minihelix model also accounts for (and predicted) the sequences of type II tRNAs with expanded V loops (Kim et al., 2018). For a recent criticism of the 3-minihelix model, see Di Giulio (2019). I reject all of Di Giulio's criticisms of my model. Di Giulio argues, as he must in order to favor a 2-minihelix model, that the anticodon stem-loop-stem and the nearly identical T stem-loop-stem are not homologs. The anticodon stem-loop-stem and the T stem-loop-stem, however, are nearly identical sequences and are clearly homologs. This point, which is obvious (Figs. 8.1, 8.2, 8.3, 8.4, and 8.5), is sufficient to demonstrate the 3-minihelix model and falsify the 2-minihelix model. From P-value statistical analysis (Pak et al., 2017), the chances that the 2-minihelix model can be correct are at least 1000:1 against. Similarly, the chances the 3-minihelix model is correct are about 1000:1 in favor.

Because tRNA arose from ordered sequences, repeats, and inverted repeats, the 31-nt minihelices that gave rise to tRNA were identified, and their sequences are known in detail. Because these sequences are ordered, some of the sequences that gave rise to minihelices are also known. tRNA world, which we now inhabit, arose from a strange prelife minihelix world. Before 31-nt minihelix world, a strange but ordered prelife polymer world must have existed. Sequences of tRNAs, therefore, provide unexpected insight into abiotic RNA sequences and chemistry. For instance, investigators who select ribozymes in vitro must select for ribozymes that generate GCG, CGC, and UAGCC repeats and CCGGGUUAAAAACCCGG inverted repeats. The 3-minihelix model, therefore, makes specific predictions about abiotic chemistry that can be challenged. The abiotic purposes of 31-nt minihelices can also be tested. I posit that 31-nt minihelices with ligated 3'-ACCA were utilized on a primitive preribosome for polyglycine synthesis, to stabilize protocells, as polyglycine is used to stabilize bacterial peptidoglycan cell walls (Pinho et al., 2013; Scheffers & Pinho, 2005; Zapun et al., 2008). Similarly, tRNA[Pri], which is most similar to tRNAGly, was likely evolved initially as an improved mechanism to synthesize polyglycine (Kim et al., 2019, 2018; Opron & Burton, 2019; Pak et al., 2018). A handful of ribozymes that can be generated in vitro can convert minihelices to tRNA (Kim et al., 2019).

It is difficult to imagine a biological model with more consistent sequence and statistical support than the 3-minihelix model for tRNA evolution. This is particularly relevant because tRNA^{Pri} evolved $\sim$4 billion years ago, but, remarkably, its initial sequence is largely preserved in existing tRNAs. So far as I can discern, no reasonable challenge to the 3-minihelix model can be made today or in the future.

The 2-minihelix model

There are two 2-minihelix models that will be considered here, one from the Di Giulio laboratory and one from the Fox laboratory (Di Giulio, 2004, 2012, 2019; Giulio, 2009; Nagaswamy & Fox, 2003; Widmann et al., 2005). As I show here, tRNA did evolve from minihelices, albeit from three not two (Fig. 8.1). So, one objection to the 2-minihelix model is that the 3-minihelix model is shown to be correct and the 2-minihelix model is, therefore, falsified. Furthermore, the 2-minihelix model cannot be reconciled with the obvious homology of the anticodon stem-loop-stem and the T stem-loop-stem (Figs. 8.1, 8.2, 8.3, 8.4, and 8.5). In a 2-minihelix model, the tRNA must be divided through the anticodon loop to separate the two minihelix sequences. With such a division, the homology of the anticodon loop and the T loop is disrupted, because the anticodon stem-loop-stem has been subdivided. Furthermore, 2-minihelix models require unimaginable convergent evolution of the anticodon stem-loop-stem and the T stem-loop-stem, because they require that the anticodon loop be formed from annealed minihelices with closing of the loop. A powerful objection to the 2-minihelix model is that the anticodon loop and the T loop must converge to identical 7-nt loops with identically placed U-turns and similar loop sequences with identical 5-nt stems. Furthermore, any 2-minihelix model that requires a relationship between the two minihelices to be joined (i.e., homology or complementarity) cannot be correct. The D loop (UAGCCUAGCCUAGCCUA) is clearly unrelated in sequence to the T loop (CCGGGUUCAAAUCCCGG; i.e., Fig. 8.4) (Pak et al., 2017). One 2-minihelix model relied heavily on the placement of a tRNA intron in the anticodon loop of a particular species of Archaea (Giulio, 2009). This is not a valid argument. Most likely, the intron was placed after the tRNA initially evolved. Introns are found in many sites within archaeal tRNAs (Kanai, 2015).

Accretion models some believe in accretion models for evolution of tRNA, in which tRNA grew a stem at a time (Caetano-Anollés & Caetano-Anollés, 2016). Accretion models have reasonably been used to describe the differences in rRNA sequences comparing archaeal, bacterial, and eukaryotic domains (Petrov et al., 2015). Because the 3-minihelix model is demonstrated, however, an accretion model for tRNA evolution cannot be correct. The accretion model is not consistent with the ordered sequences in tRNAs (repeats and inverted repeats). Nor is the accretion model consistent with anticodon stem-loop-stem and T stem-loop-stem homology. It should be noted that the 2-minihelix model is also an accretion model requiring patching with random indels.

tRNA ring model The tRNA ring model (Uroboros model) (Demongeot & Seligmann, 2019a) is also inconsistent with ordered sequences, repeats and inverted repeats in tRNAs (Figs. 8.1, 8.2, 8.3, 8.4, and 8.5). Significantly, the ring model starts with sets of ~22-nt RNA rings with anticodons, which remarkably remain stable in evolution while other sequences change. Rings must then expand into 75-nt tRNAs by random indels. Such a model cannot reasonably converge on tRNAs with closely related sequences within the proposed expansion segments, but such intermediate sequences are clearly related in tRNAs. The ring model does not attempt to describe ordered sequences that I identify. The authors make a statistical argument for their model, which I cannot evaluate. Apparently, the authors have not fully considered archaeal tRNA sequences, which are closest to tRNAomes at LUCA (the last universal common cellular ancestor) (Kim et al., 2019, 2018; Pak et al., 2018).

Predictions of the 3-minihelix model. The 3-minihelix model is highly predictive for tRNA, aminoacyl-tRNA synthetase (aaRS; i.e., GlyRS) and genetic code evolution (Kim et al., 2019). Above, we describe predictions from the 3-minihelix model for abiotic chemistry and evolution. For tRNA, the 3-minihelix model makes predictions for the early radiations of tRNAs to form tRNAomes (all of the tRNAs for an organism) (Pak et al., 2018). Specifically, in *Pyrococcus* (an ancient archaea), tRNA^{Val} (NAC) and tRNA^{Ala} (NGC) are closely related to tRNAs by sequence (Kim et al., 2019). Also, type II tRNA^{Leu} and type II tRNA^{Ser} are closely related (in many organisms). In evolution, I posit that tRNA^{Leu} and tRNA^{Ser}, which share common and neighboring genetic code rows (3rd anticodon position) but are found in different genetic code columns (2nd anticodon position), diverged initially by altering just the second position of the anticodon (Kim et al., 2019). Because tRNA^{Leu} and tRNA^{Ser} are type II tRNAs with expanded V loops, the tRNA^{Ser} V loop evolved to become a determinant for SerRS and the tRNA^{Leu} V loop evolved to become an antideterminant for SerRS to ensure specificity of amino acid attachments to similar tRNAs. Because the 2nd position of the anticodon is most important for recognition in translation, it is predicted that mutation of the 2nd anticodon position could alter amino acid specificity without many changes being necessary in other tRNA sequences. Similarly, tRNA^{Ile} (GAU and CAU) and tRNA^{Met} (CAU) (1 elongator, 1 initiator) are closely related in *Pyrococcus*, although these tRNAs subsequently diverged in more derived organisms to become more distinct. In *Pyrococcus*, tRNA^{Asp} (GUC), tRNA^{Glu} (YUC), and tRNA^{Gln} (YUG) (Y=C or U) are closely related tRNAs. With evolution from LUCA, tRNAs are expected to both diverge and converge in sequence, and this is observed. The tRNAome in *Pyrococcus*, which is an ancient Archaea, is very similar to LUCA and prelife. aaRS enzymes place an amino acid at the 3'-XCCA end of tRNAs and show a great deal of genetic diversification (Kim et al., 2019). Remarkably, closely related aaRS enzymes tend to arrange in columns within the genetic code. Columns represent the second tRNA anticodon position, which is most important for translational fidelity. Therefore, the structure of the genetic code reflects the coevolution of tRNAs, aaRS enzymes, and the code, according to predictable rules that are conserved from the initial evolution of the code. A $10 million dollar (USD) cash prize has been advanced by HeroX (https://www.herox.com/evolution2.0) to generate a functional genetic code without a deity, but this is a problem that, on paper, is largely solved (Kim et al., 2019). As noted above, tRNA evolution via the 3-minihelix model is a solved problem. aaRS evolution is largely a solved problem. Genetic code structure and evolution are largely solved problems (Kim et al., 2019). There has been some controversy about whether Archaea or Bacteria are more ancient organisms, based on the comparisons of many genes (Battistuzzi et al., 2004). For tRNAs, Archaea are clearly older and more similar to LUCA and prelife. Archaeal tRNAomes are less diverged from tRNA^{Pri}, and archaeal tRNAomes are less scrambled (i.e., by divergent and convergent evolution) than bacterial tRNAomes (Pak et al., 2018). Archaeal aaRS enzymes are less diverged from LUCA than bacterial aaRS enzymes. For instance, structurally distinct class I aaRS and class II aaRS are shown to be homologs in Archaea (Kim et al., 2019; Pak et al., 2018). In Bacteria, no such identification could be made because of the radiation of sequences. Also, in Archaea, class I ValRS-IA and IleRS-IA and class II GlyRS-IIA are shown to share a Zn-binding motif, which subsequently disappears in evolution (Kim et al., 2019, 2018; Opron & Burton, 2019; Pak et al., 2018). So, translation functions (tRNAomes and aaRS enzymes) are more ancient in Archaea than in Bacteria.

The 3-minihelix model has been considered to be just another of many models for tRNA evolution, but this assessment is not appropriate. The 3-minihelix model is more similar to a puzzle solution or a mathematical proof than a biological model. Notably, the problem of tRNA evolution was largely resolved as a puzzle solution, by inspection of typical tRNA sequences (Figs. 8.2B, 8.3, and 8.5) (Kim et al., 2019, 2018; Pak et al., 2017). Identification of anticodon and T stem-loop-stems as obvious homologs (Figs. 8.1, 8.2, 8.3, 8.4, and 8.5) solved much of problem (34/75 nucleotides). Knowledge of acceptor stems ((14 + 34)/75) improved the solution. Despite the clear conservation of the first two UAGCC repeats, solution of the D loop as a UAGCC repeat required some effort. Solution of the 5'-As* and 3'-As* (V loop) as relics of acceptor stems also required some effort and statistical analyses. This part of the solution is improved by identifying expanded V loops in type II tRNAs (i.e., tRNA^{Leu} and tRNA$^{Ser)}$ as ligated 3'-As and 5'-As sequences (Kim et al., 2018). Therefore, the same model describes evolution of type I and type II tRNA from a 93-nt tRNA precursor, and type II tRNA is an intermediate in processing to a type I tRNA (Fig. 8.1). Solution of the tRNA puzzle was much easier than a New York Times cross-word puzzle. Anyone reading this paper can now solve the puzzle of tRNA evolution and convince themselves of its validity. There are so few deviations comparing tRNA^{Pri}, tRNA^{Gly} (in ancient Archaea) and tRNA^{Typical} (in ancient Archaea) (Figs. 8.2, 8.3, 8.4, and 8.5), there can be no serious argument opposed to the 3-minihelix model or in favor of any alternate model.

References

Battistuzzi, F. U., Feijao, A., & Hedges, S. B. (2004). *BMC Evolutionary Biology, 4.* https://doi.org/10.1186/1471-2148-4-44

Caetano-Anollés, D., & Caetano-Anollés, G. (2016). *Life, 6*(4), 43. https://doi.org/10.3390/life6040043

Carter, C. W., & Wills, P. R. (2018). *Molecular Biology and Evolution, 35*(2), 269−286. https://doi.org/10.1093/molbev/msx265

Chan, P. P., & Lowe, T. M. (2016). *Nucleic Acids Research, 44*(1), D184−D189. https://doi.org/10.1093/nar/gkv1309

Chumachenko, N. V., Novikov, Y., & Yarus, M. (2009). *Journal of the American Chemical Society, 131*(14), 5257−5263. https://doi.org/10.1021/ja809419f

Demongeot, J., & Seligmann, H. (2019a). *Journal of Computational Biology, 26*(9), 1003−1012. https://doi.org/10.1089/cmb.2018.0256. Mary Ann Liebert Inc.

Demongeot, J., & Seligmann, H. (2019b). *Acta Biotheoretica, 67*(4), 273−297. https://doi.org/10.1007/s10441-019-09356-w

Demongeot, J., & Seligmann, H. (2019c). *Gene, 705,* 95−102. https://doi.org/10.1016/j.gene.2019.03.069

Demongeot, J., & Seligmann, H. (2019d). *Journal of Molecular Evolution, 87*(4–6), 152–174. https://doi.org/10.1007/s00239-019-09892-6

Demongeot, J., & Seligmann, H. (2019e). *Science of Nature, 106*(7–8). https://doi.org/10.1007/s00114-019-1638-5

Di Giulio, M. (2004). *Journal of Theoretical Biology, 226*(1), 89–93. https://doi.org/10.1016/j.jtbi.2003.07.001

Di Giulio, M. (2012). *Journal of Theoretical Biology, 310*, 1–2. https://doi.org/10.1016/j.jtbi.2012.06.022

Di Giulio, M. (2019). *Journal of Theoretical Biology, 480*, 99–103. https://doi.org/10.1016/j.jtbi.2019.07.020

Giulio, D. (2009). *Journal of Molecular Evolution, 69*, 1464–1466. https://doi.org/10.1007/s00239-009-9248-z\nDiGiulioM

Juhling, F., Morl, M., Hartmann, R. K., Sprinzl, M., Stadler, P. F., & Putz, J. (2009). *Nucleic Acids Research, 37*, D159–D162. https://doi.org/10.1093/nar/gkn772. Database.

Kanai, A. (2015). *Life, 5*(1), 321–331. https://doi.org/10.3390/life5010321

Kim, D.-E., & Joyce, G. F. (2004). *Chemistry & Biology, 11*(11), 1505–1512. https://doi.org/10.1016/j.chembiol.2004.08.021

Kim, Y., Kowiatek, B., Opron, K., & Burton, Z. F. (2018). *International Journal of Molecular Sciences, 19*(10). https://doi.org/10.3390/ijms19103275

Kim, Y., Opron, K., & Burton, Z. F. (2019). *Life, 9*(2). https://doi.org/10.3390/life9020037

McGinness, K. E., & Joyce, G. F. (2003). *Chemistry and Biology, 10*(1), 5–14. https://doi.org/10.1016/S1074-5521(03)00003-6

McGinness, K. E., Wright, M. C., & Joyce, G. F. (2002). *Chemistry and Biology, 9*(5), 585–596. https://doi.org/10.1016/S1074-5521(02)00136-9

Nagaswamy, U., & Fox, G. E. (2003). *Origins of Life and Evolution of the Biosphere, 33*(2), 199–209. https://doi.org/10.1023/A:1024658727570

Opron, K., & Burton, Z. F. (2019). *International Journal of Molecular Sciences, 20*(1). https://doi.org/10.3390/ijms20010040

Pak, D., Du, N., Kim, Y., Sun, Y., & Burton, Z. F. (2018). *Transcription, 9*(3), 137–151. https://doi.org/10.1080/21541264.2018.1429837

Pak, D., Root-Bernstein, R., & Burton, Z. F. (2017). *Transcription, 8*(4), 205–219. https://doi.org/10.1080/21541264.2017.1318811

Paul, N., & Joyce, G. F. (2002). *Proceedings of the National Academy of Sciences of the United States of America, 99*(20), 12733–12740. https://doi.org/10.1073/pnas.202471099

Petrov, A. S., Gulen, B., Norris, A. M., Kovacs, N. A., Bernier, C. R., Lanier, K. A., Fox, G. E., Harvey, S. C., Wartell, R. M., Hud, N. V., & Williams, L. D. (2015). *Proceedings of the National Academy of Sciences of the United States of America, 112*(50), 15396–15401. https://doi.org/10.1073/pnas.1509761112

Pinho, M. G., Kjos, M., & Veening, J. W. (2013). *Nature Reviews Microbiology, 11*(9), 601–614. https://doi.org/10.1038/nrmicro3088

Pressman, A. D., Liu, Z., Janzen, E., Blanco, C., Müller, U. F., Joyce, G. F., Pascal, R., & Chen, I. A. (2019). *Journal of the American Chemical Society, 141*(15), 6213–6223. https://doi.org/10.1021/jacs.8b13298

Rogers, J., & Joyce, G. F. (2001). *RNA, 7*(3), 395–404. https://doi.org/10.1017/S135583820100228X

Scheffers, D. J., & Pinho, M. G. (2005). *Microbiology and Molecular Biology Reviews, 69*(4), 585–607. https://doi.org/10.1128/MMBR.69.4.585-607.2005

Turk, R. M., Chumachenko, N. V., & Yarus, M. (2010). *Proceedings of the National Academy of Sciences of the United States of America, 107*(10), 4585–4589. https://doi.org/10.1073/pnas.0912895107

Widmann, J., Di Giulio, M., Yarus, M., & Knight, R. (2005). *Journal of Molecular Evolution,*
    61(4), 524–530. https://doi.org/10.1007/s00239-004-0315-1
Zapun, A., Vernet, T., & Pinho, M. G. (2008). *FEMS Microbiology Reviews, 32*(2), 345–360.
    https://doi.org/10.1111/j.1574-6976.2007.00098.x
Zhang, B., & Cech, T. R. (1998). *Chemistry and Biology, 5*(10), 539–553. https://doi.org/
    10.1016/S1074-5521(98)90113-2

CHAPTER 9

Methylating agents as adjunct therapy to chemotherapeutic alkylating medications for improved outcomes in chronic lymphocytic leukemia: a case study

Bruce K. Kowiatek
Allied Health Sciences, Blue Ridge Community and Technical College, Martinsburg, WV, United States

Introduction

Although the nonenzymatic methylation of cytidine (C) by S-adenosylmethionine (SAM) (Fig. 9.1) and methylcobalamin (Fig. 9.2) to 5-methylcytidine and its subsequent spontaneous deamination to thymidine (T) in DNA at physiologic pH and temperature has been known since the early 1980s and has been implicated in some C to T point mutagenic cancers (Bolton, 1982), cancers in general, and chronic lymphocytic leukemia (CLL) in particular, display a global and promoter hypomethylation respectively of their DNA (Ehrlich, 2009; Upchurch et al., 2016); SAM, therefore, sold as the over-the-counter (OTC) supplement SAM-e, as well as the OTC methylcobalamin precursor cyanocobalamin may still possibly

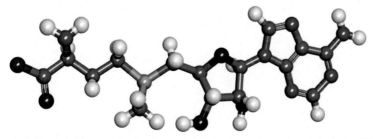

Figure 9.1 Methylating agents: S-adenosylmethionine; primary methylator in all living organisms and viruses.

The Makings of a Clinical Protocol
ISBN 978-0-323-95749-6
https://doi.org/10.1016/B978-0-323-95749-6.00009-0

179

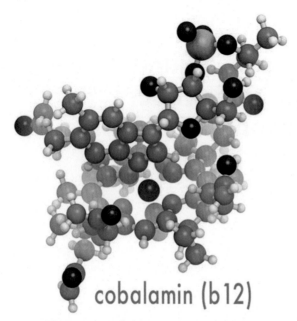

Figure 9.2 Methylating agents: cobalamin.

play a role as potential adjunct therapies in certain cancers, particularly those treated with alkylating chemotherapeutic agents used in CLL, much in the same way that folic acid (Fig. 9.3) is used as a rescue adjunct therapy when using antifolate chemotherapeutic agents (Hagner & Joerger, 2010). Clinical trials in support of this proposal were set to begin just prior to the COVID-19 pandemic but have been indefinitely postponed; however, an ongoing case study employing this methylation protocol has led to its

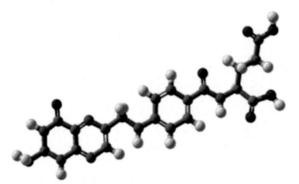

Figure 9.3 Methylating agents: folic acid (folate).

national implementation in the U.S. is currently underway with extremely positive results.

Methylating agents

- All are involved in DNA and RNA synthesis.
- All address cancer hypomethylation.
- All are antiviral by their nature via gene silencing (which can also be involved in gene activation) (Hancock, 1984).
- Possible antiSARS-CoV-2 activity? Studies may be warranted!
- Antiviral activity may complement anticancer mechanism of action; both Human T-cell lymphotropic virus type 1 (HTLV-1) and Epstein—Barr virus (EBV) has been implicated in the progression of CLL (Black & Microbiology, 1993).

The Kowiatek protocol

First Line: SAM-e 400 mg po qd; d/c if patient's SSRI exceeds 10 mg daily and switch to.

Second Line: cobalamin as cyanocobalamin 1000 mcg sl or po qd.

Third Line: Folic acid (folate) 800—1000 mcg po qd; not recommended in antifolate chemotherapies unless as a rescue.

Case study involving the Kowiatek protocol

DS, 52-yo Caucasian female diagnosed with Chronic Lymphocytic Leukemia (CLL).

Administered Radiation Therapy plus Chemotherapy (RTC): bendamustine (Treanda), an alkylating agent, and rituximab (Rituxan), a monoclonal antibody.

Administered Packed Red Blood Cells (PRBCs).

Administered SAM-e 400 mg po qd immediately postchemo.

Tolerated radiation and chemo extremely well with minimal side effects: negative for alopecia and minimal n/v.

CBC highlights immediately post-chemo (12/13/2019)

WBC 4.82 (3.98—10.04)

RBC 4.14 (3.93—6.08)

MCV 96.10 (79.0–94.8) High.
MCHC 31.90 (32.2–36.5) Low.
PLT 228.0 (150.0–369.0)

03/30/2020: change in methylating agent

Although still doing physically well with minimal side effects (still alopecia negative and now n/v negative), the patient began experiencing increased anxiety and depression (most likely COVID-19-related); her SSRI escitalopram (Lexapro) was increased from 10 to 20 mg qd, SAM-e was d/c'd and changed to cobalamin as cyanocobalamin 1000 mcg sl qd, illustrating the only main limitation to SAM-e (Patent & Services, 2006).

CBC highlights of first follow-up allowed by easing of COVID-19 restrictions (04/29/2020)

WBC 4.67 (3.98–10.04)
   RBC 4.12 (3.93–6.08)
   MCV 93.40 (79.0–94.8)
   MCHC 32.20 (32.2–36.5)
   PLT 164.0 (150.0–369.0)

Statistical analysis of 4/29/2020 CBC highlights

WBC decreased by 6% (statistically significant)
   RBC decreased by 0.5% (not statistically significant)
   MCV decreased by 3% (not statistically significant, but normalized)
   MCHC increased by 0.9% (not statistically significant, but normalized)
   PLT 164.0 decreased by 28% (statistically significant)

Discussion of results

Statistically significant reduction in WBC and PLT, the latter reducing the risk of thromboses and emboli, also a significant risk factor seen in sedentary COVID-19 patients.
   Normalization of other pertinent lab values.
   Attenuation of postchemo side effects.
   Well-tolerated by the patient.
   Case study goal of improved outcomes met.

CBC highlights 4/13/2021

WBC 6.0 (4.0–11.0)
   RBC 4.53 (3.80–5.00)
   Hgb 13.7 (12.0–16.0)
   Hct 43.6 (36.0–48.0)
   MCV 96 (80–100)

4/13/2021 CBC analysis

All CBC values remain normalized.

Zero signs and symptoms of relapse with only two more scheduled follow-ups before discharge from oncology care if remission continued.

Patient back to light-duty work and only experiencing mild pain and discomfort postbroken foot.

Further demonstrates the utility and success of the protocol.

Updates: 8/23/2021 and 3/14/2022

The patient returned to full-time employment 10/05/2020.

The patient has discontinued protocol and continues to remain in full remission for 2+ years as per her oncologist and labs.

The patient broke two metatarsal bones in the right foot 12/26/2020.
Underwent period of convalescence and recovery until 04/01/2021.
All CBC values remain normalized as of 3/14/2022.
The protocol is now being widely implemented nationwide in the U.S.

Initial protocol prospectus, July/August 2019

Fig. 9.4.
Fig. 9.5.

U.S. nationwide implementation of protocol first-line as per the U.S. FDA https://www.ehealthme.com/compare-drugs/3615/

10,401 females aged 54 (±5) who take Sam-E, Bendamustine, and Rituxan for CLL are studied.

A number of reports submitted per year with Very High Effectiveness.
Fig. 9.6.

J Biotechnol Biomed 2019; 2 (2): 048-056   DOI: 10.26502/jbb.2642-9128009

Commentary

Non-enzymatic Methylation of Cytosine in RNA by S-adenosylmethionine and Implications for the Evolution of Translation

Bruce K Kowiatek[*]

Blue Ridge Community and Technical College, Martinsburg, WV, USA

*Corresponding Author: Bruce K. Kowiatek, Blue Ridge Community and Technical College, 13650 Apple Harvest Dr, Martinsburg, WV, USA, E-mail: bkowiatek@yahoo.com

Received: 14 June 2019; **Accepted:** 01 July 2019; **Published:** 23 July 2019

Abstract

The non-enzymatic methylation of cytosine (C) to form 5-methylcytosine (5-mC) in deoxyribonucleic acid (DNA) by the intracellular methyl group donor S-adenosylmethionine (SAM), resulting in S-adenosylhomocysteine (SAH) and minor thymine (T) via spontaneous deamination, implicated in certain point mutagenic cancers, has been widely known since the 1980s, as has the proposed Watson-Crick mechanism of the adenine (A) moiety of SAM base-pairing with T or uracil (U). Such analogous base-pairing and non-enzymatic methylation in ribonucleic acid (RNA), however, has not been as widely addressed, particularly with respect to the origins and evolution of the process of translation initiation in the context of the hypothesized RNA world that preceded the current DNA-

Figure 9.4 Initial protocol prospectus, July/August 2019.

U.S. nationwide implementation of protocol second-line as per the U.S. FDA https://www.ehealthme.com/drug-interactions-checker/240576/

94 females aged 54 (±5) who take Cyanocobalamin, Bendamustine, and Rituxan for CLL are studied.

A number of reports submitted per year with Very High Effectiveness. Fig. 9.7.

U.S. nationwide implementation of protocol third-line as per the U.S. FDA https://www.ehealthme.com/drug-interactions-checker/240514/

339 females aged 54 (±5) who take Folic Acid, Bendamustine, and Rituxan for CLL are studied.

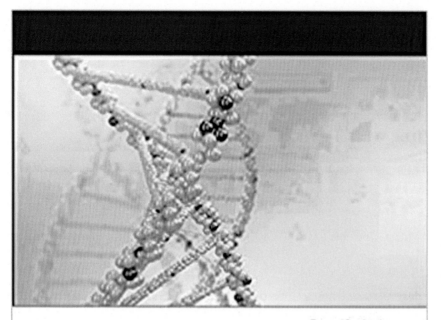

Bruce Kowiatek

Synthesis:

The Biochemical Basis of Life

Figure 9.5 Textbook published on protocol prospectus.

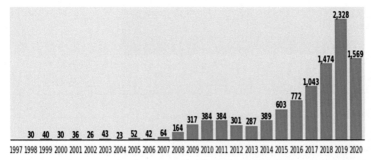

Figure 9.6 U.S. nationwide implementation of protocol first-line as per the U.S. FDA.

Figure 9.7 U.S. nationwide implementation of protocol second-line as per the U.S. FDA.

A number of reports submitted per year with Very High Effectiveness. Fig. 9.8.

Nationwide implementation of protocol first-line as per the U.S. FDA for breast cancer stage 1 https://www.ehealthme.com/drug-interactions-checker/248222/

356 females aged 59 (±5) who take Methotrexate, S–Adenosyl-L-Methionine, Cyclophosphamide, and Fluorouracil are studied.

A number of reports submitted per year with Very High Effectiveness. Fig. 9.9.

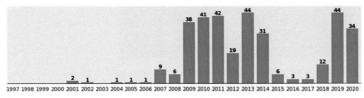

Figure 9.8 U.S. nationwide implementation of protocol third-line as per the U.S. FDA.

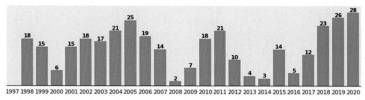

Figure 9.9 Nationwide implementation of protocol first-line as per the U.S. FDA for breast cancer stage 1.

Latest research August 2021

Fig. 9.10.

Self-Dividing Micelles: A Mechanistic Look with Evolutionary and Clinical Implications

DOI: https://doi.org/10.36811/ijbm.2021.110023   IJBM: August: 2021: Page No: 08-20

International Journal of Biology and Medicine

Research Article   Open Access

Self-Dividing Micelles: A Mechanistic Look with Evolutionary and Clinical Implications

Bruce K Kowiatek

Blue Ridge Community and Technical College, USA

*Corresponding Author: Bruce K Kowiatek. Blue Ridge Community and Technical College, 13650 Apple Harvest Dr, Martinsburg, WV 25403, USA. E-mail: bkowiate@blueridgectc.edu

Received Date: Jul 26, 2021 / Accepted Date: Aug 03, 2021/ Published Date: Aug 05, 2021

Abstract

Micellar therapy has become a usefully viable treatment arm in various fields, ranging from oncology to bioimaging. As such, research leading to any improvements or adaptations in administration and techniques can have far-reaching consequences. Potential aspects of prebiotic chemistry may also be explored in such research as well. To that end, proof-of-concept experiments were performed to elucidate a possible mechanism of action for prebiotic protocell division. Representative potentially prebiotically plausible biomolecules, i.e., a fatty acid, amino acid, and nucleotide were mixed and heated in water and subjected to microscopic examination for observation of possible self-division and laboratory testing for the presence of polypeptides and polynucleotides (Biuret, MALDI mass-spec, etc.) with and without the presence of nucleotide. The results are presented for the first time here and a mechanism is proposed that best fits the data obtained. The evolutionary, e.g., prebiotic biomolecular cooperativity, and clinical, e.g., potential antineoplastic micellar/vesicular therapy, ramifications are discussed as well.

Keywords: Micelle; Liposome; Protocell; MRNA; Self-division; Mechanism; Solid tumors

Figure 9.10 Latest research August 2021.

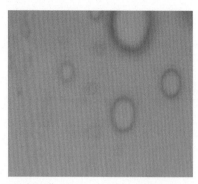

Figure 9.11 Micelle prenucleotide addition; no division.

Micelle self-division

100× oil immersion magnifications of methylene-blue-dyed decyl-maltoside and polyhistidine micelles dividing with the introduction of the nucleotide adenosine monophosphate (AMP).

The full 1:30 recording of representative micelle divisions can be found at https://www.youtube.com/watch?v=2zjDAdG UeZM.

Fig. 9.11.
Fig. 9.12.
Fig. 9.13.
Fig. 9.14.

On the horizon

The latest research now focuses on the development of self-dividing micelles and vesicles (liposomes) to potentially deliver and distribute antineoplastic chemotherapeutic agents in solid tumors.
- Monolayer micelles can deliver and distribute lipophilic agents.
- Bilayer liposomes can deliver and distribute hydrophilic agents.

Also exploring the prospect of mRNA cancer vaccines utilizing such self-dividing micelles and vesicles to deliver and distribute only stop mRNA codons for the treatment and destruction of rapidly-dividing solid tumors.

Figure 9.12 Micelle postnucleotide addition; predivision.

Figure 9.13 Micelle postnucleotide addition; division.

Figure 9.14 Micelle postnucleotide addition; postdivision.

Acknowledgments

Special thanks to:
Patient DS for her participation in this Case Study.
Shenandoah Oncology of Valley Health System, Winchester, VA.

References

Black, J. (1993). Microbiology. *Journal of Applied & Environmental Microbiology, 2*(6), 303–308. https://doi.org/10.12691/jaem-2-6-6

Bolton, S. (1982). (pp. 211–216). New York: Marcel Dekker.

Ehrlich, M. (2009). *Epigenomics, 1*(2), 239–259. https://doi.org/10.2217/EPI.09.33

Hagner, N., & Joerger, M. (2010). *Cancer Management and Research, 2*(1), 293–301. http://www.dovepress.com/getfile.php?fileID=8156%20.

Hancock, R. L. (1984). *Medical Hypotheses, 15*(3), 323–331. https://doi.org/10.1016/0306-9877(84)90022-7

Patent, I. C., & Services. (2006).

Upchurch, G. M., Haney, S. L., & Opavsky, R. (2016). *Frontiers in Oncology, 6.* https://doi.org/10.3389/fonc.2016.00182

CHAPTER 10

Chaos, order, and systematics in evolution of the genetic code

Lei Lei[1] and Zachary F. Burton[2]

[1]School of Biological Sciences, University of New England, Biddeford, ME, United States; [2]Department of Biochemistry and Molecular Biology, Michigan State University, East Lansing, MI, United States

Introduction

Eukaryotic cells divide into compartments. Some compartments are set aside by membranes but others are membraneless and divided instead by liquid phases. Components of membraneless compartments concentrate through local interactions and selection and exclusion of defining components. Hydrogels and liquid-liquid phase separation (LLPS) form these functional units. In human neurological disease and cancer, hydrogel compartments can disassemble and, in some cases, lead to generation of amyloid accretions. Although not yet as extensively studied, hydrogels and LLPS are also becoming recognized in prokaryotic systems. In this paper, we explore hydrogel and LLPS compartments as drivers of the establishment of the genetic code.

Evolution of life on Earth required a small number of key transitions (Fig. 10.1). In this paper, we concentrate on the prelife to cellular life transition and the evolution of coding systems, but we use examples from later evolution to highlight very early events. We consider evolution of life on Earth to be a fairly simple outline with overwhelming relevant detail. The first cellular life on Earth is described as LUCA (the last universal common cellular ancestor), which we consider to be the first organisms with an intact DNA genome and an intact cell (Di Giulio, 2011; Koonin & Novozhilov, 2009, 2017; Weiss et al., 2018). The second major transition is the great divergence of Archaea and Bacteria (Burton & Burton, 2014; Burton et al., 2016; Harish, 2018). Based on our analyses and those of others, we consider Archaea to be most similar to LUCA and Bacteria to be more diverged. The third major transition is the genetic fusion of multiple Archaea and multiple Bacteria to generate Eukaryota (Brueckner & Martin, 2020; Eme et al., 2017; Furukawa et al., 2017; Zachar & Boza, 2020). Endosymbiosis of an α-proteobacterium taken up by an Asgard Archaea

The Makings of a Clinical Protocol
ISBN 978-0-323-95749-6
https://doi.org/10.1016/B978-0-323-95749-6.00014-4

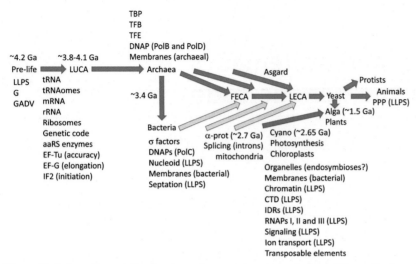

Figure 10.1 A working outline for the main transitions in evolution of life on Earth and the involvement of hydrogels (i.e., LLPS). *PPP*, promoter-proximal pausing; *cyano*, cyanobacterium; *α-prot*, α-proteobacterium; *TBP*, TATA box-binding protein; *TFB*, transcription factor B; *TFE*, transcription factor E; *FECA*, first eukaryotic common ancestor; *LECA*, last eukaryotic common ancestor; *RNAP*, RNA polymerase.

was a key step in eukaryogenesis (Long et al., 2019). The tortured evolutionary path of Eukarya describes the evolution of eukaryote complexity. Subsequently, another endosymbiotic event involving a cyanobacterium invading an alga gave rise to plants. Evolution of animal complexity required evolution of a promoter-proximal pausing mechanism of RNA polymerase II and the RNA polymerase II CTD (carboxy–terminal domain) (Boehning et al., 2018; Guo et al., 2020; Lu et al., 2018). The best descriptions of the key stages relate to evolution of biological coding, translation, and transcriptional mechanisms. We discuss the importance of hydrogels (LLPS) in transitions. In this paper, when we refer to hydrogels or LLPS, we consider these features in all their complexity, including membraneless organelles and associated amyloids (Harmon et al., 2017; Portz & Shorter, 2020; Yoshizawa et al., 2020). It is our opinion that peptide disorder compartmentalized and regulated hydration and caused separation and concentration of reactants in protocells and was a major driving force in the early evolution of life and, specifically, in the evolution of the genetic code, which is the most central feature of evolution of complex life on Earth.

To understand the prelife to life transition, requires bottom-up and top-down approaches (Chatterjee & Yadav, 2019; Kunnev & Gospodinov, 2018; Lei & Burton, 2020; Mariscal et al., 2019). In a bottom-up approach, a goal is to develop plausible prebiotic, coacervate, and selfreplicating polymerization systems. The top-down approach, by contrast, is intended to infer some major pathways in the prelife world often from analyses of conserved sequences. The advantage of the bottom-up approach is that many prebiotic reactions are interesting and potentially on-pathway. The potential disadvantage of a bottom-up approach is that too many pathways are possible and too many plausible pathways may be dead ends or may not result in dominant pathways. The potential advantage of the top-down strategy is that inferences based on sequence are likely to reflect dominant and successful pathways. The limitation of a top-down strategy is that many important pathways may not be represented or may not be recognized in existing sequence data sets. Because authors of this manuscript are molecular biologists, our approach has been sequence-based and top-down. We find that top-down strategies describe the early evolution of translation systems and transcription systems and the divergence of Archaea and Bacteria. Top-down approaches also enrich bottom-up views. For instance, based on top-down methods, we posit models for prebiotic chemistry (see below). Of course, when top-down approaches and bottom-up approaches meet, a richer analysis of the prelife to life transition has been achieved.

We posit that the major event in the divergence of Archaea and Bacteria was the evolution of bacterial α-transcription factors (Fig. 10.1) (Burton, 2014; Burton & Burton, 2014). In Bacteria, σ factors bind to RNA polymerase to facilitate binding to the promoter. σ helix-turn-helix (HTH) factors are homologs of archaeal TFB (Iyer & Aravind, 2012). Comparing bacterial σ factors to archaeal TFB, however, σ factors alter bacterial promoter recognition and transcriptional control in fundamental ways. This radical shift in core transcriptional mechanisms, promoters, and control caused Bacteria to become significantly different from Archaea, while Archaea remained very similar to LUCA. Bacteria also adopted a new replicative DNA polymerase (PolC) relative to Archaea (PolB and PolD), so this is another fundamental difference comparing Bacteria and Archaea (Koonin et al., 2020).

Much is not yet known about the evolution of Eukaryota. We view eukaryotes as genetic fusions of multiple Archaea and multiple Bacteria without a very clear model for how this transition occurred. We view the transition as a multistage process of endosymbiotic or other large horizontal gene transfer events (i.e., by feeding: referred to as "foodchain gene adoption") (Eme et al., 2017; Fournier & Poole, 2018; Long et al., 2019; O'Malley et al., 2019; Pittis & Gabaldón, 2016). It is our opinion that horizontal transfers of small packets of genes are generally less successful than larger transfers. In Fig. 10.1, we indicate a few possible events in the FECA (first eukaryotic common ancestor) to LECA (last eukaryotic common ancestor) transition, focusing on evolution of cell architectures and transcriptional mechanisms. Splicing appears to have evolved near the time of LECA (Koonin & Novozhilov, 2009). It appears that Eukarya evolved a new use for hydrogels (LLPS) involving intrinsically disordered regions (IDRs) of proteins (Boehning et al., 2018; McSwiggen et al., 2019; Yoshizawa et al., 2020; Zhou et al., 2018). Histone tails and the CTD of RNA polymerase II are IDRs with regulated interactions and protein readers (Boehning et al., 2018; Guo et al., 2020; Lu et al., 2018). Other factors with IDRs cooperate in the transcription cycle, sequestering complexes in LLPS compartments. So far as we can ascertain, prokaryotes utilize LLPS probably utilizing short disordered protein regions.

There is some redundancy in this paper compared to previously published work from our laboratory on evolution of tRNAs, tRNAomes, aminoacyl tRNA synthetases (aaRS enzymes), and the genetic code. Because the paper describes and combines multiple complex subjects, some redundancy was inevitable. The current paper provides refined models, insights, and perspectives. A dominant theme of this review is that hydrogels, amyloids, and LLPS drove the early evolution of the genetic code. Specifically, we posit that hydrogels and related assemblies provided the main Darwinian driving force behind genetic code evolution. We provide highly detailed and connected models for key intermediates in the prelife⇒life transition. Specifically, our tRNA evolution model, aaRS evolution model, and genetic code evolution model are mutually reinforcing and highly predictive.

Artificial intelligence in evolution of life

At some level, evolution of life on Earth can be described according to principles of artificial intelligence (Kim et al., 2019; Lei & Burton, 2020). A system is capable of "learning" (teaching itself) if it can build up intellectual property that enhances its subsequent capabilities. In a biological system, evolution must solve the coding problem, because, without sophisticated biological coding, no life as recognized on Earth is possible. So evolution of the genetic code is the core feature of evolution of life on Earth. tRNA was the central driver to evolve biological coding. Therefore, evolution of tRNA is a central story. We consider tRNA to be the core intellectual property in evolution of translation systems, including (1) tRNAomes (all of the tRNAs in an organism) (Pak, Du, et al., 2018); (2) the genetic code (Kim et al., 2019; Lei & Burton, 2020); (3) mRNA; (4) aminoacyl-tRNA synthetases (aaRS; i.e., GlyRS-IIA; IIA indicates the aaRS I or II structural class and A-E subclass) (Kunnev & Gospodinov, 2018; Pak, Kim, et al., 2018); (5) rRNA; and (6) ribosomes (Opron & Burton, 2019). The system taught itself to code, centered on tRNAs, and then vastly enriched the code and its expression by coevolving proteins. Biological coding expands the capacity of the system to create highly functional proteins and protein assemblies and then to evolve complex organisms.

Evolution of tRNA is also a story of artificial intelligence. tRNA evolved from ligation of 3—31-nt minihelices, as described below. Furthermore, the minihelices were comprised of repeating sequences and inverted repeats, so minihelices and tRNAs were constructed from highly ordered ancient sequences from about 4 Ga ago. Many have considered earliest evolution from random biopolymers, but tRNA did not evolve from random sequences, and evolution of tRNA is the central issue in evolution of translation systems and the genetic code. For computational studies, it remains an important question of why and how evolution from ordered sequences can give rise to some of the most central biomolecules.

The prelife to life transition: evolution of translation

According to our vision for evolution of life on Earth, tRNA was the core intellectual property (Damer & Deamer, 2015; Lei & Burton, 2020). Evolution of tRNA directed evolution of mRNA, rRNA, and the genetic code, which evolved around the tRNA anticodon. Evolution of aminoacyl-tRNA synthetases (aaRS), which are the enzymes that attach amino acids to tRNA, tracked the evolution of the genetic code. At least in part, some of the oldest rRNA segments appear to have evolved from amalgamations of tRNAs and tRNA-like sequences, indicating that tRNA predates rRNA, at least in rRNA's current form (de Farias et al., 2019; Root-Bernstein & Root-Bernstein, 2015, 2016, 2019; Root-Bernstein et al., 2016). The primitive ribosome we consider to be a decoding center scaffold (pre-16S rRNA) and a mobile and separate peptidyl transferase center (PTC; pre-23S rRNA) (Opron & Burton, 2019). The PTC appears to be a dehydration chamber for formation of peptide bonds utilizing diverse amino acid substrates (Bernier et al., 2018; Gulen et al., 2016). We, in part, describe how ribosomes could have evolved to prokaryotic forms before LUCA.

Evolution of tRNA

A number of models have been advanced to describe tRNA evolution. We favor the 3-minihelix model advanced by our laboratory, which we find to be best supported by sequence and statistical analyses and also most predictive (Burton, 2020; Kim et al., 2018; Pak, Du, et al., 2018; Pak et al., 2017; Root-Bernstein et al., 2016). The model fully accounts for the sequences of type I and type II tRNAs. Most tRNA are type I with a 5-nt V loop (V for variable). Type II tRNAs (i.e., tRNA^{Leu} and tRNA^{Ser} in Archaea) have an expanded V loop that initially was 14-nt, although type II V loops have expanded and contracted in evolution. The 3-minihelix model is strongly predictive to describe evolution of the genetic code and evolution of ribosomes. We identify a tRNA primordial sequence (tRNA^{Pri}) from which tRNAs radiate. We showed that in Archaea, tRNA^{Pri} is very close in sequence to tRNA^{Gly}, indicating that glycine was the first encoded amino acid (Pak, Du, et al., 2018). Remarkably, tRNA^{Pri} is very close in sequence to a typical tRNA sequence (similar to a consensus sequence) from Archaea (Juhling et al., 2009).

Evolution of type I tRNA

tRNA^{Pri} was formed by ligation of three 31-nt minihelices of almost completely known sequence (Figs. 10.2 and 10.3). Fig. 10.2A shows the structure of a type I tRNA colored according to the model. Fig. 10.2B shows a typical type I tRNA (similar to a consensus sequence) from Pyrococcus furiosis, an ancient Archaea (Juhling et al., 2009). Fig. 10.3C outlines the 3-minihelix model for tRNA evolution. In the model (Fig. 10.3C), a 93-nt tRNA precursor was formed by ligation of three 31-nt minihelices. The 93-nt precursor was then processed into type I and type II tRNAs. The 31-nt minihelices that became the anticodon stem-loop-stem and the T stem-loop-stem were initially identical (~GCGGCGGCCGGGUU/AAAAACCCGGCCGCCGC; stem-loop-stem microhelix core: ~CCGGGUU/AAAAACCCGG; there is slight sequence ambiguity in the primordial ~UU/AAAAA (/indicates a U-turn) loop; there is no ambiguity in the 5-nt stems). The minihelix that became

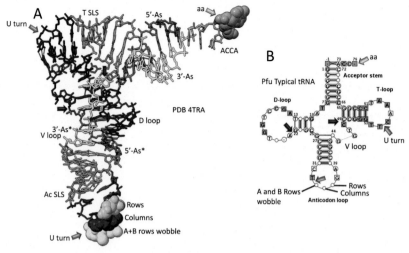

Figure 10.2 The three 31-nt minihelix model for tRNA evolution. (A) A type I tRNA. (B) A typical type I tRNA from *Pyrococcus furiosis* (Pfu). *Arrow* colors in A—C: *red*) internal deletion endpoints; *blue*) U-turns; *yellow*) amino acid placements. (For interpretation of the references to colour in this figure legend, the reader is referred to the web version of this article.)

the D loop sequence is distinct because it has a 17-nt microhelix core based on a UAGCC repeat (initially GCGGCGGUAGCCUAGCCUAGC-CUACCGCCGC; 17-nt UAGCC repeat microhelix core: UAGC-CUAGCCUAGCCUA). Remarkably, the UAGCC repeat in the D loop is apparent in typical tRNAs from ancient Archaea (Fig. 10.2B; UAGC-NUAGCCUGGUNNA). To generate a type I tRNA^{Pri} requires two internal 9-nt deletions in the 93-nt precursor surrounding the anticodon stem-loop-stem within ligated acceptor stems (type I tRNAs missing 3'-ACCA were initially 75-nt) (Fig. 10.3C). In type I tRNA in Archaea, only a few small D loop deletions (i.e., 1—4 nt) and 1-nt deletions in the 5-nt V loop were tolerated (Juhling et al., 2009). To form a functional tRNA that could attach an amino acid, ligation (or genetic and/or enzymatic attachment) of 3'-ACCA was necessary.

Archaeal tRNAs radiated from tRNA^{Gly}

In support of the three 31-nt minihelix model for tRNA evolution, we show that tRNA^{Pri}, tRNA^{Gly}, and tRNA^{Typical} are closely related sequences (Fig. 10.4). Fig. 10.4A shows a typical tRNA^{Gly} from three Pyrococcus species. Fig. 10.4 shows an annotated multiple sequence

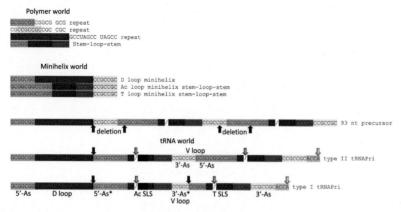

Figure 10.3 The three 31-nt minihelix model for tRNA evolution. (C) The three 31-nt minihelix model. Primordial type I and type II tRNAs were derived from a 93-nt precursor that was formed by ligation of three 31-nt minihelices of mostly known sequence. A polymer world preceded minihelix world. Currently, we occupy a tRNA world that, because of its success, has persisted for ∼4 Ga (1 Ga = 1 billion years) on Earth. Colors in (A) and (C): *green*) 5′-acceptor stems and 5′-acceptor stem remnants (5′-As*); *magenta*) D loop 17-nt microhelix; *yellow*) 3′-acceptor stems and 3′-acceptor stem remnants (3′-As*); *cyan*) 5′ anticodon and T loop stems; *red*) U-turn loops (anticodon and T loops); and *cornflower blue*) 3′- anticodon and T loop stems. SLS indicates stem-loop-stem. Molecular graphics was done using the program UCSF ChimeraX. (For interpretation of the references to colour in this figure legend, the reader is referred to the web version of this article.)

alignment of tRNA^{Pri}, tRNA^{Gly}, and tRNA^{Typical}. Despite ∼4 Ga of evolution, the three sequences are nearly identical (Pak, Du, et al., 2018). Sequence deviations from tRNA^{Pri} can be explained based on tRNA folding (Burton, 2020).

Evolution of type II tRNAs with an expanded V loop

The same model describes evolution of type I tRNA and type II tRNA with an expanded V loop, indicating that both models are correct (in Archaea, tRNA^{Leu} and tRNA^{Ser} are type II tRNAs) (Kim et al., 2018). To generate a type II tRNA^{Pri} required a single internal 9-nt deletion corresponding precisely to the more 5′-deletion in generation of type I tRNA^{Pri} (type II tRNAs were initially 84-nt without 3′-ACCA) (Figs. 10.2C and 10.5). Fig. 10.5A shows a structure of tRNA^{Leu}. The expanded V loop was generated from a 3′-acceptor stem ligated to a 5′-acceptor stem, as indicated in the model (Fig. 10.2C). Fig. 10.5B shows a typical tRNA^{Leu} from three ancient archaeal Pyrococcus species (Juhling et al., 2009). In ancient

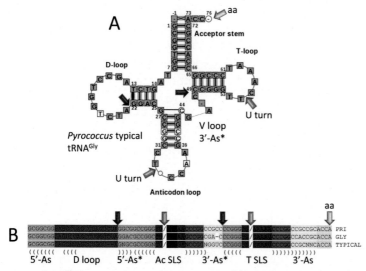

Figure 10.4 tRNA^{Gly} was the primordial tRNA (tRNA^{Pri}) from which other tRNAs radiated. (A) A typical tRNA^{Gly} from three Pyrococcus species. (B) An annotated sequence alignment. PRI) tRNA^{Pri}; GLY) tRNA^{Gly} (as in A); and TYPICAL) tRNA^{Typical} from *Pyrococcus furiosis*. *Purple* indicates the anticodon/indicates a U-turn. (For interpretation of the references to colour in this figure legend, the reader is referred to the web version of this article.)

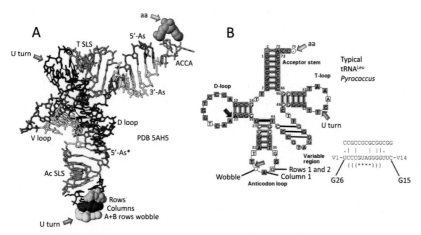

Figure 10.5 Type II tRNAs. (A) A type II tRNA structure (tRNA^{Leu}). 4-nt in the anticodon loop were missing from the structure, so the anticodon loop shown is from PDB 4TRA (tRNA^{Phe}). (B) A typical tRNA^{Leu} from three Pyrococcus species. The sequence alignment shows a comparison of the typical tRNA^{Leu} V loop to the primordial sequence. (C) A color key for a type II tRNA^{Pri} and description for the image in (A).

Archaea, tRNA^{Leu} is closer in sequence to a type II tRNA^{Pri} (Fig. 10.3C) than tRNA^{Ser} (Pak, Du, et al., 2018). Fig. 10.5C describes the tRNA segments and coloring in Fig. 10.5A. In Fig. 10.5B an alignment of the primordial type II tRNA V loop and the typical tRNA^{Leu} V loop is shown. The length of the tRNA^{Leu} V loop is 14-nt, as predicted from the model (Fig. 10.3C). The V loop (numbered V1−V14) is selected to form a G26~UV1 wobble pair and a G15=CV14 reverse Watson-Crick base pair (referred to as the Leavitt base pair), as indicated (Kim et al., 2018). The primordial V loop would pair along its entire length. In type II tRNAs, by contrast, the V loop is evolved to form a loop with a short stem. Also, the sequence of the tRNA^{Leu} V loop has diverged from the tRNA^{Ser} V loop, which is a direct determinant for SerRS-IIA serine addition, to avoid tRNA charging errors (Perona & Gruic-Sovulj, 2014). The tRNA^{Leu} V loop is evolved to be an antideterminant for SerRS-IIA Fig. 10.2A. To attach amino acids to tRNAs, we posit that, initially, ACCA was ligated to tRNAs, minihelices, microhelices, and other RNAs, utilizing a ribozyme ligase. In the ancient world, RNAs covalently linked to amino acids must have been frequently utilized as substrates in catalysis (Gospodinov & Kunnev, 2020; Kunnev & Gospodinov, 2018). This mode of catalysis remains apparent today (see below).

Demonstration of the model

The evidence for the three 31-nt minihelix model is compelling. For instance, statistical analysis shows P-values of .001 (highest indication of homology) for: (1) homology of the anticodon and T stem-loop-stems (17-nt microhelix segments); (2) homology of the last 5-nt of the D loop (5′-As*; As for acceptor stem) and the last 5-nt of the 5′-acceptor stem; and (3) homology of the 5-nt V loop (3′- As*) and the first 5-nt of the 3′-acceptor stem (Burton, 2020; Pak et al., 2017). Inspection of the typical tRNA (Fig. 10.2B) is sufficient to confirm the homology of the anticodon stem-loop-stem (27- CCGGNCUNNNGANCCGG-43) and the T stem-loop-stem (49-CCGGGUUCAAAUCCCGG-65) (see also Fig. 10.4). Standard tRNA numbering does not match tRNA^{Pri} because standard numbering is based on tRNAs with a 3-nt deletion in the D loop (based on eukaryotic tRNAs). In some ancient Archaea, tRNA^{Gly} (Fig. 10.4A) and tRNA^{Leu} (Fig. 10.5B) have full-length D loops. Homology of the anticodon and T stem-loop-stems, which is obvious from inspection, is sufficient to confirm the three 31-nt minihelix model (Figs. 10.2, 10.3, 10.4, and 10.5). We showed that the expanded V loop of type II tRNAs was initially a 3′-acceptor stem ligated to a 5′-acceptor stem, as predicted by the model for processing of the 93-nt tRNA precursor (Figs. 10.3C and 10.5) (Kim et al., 2018).

We consider these analyses to prove our tRNA evolution model is correct and to falsify alternate models (Burton, 2020). We showed that tRNAomes in ancient Archaea cluster tightly around tRNA^{Pri} (Pak, Du, et al., 2018). Ancient Bacteria (i.e., Thermus thermophilus) have fairly compact tRNAomes centered on tRNA^{Pri}. More derived Bacteria (i.e., Escherichia coli) have more diverged tRNAomes centered on tRNA^{Pri}. Because of internal homologies in tRNA sequences, no accretion model (involving random insertions-deletions; indels) can be correct for tRNA evolution (Pak, Du, et al., 2018). Other tRNA evolution models (i.e., 2-minihelix and Uroboros) are accretion models with random indels (Demongeot & Seligmann, 2020; Di Giulio, 2019). In both the 2-minihelix and Uroboros models, random indels lead perplexingly to ordered and repeated sequences in tRNAs (Pak, Du, et al., 2018). Given rules of genetics, we do not know how this is possible. By contrast, we do not think the three 31-nt minihelix model (our model) can be falsified.

We consider the three 31-nt minihelix model to be highly robust and predictive. Given archaeal tRNA sequences, we are unsure how anyone can seriously question the three 31-nt minihelix model (Figs. 10.2, 10.3, 10.4, and 10.5) (Burton, 2020; Kim et al., 2018, 2019; Pak, Du, et al., 2018; Pak et al., 2017). So far, every prediction of the model has been justified (see above). The model describes type I and type II tRNA sequences (Figs. 10.2 and 10.5) (Burton, 2020; Kim et al., 2018; Pak et al., 2017). The model accurately describes internal tRNA homologies including sequence repeats and stem-loop-stems (Figs. 10.2, 10.3, 10.4, and 10.5). The model supports U-turn loop (anticodon and T loop) structural similarities (Burton, 2020; Kim et al., 2018; Pak et al., 2017). The model has also been used to root tRNAome structures (Pak, Du, et al., 2018).

Because tRNA evolution indicates an ancient polymer world and minihelix world preceding the current tRNA world (Fig. 10.3C), about 200−300 million years of prelife evolution are described by the top−down generated three 31-nt minihelix model. Surprisingly, polymer, microhelix, minihelix, and tRNA sequences were derived from ordered sequences: repeats and inverted repeats (Figs. 10.2, 10.3, 10.4, and 10.5). The ancient, prelife world, therefore, included ordered polymers from which tRNAs evolved. Features of the model derive from tRNA sequences and structure.

Evolution of the genetic code (overview)

We posit that the genetic code evolved around the tRNA anticodon following a simple set of rules, which appear never to have been violated (Damer & Deamer, 2015; Lei & Burton, 2020). For the 2nd and 3rd positions of the anticodon, the rules are C>G>U>>A. Preferences are much stronger for the 3rd anticodon position than for the 2nd anticodon position, because the 2nd anticodon position is most central and, therefore, the easiest to read (Opron & Burton, 2019). Consistent with the rule, however, C is strongly preferred in the 2nd position, just as it is in the 3rd position, as evidenced by the position of glycine in the code (see below). We posit that the genetic code initially sectored on the 2nd anticodon position, because the 2nd position was easiest to read on a primitive ribosome. Essentially, the system was teaching itself to encode proteins by accurately matching and reading codons and anticodons. Furthermore, we posit that, on a primitive ribosome, the 1st and 3rd anticodon positions were initially wobble positions. At a wobble position, only pyrimidine-purine discrimination was initially possible, so tRNA wobbling in translation limited the size of the code. Because of wobbling, tRNA, not mRNA, limited the final size of the genetic code. Considering a genetic code of 64 assignments in mRNA, therefore, is not reasonable. Because of wobbling in the 1st anticodon position, the genetic code has a maximum complexity in tRNA of 32 assignments ($2 \times 4 \times 4$). Because some genetic code sectors cannot easily be split, the standard genetic code evolved to 20 amino acids plus stops (21 assignments) rather than encoding additional amino acids (up to 32 assignments).

At the 3rd anticodon position, wobbling was abolished by evolution of the elongation factor (EF)-Tu "latch" (also referred to as conformational closing of the 30S ribosome subunit) (Lei & Burton, 2020; Loveland et al., 2017, 2020; Rozov, Demeshkina, et al., 2016; Rozov, Westhof, et al., 2016; Rozov et al., 2018). tRNA enters the ribosome bound to the GTPase chaparonin EF-Tu. On the ribosome, EF-Tu holds the tRNA until GTP is hydrolyzed and the 30S ribosome subunit tightens its conformation and the EF-Tu GTPase latch is set. Then EF-Tu dissociates, allowing the verified tRNA with its tightened mRNA codon attachment to rotate its $3'$-aa end into the ribosome PTC A site (addition or aminoacyl site). Setting the latch allows 4-base discrimination at the 3rd anticodon position. 4-base resolution was readily achieved at the 2nd anticodon position, because the 2nd anticodon position is most central and the easiest to read. 4-base resolution, therefore, was obtained at the 3rd anticodon position through evolution of the EF-Tu latch. The latch includes Thermus thermophilus (Tth) rRNA positions 16S rRNA G530, A1492, and A1493 and 23S rRNA A1913. The latch checks for Watson-Crick pairing to the mRNA codon at the anticodon 2nd and 3rd positions. The latch also checks the accuracy of pairing at the wobble position. Wobbling is necessary to evolve a genetic code based on RNA, and wobbling is a major story in the evolution of the code. The EF-Tu latch was a major determinant of translational accuracy and an essential evolutionary advance in building the code.

At the wobble 1st anticodon position, the sequence preference rule is $G > (U \sim C) >>>>> A$ (Damer & Deamer, 2015; Lei & Burton, 2020). Only purine versus pyrimidine discrimination is initially possible at a wobble position. Wobble G appears to be favored over $U \sim C$, because Asp (wobble G) appears to enter the code before Glu (wobble U/C) (see below). A is seldom or never used in the wobble anticodon position in Archaea. When wobble A is encoded in Bacteria and Eukarya, A is modified by deamination to inosine (Pak, Kim, et al., 2018; Saint-Léger et al., 2016). Essentially, A is not tolerated in the tRNA anticodon wobble position. Partly, A is not tolerated because A in the tRNA wobble position does not pair well with U in the mRNA wobble position. As noted above, before evolution of the EF-Tu latch, A was also poorly tolerated in the anticodon 3rd position. A is not necessary in the anticodon wobble position because G pairs with C (Watson-Crick pairing) and with U (wobble pairing) (Agris et al., 2018).

In the wobble anticodon position, U and C are read degenerately. Initially, one might expect anticodon wobble C to show reasonable specificity for codon G. Similarly, anticodon U might be expected to read codon A (Watson-Crick pairing) and codon G (wobble pairing), resulting in anticodon wobble ambiguity. Generally, Archaea use both anticodon wobble C and U tRNAs to encode the same amino acid, indicating that anticodon wobble ambiguity was too high a barrier in evolution to easily separate wobble C and U tRNAs to encode two different amino acids. In principle, such separation of functions might be achieved by tRNA wobble modifications (Agris et al., 2017, 2018). To encode tryptophan, the anticodon CCA is used. The UCA anticodon, however, is not generally utilized because UCA corresponds to the UGA stop codon, which is recognized in mRNA by a protein release factor (Burroughs & Aravind, 2019). To encode methionine, anticodon CAU is utilized to read AUG codons. In Archaea, isoleucine also utilizes CAU with C modified to agmatidine to read only codon AUA (Ile) and not AUG (Met) (Köhrer et al., 2014; Mandal et al., 2010; Satpati et al., 2014; Voorhees et al., 2013). To avoid ambiguity in coding, anticodon UAU is rarely utilized in Archaea and Bacteria (Pak, Kim, et al., 2018). With very few exceptions, tRNA wobble modifications cause U and C to be read with more ambiguity than expected for an unmodified base. Generally, tRNA wobble modifications support broader reading of synonymous codons rather than evolving higher tRNA specificity in coding (Agris et al., 2017, 2018).

We strongly support the concept that the genetic code evolved as a 32-assignment code, primarily around the tRNA anticodon (Demongeot & Seligmann, 2020; Kim et al., 2019; Lei & Burton, 2020; Opron & Burton, 2019). To make sense out of the genetic code, therefore, requires a view centered on tRNA and the tRNA anticodon. By contrast, 64-assignment codes, based on mRNA codons, are not reasonable nor descriptive of the evolutionary process. Below, we describe a detailed pathway for evolution of the genetic code based on these ideas.

Evolution of ribosomes

We support the model that rRNA arose from tangles of ligated RNAs that included amalgamations of tRNAs, as also has been proposed by others (de Farias et al., 2019; Root-Bernstein & Root-Bernstein, 2015, 2016, 2019). We imagine an ancient world in which RNAs were replicated by ligation, catalyzed by a ribozyme ligase, followed by complementary replication catalyzed by a template-dependent ribozyme replicase. Attaching a snap-back primer to RNAs would prime their complementary replication. 31-nt minihelices can function as snap-back primers. 17-nt microhelices (i.e., anticodon and T stem-loop-stems) can also function as snap-back primers (Fig. 10.3C). Minihelices and microhelices can be removed from larger RNAs via endonucleolytic cleavage of the RNA catalyzed by a ribozyme (i.e., cutting at the base of stems). In such a world, long RNAs with diverse sequences were generated, and some of these could function as a primitive decoding center scaffold and others as a mobile PTC (Opron & Burton, 2019; Zhang & Cech, 1998). Such tangled RNAs were also an incubator for evolution of novel ribozymes.

The patterns of rRNAs were established before LUCA. One indication of this conclusion is that 16S and 23S rRNAs in Archaea and Bacteria are very similar in sequence and have similar functional RNA motifs (Bernier et al., 2018; Gulen et al., 2016). Archaeal and bacterial rRNA sequences align essentially over their entire lengths without frequent insertion-deletion. As some examples, in 16S rRNA, both Archaea and Bacteria have similar sequences for: (1) the decoding center; (2) the EF-Tu latch; and (3) the ribosome attachment site. In 23S rRNA, Archaea and Bacteria have similar sequences for (1) the A-site (addition or aminoacyl site); (2) the P-site (peptidyl site); (3) the EF-Tu latch; and (4) the SRL (sarcin–ricin loop). We conclude, therefore, that 16S and 23S rRNAs were largely established before LUCA and persisted in Archaea and Bacteria with only minor changes and few large insertions–deletions.

rRNA may be derived in part from amalgamated tRNAs

In support of the idea that segments of rRNAs may initially have been generated from ligated tRNAs, we show Fig. 10.6. We searched an aligned region of the archaeal and bacterial PTC, located between the tRNA 3′-end CCA-binding segments named the P-loop and the A-loop, using the *Pyrococcus furiosis* (Pfu) tRNAome, which is very similar to a LUCA tRNAome (Pak, Du, et al., 2018). We searched aligned segments of archaeal (*Methanocaldococcus infernus*; Min) and bacterial (*Thermus thermophi-lus*; Tth) PTCs (Fig. 10.6A). We find tRNA-like sequences that were identified using multiple tRNA probes that align in both archaeal and bacterial sequences. For this search, the smallest (most likely homologous) e-value obtained was 7×10^{-4} (~ 1 chance in 1400 of being due to random chance) for an alignment of a Pfu tRNA (Arg (TCT)) to the Tth PTC (Fig. 10.6B). The same region is detected as tRNA-like with aligned tRNA segments in the archaeal Min PTC using multiple Pfu tRNA probes. The alignment appears to extend in the plus/plus orientation from the tRNA D loop across the anticodon stem-loop-stem and a 5-nt (type I tRNA) V loop to the first base of the T loop of the tRNA, indicating that full-length tRNAs rather than minihelices or microhelices (Fig. 10.3C) were present for evolution of the PTC. Probably, the homology is to a type I tRNA because it appears to extend over a 5-nt V loop. This same alignment can be obtained using a search with a typical type I tRNA sequence from ancient Archaea. We conclude, therefore, that type I tRNAs probably evolved prior to the 23S rRNA PTC, and tRNA sequences probably contributed to PTC evolution. We have done similar analyses with 16S rRNA and other segments of the 23S rRNA with similar results. We detect both plus/plus and plus/minus alignments to tRNAs, indicating

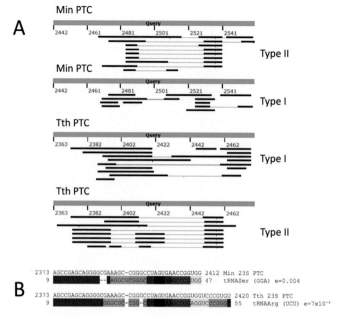

Figure 10.6 A tRNA-like segment of the PTC of 23S rRNA. (A) Alignments of Pfu tRNA sequences (*black bars*) to aligned archaeal Min (*top*) and bacterial Tth (*bottom*) PTC fragments. Type I and type II tRNAs were searched separately. (B) A top alignment in this search that was identified using multiple probes. tRNA Colors: *magenta*) D loop; *green*) 5′-acceptor stem remnant; *cyan*) 5′-anticodon and T stem; *red*) anticodon and T loop; *purple*) anticodon; *cornflower blue*) 3′-anticodon stem; and *yellow*) 3′-acceptor stem remnant (V loop). The e-value is dependent on the size of the PTC fragment used in the search, which in this case is short (∼117 nt), decreasing the e-value compared to longer PTC fragment searches. (For interpretation of the references to colour in this figure legend, the reader is referred to the web version of this article.)

that complementary replication predates evolution of rRNAs. In Fig. 10.5, plus/plus alignments are prominent (Fig. 10.6B). Others have reported similar findings using other bioinformatics approaches (de Farias et al., 2019; Root-Bernstein & Root-Bernstein, 2015, 2016, 2019). We note that in the tRNA-aligned segment of the PTC no tRNA-like stem-loop-stems were detected (not shown). RNAs tend to fold according to longer range RNA contacts, so this result was not unexpected.

In addition to the decoding center of the ribosome (16S rRNA; 30S subunit), which forms a scaffold on which to run the mRNA, and the PTC (23S rRNA; 50S subunit), at which amino acids are joined to a peptide chain, the ribosome has additional features, which we consider to be subsequent evolutionary add-ons (Opron & Burton, 2019). Remarkably, the ribosome must be coevolved with the genetic code, and a model for evolution of the code parallels these advances, most of which occurred prior to LUCA. Ribosomes, of course, continue to evolve in Eukarya, but these enhancements are generally regulatory to support cell- and organism-specific functions (Petrov et al., 2014, 2015).

The prokaryotic ribosome

A recent cryo-electron microscopy paper reveals the Thermus thermophilus ribosome and its dynamics and fidelity in amazing detail (Loveland et al., 2020). We highly recommend this paper to any with an interest in general translational mechanisms and fidelity. Here, we provide a general description of the translational mechanism with particular attention to evolution of the EF-Tu GTPase "latch," which we consider to be the fundamental advance in evolution of translation systems and the genetic code. In the paper referenced above, the "latch" is described as conformational closing of the 30S ribosome subunit. Also, of importance is the recognition that IF2, EF-Tu, and EF-G are ancient homologous GTPases that function in translation.

So far as we can discern, the prokaryotic ribosome was evolved before LUCA and the same basic functional design was maintained in Archaea and Bacteria. Initiation occurs on the small 30S subunit aided by initiation factors (IF1, IF2 and IF3). IF2, elongation factor (EF)-Tu and EF-G are homologous GTPases that function as chaparonins in the translation process (Opron & Burton, 2019). Many Archaea and Bacteria have a UCCU sequence near the 3′-end of the 16S rRNA to orient the sequence - ~AGGA on mRNA (the ribosome attachment site) relative to the AUG start codon sequence, which must be positioned in the ribosome P site for translation initiation. Incoming tRNAs first associate with EF-Tu before binding to mRNA. The 16S rRNA (30S subunit) has a "head," "neck," and "body." The head can adjust its rotation to orient the mRNA for initiation and to help by reversible swiveling and mRNA sliding with forward translocation during elongation. The mRNA runs along the neck where it is ratcheted forward via reversible swiveling of the head. The mRNA is held forward in part by bound tRNAs, maintaining the translation register.

For elongation, the 23S rRNA (50S subunit) associates with the 16S rRNA (30S subunit) with bound mRNA and the IFs then dissociate. The aa-tRNA-EF-Tu complex enters, GTP is hydrolyzed, the ribosome latch tightens (the 30S subunit closes), and the aa-tRNA rotates its 3′-XCCA-aa end (X is the discriminator base for aaRS discrimination and amino acid placement on tRNA) into the A-site (addition or aminoacyl site), also aligning peptide-tRNA in the P-site (peptidyl site). During tRNA rotation, EF-Tu dissociates from the A-site aa-tRNA and EF-G binds to the same ribosome site that had been occupied by its homolog EF-Tu during previous steps. EF-G hydrolyzes GTP and stimulates forward translocation. There is limited reversible rotation of the 23S rRNA (50S subunit) versus the 16S rRNA (30S subunit), facilitating forward translocation. tRNAs advance from the A-site to the P-site to the E-site (exit site). Having 2-3 tRNAs bound to the mRNA during elongation helps to maintain the translation frame.

EF-Tu tightens the ribosome "latch," which is a central feature of ribosome evolution (Loveland et al., 2017; Rozov, Demeshkina, et al., 2016; Rozov et al., 2018; Rozov, Westhof, et al., 2016). The latch closes around the aa-tRNA-mRNA helix bound in the ribosome A-site. Enclosure includes interactions with 16S rRNA G530, A1492 and A1493, and 23S rRNA A1913 (Tth numbering). The closed conformation of the ribosome confirms 4-base recognition at the 2nd and 3rd tRNA anticodon positions. Evolution of the latch, therefore, allowed evolution of the genetic code to advance beyond ~ 8 amino acids (i.e., 2×4 assignments; we posit that only a single wobble position can be read at one time on the primitive ribosome) (Kim et al., 2019; Lei & Burton, 2020). Prior to evolution of the EF-Tu GTPase latch, both the 1st and 3rd anticodon positions were wobble positions, limited to pyrimidine versus purine resolution. Evolution of the EF-Tu GTPase latch, therefore, "teaches" the ribosome to potentially read a 32-assignment code, which froze at a 20-amino acid + stop codon standard code (Koonin, 2009, 2017; Koonin & Novozhilov, 2009, 2017). In order for the A-site tRNA to advance to the P-site, the latch must open. Because tRNAs bound in the A-site, P-site, and E-site have anticodon interactions paired at the mRNA, associated with the 16S rRNA (30S subunit), and also 3′-end interactions with 23S rRNA (50S subunit), multiple intermediate structures (referred to as hybrid states) are possible (Loveland et al., 2020). Hybrid states appear to rotate around the EF-Tu latch. Setting of the latch results in dissociation of EF-Tu and is followed by a large rotation of the 3′-end of the verified aa-tRNA into the PTC A-site, a step referred to as "accommodation." Rotation of the deacylated P-site tRNA into the E site, by contrast, is associated with rotation of the 3′-end of the tRNA associated with opening of the latch and forward translocation. Depending on the step, therefore, tRNAs ratchet independently at their 3′-ends and anticodon ends, creating the hybrid states.

For chemistry, a P-site tRNA has XCCA-peptide at its 3′-end (X = the discriminator base). The A-site tRNA has XCCA-aa at its 3′-end. In the 23S rRNA (50S subunit) P-site, the sequence 2248- CUGGGGCGG-2256 presents 2251-GG-2252 to form Watson-Crick pairs with the 3′-CC of the P-site peptide-tRNA. In the 23S rRNA (50S subunit) A-site, the sequence 2548-GGGCUGUUCGCCC-2560 presents 2553-G to pair with 3′-CC (the 2nd C), to orient the A-site aa-tRNA. It appears that proximity of P-site and A-site tRNAs in the dehydrating environment of the PTC may be sufficient to form the next peptide bond (Bernier et al., 2018).

The peptide chain elongates by its transfer to the A-site tRNA, resulting in deacylation of the P-site tRNA. After deacylation, the P-site tRNA can advance its 3′-end to the E-site. Because the peptide chain is transferred to the A-site tRNA from the P-site tRNA, the peptide was lengthened by one amino acid. Once the P-site tRNA releases the peptide chain to the A-site tRNA, the deacylated P-site tRNA can then translocate to the E-site, displacing and releasing the E-site tRNA. So the march of tRNAs aided by EF-G through a compact tRNA-shaped tunnel in 23S rRNA helps to ensure forward translocation and maintenance of the translation frame. Because the 23S rRNA apparently evolved to match the shapes of advancing tRNAs, we posit that the final confirmation of 23S rRNA and the 50S subunit evolved around tRNAs and that tRNAs evolved prior to the final evolved shape of the prokaryotic ribosome.

The exiting peptide chain extends from the active site A-peptide (before translocation) or P-peptide site (after translocation), through a channel in the ribosome. When the peptide exits the ribosome, it can begin to fold or it can be targeted to a membrane for transport or excretion. Translation termination occurs when protein release factors bind to stop codons in the mRNA. Because there are no tRNAs corresponding to stop codons, anticodons that are complementary to stop codons are not represented in the tRNA-centric standard genetic code.

The genetic code expanded by two mechanisms we can identify: (1) tRNA charging errors; and (2) modification of amino acids bound to tRNAs (e.g., Asp⇒Asn, Glu⇒Gln, and pSer⇒Cys (pSer for phosphoserine)). We posit that the first amino acids to enter the code filled large sectors of the code, and these sectors were then invaded by other amino acids. Invasion follows a strict set of rules that we describe in more detail above and below. Significantly, because invasion by incoming amino acids required tRNA charging errors or amino acid modifications, translational fidelity is very important for the eventual freezing of the code. Ribosome fidelity, for instance, evolution of the EF-Tu GTPase latch, was fundamental to first expand and then to freeze the code.

We posit that hydrogels and related LLPS compartments are very important in prokaryotic translational functions, transcription-translation coupling, protein folding, and chaparonin functions, although we were unsuccessful at finding specific references. In prokaryotic systems, hydrogels are small and hydrogels are disordered by their nature, making hydrogels difficult to visualize and analyze using current imaging methods.

Aminoacyl-tRNA synthetases (aaRS)

We posit an updated model for aaRS evolution (Damer & Deamer, 2015; O'Donoghue & Luthey-Schulten, 2003; Opron & Burton, 2019; Pak, Du, et al., 2018; Pak, Kim, et al., 2018). Remarkably, the pattern of aaRS evolution matches the pattern of genetic code evolution, providing a pathway for evolution of the genetic code. Also, apparent coevolution of the genetic code and aaRS enzymes indicates that the models we present for aaRS enzyme evolution and genetic code evolution are mutually reinforcing, reliable and predictive. Remarkably, the evidence we cite of coevolution has been maintained during ~ 4 Ga of evolution with significant potential for divergence. aaRS evolution patterns show coevolution with genetic code columns, which represent the 2nd tRNA anticodon position, the most important position for translational accuracy. As we discuss below, amino acids appear to add into the genetic code by rows, which represent the 3rd anticodon position. Recently, our laboratory clarified aaRS evolution using the Phyre2 protein-fold recognition server, which utilizes sequence and structure to align sequences available in the Protein Data Base with a seed sequence (Kelley et al., 2015). Phyre2 provides evolutionary relationships of all (or most) aaRS enzymes in a structural class at once. The results provide a road map for evolution of the genetic code. Here, we provide an explanation for the radiation of the aaRS enzymes according to models for genetic code and tRNAome evolution.

aaRS structural classes

There are two structural classes of aaRS (class I and class II) with multiple structural subclasses (i.e., A—E) (Perona & Gruic-Sovulj, 2014). Class I and class II aaRS have incompatible folds but are homologs by sequence (see below). Class I aaRS enzymes have an active site arranged on a set of parallel β-sheets. As a result, class I aaRS have been referred to as a "Rossmann-like" fold. Class I aaRS, however, are not homologs of Rossmann fold proteins. By contrast, class II aaRS mount their active sites on a set of antiparallel β-sheets. Both class I and class II aaRS enzymes are among the very first proteins to evolve on Earth, so aaRS are ancient proteins that evolved before LUCA and coevolved with the genetic code. Both class I and class II aaRS enzymes can have an extra editing domain to remove an inappropriately attached amino acid from a tRNA. Remarkably, in Archaea, only amino acids found in the left half of the genetic code (columns 1 and 2) have editing active sites (see below).

Archaeal GlyRS-IIA was the first aaRS enzyme to evolve. Identifying GlyRS-IIA as the primordial aaRS indicates, once again, that glycine may have been the first encoded amino acid and that glycine maintained the dominant position in the evolving code. Of course, GlyRS-IIA is a product of protein encoding, so significant evolution of the genetic code must have been supported initially by ribozymes charging tRNAs (i.e., GlyRS-RBZ; RBZ for ribozyme) (Illangasekare & Yarus, 2012; Turk et al., 2010; Yarus, 2011). We posit that, because of coevolution, divergence of aaRS enzymes followed the pattern of evolution of the evolving tRNAome, and tRNA^{Gly} held the most favored position in the code. GlyRS-IIA, therefore, evolved as the first protein aaRS, and all class I and class II aaRS diverged from GlyRS-IIA.

Folding of primitive GlyRS-IIA was directed by a Zn-finger near the protein N-terminus (Pak, Kim, et al., 2018). All class II aaRS enzymes derive in lineage from GlyRS-IIA. To form class IA aaRS enzymes (i.e., IleRSIA and ValRS-IA), a primitive GlyRS-IIA was extended at its N-terminus and then refolded. Because of the N-terminal extensions, class IA enzymes are about twice as long as GlyRS-IIA. Other class I aaRS enzymes were derived from a class IA aaRS (probably ValRS-IA) (Kim et al., 2019; Lei & Burton, 2020; Opron & Burton, 2019; Pak, Kim, et al., 2018). Initially, class IA enzymes folded around the N-terminal protein extension and two Zn-fingers (Pak, Kim, et al., 2018). The more C-terminal Zn-finger found in IleRS-IA and ValRS-IA, in some ancient

Archaea, corresponds to the single Zn-finger in ancient archaeal GlyRS-IIA. In Fig. 10.7, sequence alignments are shown comparing a common segment of archaeal GlyRS-IIA enzymes and IleRS-IA and ValRS-IA enzymes. To succeed, these searches must be done using ancient archaeal species. The e-values for the alignments are 3×10^{-13} (GlyRS-IIA to IleRS-IA) and 6×10^{-11} (GlyRS-IIA to ValRS-IA). The chances of these independent alignments of homologous GlyRS-IIA regions being due to random events (i.e., convergent evolution) would be ~1 to 1023 against. We conclude that class IIA and class IA aaRS enzymes are homologs by sequence that have subsequently diverged in evolution to maintain the fidelity of translation. Other models (i.e., Carter-Ohno-Rodin) for class I and class II aaRS evolution have been published (Carter et al., 2014; Chandrasekaran et al., 2013; Martinez-Rodriguez et al., 2015; Rodin et al., 2009), but these models are not correct. In the Carter-Ohno-Rodin model, "urzymes" for class I and class II aaRS were posited to be generated from both strands of a primordial bidirectional gene. Such a model is inconsistent with simple homology of class I and class II aaRS, as we demonstrate (Fig. 10.7) (Pak, Kim, et al., 2018).

We posit that hydrogels sequestering tRNAs may have been involved in the earliest folding and class divergence of aaRS enzymes. Class IA and class IIA aaRS folding were initially directed by Zn-binding and the N-terminal extension of class IA enzymes, which comprises part of the class I aaRS active site. Because class I and class II aaRS bind opposite faces of their cognate tRNAs, tRNA binding might have also promoted appropriate aaRS folding. Hydrogels can sequester RNAs and could promote early aaRS class I and class II folds.

```
      Score    Expect        Method          Identities  Positives  Gaps
   52.0 bits(123) 6e-11   Compositional matrix adjust. 25/61(41%) 38/61(62%) 4/61(6%)
   Query  359  TVKPNMGILGPRFKGKAAKIANALKALKPEELSGDV----IELTIDGEKITIEKDAVAFEK  415 Cmni GlyRS-IIA
                VKPNM I+GP+F+ +A  I    L ++ P E++  +    I + IDGE I +E ++VA EK
   Sbjct  782  NVKPNMAIIGPKFRKQAGAIIKTLTSMDPVEVANIISKGNININIDGEDIELEPESVAIEK  842 Mbu  ValRS-IA
```

```
      Score    Expect        Method          Identities  Positives  Gaps
   60.1 bits(144) 3e-13   Compositional matrix adjust. 26/63(41%) 43/63(68%) 0/63(0%)
   Query  357  KVVAKPDMKKFGPLFKGDSPKIKAVLDETDATIIKNAFEADGTFKVEVEGKEYELTEDLVSFQ  419 Mco GlyRS-IIA
                K++AKP++K  GP  +GD+PK+    L E D + IK+  +A+G++ VEV+G+  EL  D + F+
   Sbjct  872  KIIAKPNLKTLGPRLRGDAPKVMKHLTEADGSEIKSILDAEGSYSVEVDGRSIELGVDDILFE  934 Mbr IleRS-IA
```

Figure 10.7 Class IIA and class IA aaRS enzymes with incompatible folds are homologs by sequence. Abbreviations: (Cmni) Candidatus Methanoperedens nitroreducens; (Mbu) Methanococcoides burtonii; (Mco) Methanobacterium congolense; (Mbr) Methanobacterium bryantii. These alignments can be extended.

The pattern of aaRS evolution gives the pattern of amino acid placements in the standard genetic code

Fig. 10.8 shows divergence of aaRS enzymes as they relate to the standard genetic code in Archaea. The graph represents the closest homologs in the Protein Data Base identified using the Phyre2 protein-fold recognition server (Fig. 10.8A), so alignments and homology models represent sequence similarity and structural modeling (Kelley et al., 2015; Kim et al., 2019). Distances in the map represent evolutionary distances, so clustered aaRS are closely related. Remarkably, all class I aaRS enzymes were connected using the Phyre2 server. Many of these connections would not have been detected using sequence alignments. By contrast, some of the nodes in the class II aaRS map could not be connected using Phyre2. For instance, no relevant connection of class IIA and class IID enzymes could be obtained.

Other approaches to aaRS evolution have not provided as clear a picture or one so clearly correlated with genetic code evolution (O'Donoghue & Luthey-Schulten, 2003). Fig. 10.8B shows how the tRNA anticodon

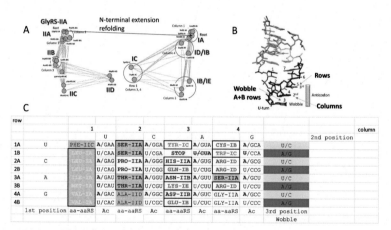

Figure 10.8 aaRS and standard genetic code coevolution in Archaea. (A) aaRS evolution. Distances between nodes indicate evolutionary distance. The *red arrow* indicates that GlyRS-IIA is homologous to ValRS-IA and IleRS-IA (Fig. 10.6). Structural classes and subclasses of aaRS enzymes are indicated. (B) Relationship between the tRNA anticodon and the genetic code. (C) The standard genetic code as a codon-anticodon table. Coloring of aa-aaRS is meant to stress evolutionary relatedness mostly in genetic code columns. *Gray* shading indicates aaRS enzymes with editing active sites in Archaea. Versions of this figure were previously published and the figure is reprinted here with permission. (For interpretation of the references to colour in this figure legend, the reader is referred to the web version of this article.)

relates to the genetic code. The anticodon 2nd position relates to code columns. The anticodon 3rd position relates to code rows (1—4). The anticodon 1st wobble position relates to A and B rows. In Fig. 10.7C, the standard genetic code (codon-anticodon table) is shown for Archaea with coloring for closely related aaRS enzymes, strongly indicating genetic code evolution within code columns (anticodon 2nd position). Because the genetic code evolved around the tRNA anticodon, and because genetic code evolution is tracked by aaRS evolution, we strongly advocate presenting the code as a codon-anticodon table including aaRS evolutionary data.

A model for code sectoring based on aaRS coevolution

We posit that the genetic code coevolved with aaRS enzymes and tRNAomes and that a record of that coevolution is maintained in the pattern of aaRS divergence and the distributions of amino acids in the code. Here, we posit models for the sectoring of the genetic code correlated with aaRS evolution in Archaea and for modifications of the model in Bacteria (Kim et al., 2019; Lei & Burton, 2020). In Fig. 10.1, we indicate that Bacteria may have been derived from Archaea (Long et al., 2019; Marin et al., 2017). As a first consideration, most genetic code sectoring is within code columns, indicating powerful coevolution of the genetic code, aaRS enzymes and the 2nd anticodon position of tRNA. In column 1 (2nd anticodon position A), valine, leucine, and isoleucine are hydrophobic amino acids, and ValRS-IA, MetRS-IA, IleRS-IA, and LeuRS-IA are all closely related class IA aaRS enzymes (Fig. 10.8). From metabolic pathways, valine can be converted to leucine in five enzymatic steps. We posit that this conversion may have initially occurred with valine bound to tRNA. So, Val-tRNA^{Leu} ⇒ Leu-tRNA^{Leu} (catalyzed by five enzymes) prior to evolution of LeuRS-IA, which we posit was derived from ValRS-IA after duplication. PheRS-IIC may have been derived from a similar enzyme to pSerRS-IIC (pSer for phosphoserine) (Hauenstein & Perona, 2008; Mukai et al., 2017). pSerRS-IIC was probably the route by which cysteine was first introduced into the genetic code (see below). To suppress translation errors, the aaRS enzymes in genetic code column 1 have separate editing active sites to remove an inappropriately attached amino acid (Perona & Gruic-Sovulj, 2014), further demonstrating their similarity and their evolution in genetic code columns. Significantly, in Archaea, only amino acids found on the left half of the genetic code (columns 1 and 2) utilize aaRS enzymes with editing active sites (Fig. 10.8C) (Damer & Deamer, 2015; Pak, Du, et al., 2018; Pak, Kim, et al., 2018).

In column 2 (2nd anticodon position G), serine and threonine are similar amino acids, and SerRSIIA, ProRS-IIA, and ThrRS-IIA are closely related class IIA aaRS enzymes. We posit that AlaRS-IID may have been derived from a similar enzyme to pSerRS-IIC, that is, by duplication, mutation, and repurposing. Probably, AlaRS-IIA was replaced by AlaRS-IID early in code evolution (i.e., before LUCA), in order to enhance the fidelity of tRNA charging. SerRS-IIA, ThrRS-IIA, and AlaRS-IID have editing active sites. In Archaea, AlaX is a tRNA editing function homologous to AlaRS-IID but without a synthetic active site to add alanine to $tRNA^{Ala}$. We consider these observations to strongly support coevolution of amino acids, aaRS enzymes, and the genetic code within column 2.

In column 3 (2nd anticodon position U), aspartate and asparagine are related amino acids, and AspRS-IIB, AsnRS-IIB, and HisRS-IIA are reasonably closely related enzymes. We posit that AspRSIIB was initially AspRS-IIA from which HisRS-IIA was derived (see below). AsnRS-IIB was derived from AspRS-IIB. In some ancient Archaea, Asp-tRNA^{Asn} is converted enzymatically to Asn-tRNA^{Asn} by Asp-tRNA^{Asn} amidotransferase, indicating an important mechanism for evolution of the genetic code through modification of amino acids bound to tRNAs (Feng et al., 2005; Kim et al., 1996; Rogers & Söll, 1995; Schön et al., 1988; Sheppard et al., 2007). Of course, aspartate and glutamate are closely related amino acids, drawing a further linkage of most of the amino acids in column 3. In column 3, glutamate and glutamine are closely related amino acids and GluRS-IB, LysRS-IE, and GlnRS-IB are closely related enzymes. The structural classification of LysRS-IE is deceptive (Fig. 10.8A). Despite the different subclassifications, LysRS-IE is very similar to GluRS-IB and GlnRS-IB. In some ancient Archaea, Glu-tRNA^{Gln} is converted to Gln-tRNA^{Gln} by modification utilizing a Glu-tRNA^{Gln} amidotransferase, indicating an evolutionary intermediate leading to replacement with GlnRS-IB (Feng et al., 2005; Kim et al., 1996; Rogers & Söll, 1995; Schön et al., 1988; Sheppard et al., 2007). Tyrosine and TyrRS-IC are late additions to the genetic code. No column 3 aaRS enzymes in Archaea have editing active sites. In Bacteria, LysRS-IIB replaced archaeal LysRS-IE. In Bacteria, LysRS-IIB edits (Perona & Gruic-Sovulj, 2014). Bacterial LysRS-IIB appears to be derived from AspRSIIB, further indicating evolution within code columns, even when an aaRS appears to have been replaced in evolution. Surprisingly, there appears to be very little chaos in evolution of the genetic code.

In column 4 (2nd anticodon position C), a jumble of amino acids and aaRS enzymes is found. Archaeal GlyRS-IIA is the aaRS enzyme from which all aaRS enzymes class II and class I were derived (Fig. 10.7A). Class I aaRS enzymes were generated by refolding an ancient GlyRS-IIA probably to ValRS-IA (Fig. 10.7). In Bacteria, GlyRS-IID replaces archaeal GlyRS-IIA. GlyRS-IID, in Bacteria, is probably derived from AlaRS-IID, which arose prior to LUCA (Fig. 10.8A). Using Phyre2, no direct homology was detected linking class IIA and class IID aaRS enzymes (without intermediates), indicating that GlyRS-IID and AlaRS-IID were reinvented. We posit that in Bacteria, archaeal GlyRSIIA and, before LUCA, AlaRS-IIA were replaced in evolution to increase translational fidelity. ArgRSID and subclass IA enzymes are closely related despite the subclassification of ArgRS-ID. ArgRS-ID is also closely related to CysRS-IB, and cysteine and arginine are found in nearby sectors, in column 4. In some ancient Archaea, pSer-tRNA^{Cys} is converted to Cys-tRNA^{Cys} by Sep-tRNA:Cys-tRNA synthase (Sep for phosphoserine), indicating how cysteine first entered the genetic code and how CysRS-IB arose (Hauenstein & Perona, 2008; Mukai et al., 2017). Cysteine was needed in proteins from an early time in evolution to ligate metals (Weiss et al., 2018). Subsequently, CysRS-IB could have evolved from ArgRS-ID to charge tRNA^{Cys} directly. Tryptophan and TrpRS-IC are posited to be the final additions to the genetic code (Fournier & Poole, 2018; Mukai et al., 2017). TrpRS-IC was probably derived from TyrRS-IC. We posit that serine invaded column 4 of the genetic code by jumping from column 2 (see below).

aaRS accuracy

The genetic code evolved around the tRNA anticodon. To an extent, this coevolution relates to aaRS enzymes recognizing the tRNA anticodon as a direct determinant to accurately place an amino acid on the cognate tRNA. Exceptions to this general rule, however, are also of interest for understanding evolution of the code (see below). The accuracy of amino acid placement by aaRS enzymes is a complicated issue that is only addressed briefly here. Because tRNAs are so similar in form and sequence, subtle determinants and antideterminants are recognized by aaRS enzymes. As examples, aaRS enzymes may recognize some of the following determinants and antideterminants in tRNAs: (1) the discriminator base; (2)

the acceptor stems; (3) the anticodon; (4) the V loop; and (5) the D loop (Perona & Gruic-Sovulj, 2014). Generally, class I and class II aaRS enzymes recognize opposite faces of the tRNA, so they may recognize different features as determinants and/or antideterminants for discrimination and amino acid placement. The active site of the aaRS also has particular properties that accept the appropriate amino acid and reject incorrect substrates. For instance, the size of the active site pocket is appropriate for the amino acid substrate rejecting larger substrates. Amino acids with greater character, that is, charge, hydrogen bonding and flexibility or rigidity, tend to more easily be discriminated in the aaRS active site. Hydrophobic and neutral amino acids, by contrast, are associated with aaRS enzymes with editing active sites. Remarkably, aaRS enzymes that edit were largely restricted to the left half of the genetic code (columns 1 and 2). Also, hydrophobic and neutral amino acids that, generally, require aaRS editing are found in columns 1 and 2.

The aaRS enzymes that lack tRNA anticodon recognition include AlaRS-IID, LeuRS-IA, and SerRS-IIA (Perona & Gruic-Sovulj, 2014). Significantly, these aaRS enzymes that lack anticodon recognition have editing active sites to suppress charging errors. In ancient Archaea, the editing function of AlaRS-IID is supplemented by AlaX enzymes that edit inappropriately aminoacylated tRNA^{Ala} but lack an active site to add alanine. Apparently, a different evolutionary route was taken to support the accuracy of alanine charging on tRNA^{Ala}. Probably, AlaRS-IID evolved to replace a now extinct AlaRS-IIA to reduce tRNA^{Ala} charging errors. Remarkably, tRNA^{Leu} and tRNA^{Ser} are the only type II tRNAs in the archaeal standard code. Furthermore, leucine, serine, and arginine are the only amino acids with 6-codon sectors. Below, we propose a model to explain the evolution of the 6-codon sectors. In 6-codon sectors, multiple columns (serine) and rows (leucine, serine and arginine) are crossed. Because this causes ambiguities reading the anticodon, other strategies for tRNA discrimination became necessary. Recognition of the expanded V loops, for instance, aids tRNA^{Leu} and tRNA^{Ser} discrimination. Arginine is a large, stiff amino acid with fairly unique hydrogen-bonding potential, so ArgRS-ID active site specificity for arginine and other tRNA^{Arg} determinants largely describes the specificity of tRNA^{Arg} charging. ArgRS-ID does not edit. Also, ArgRS-ID does recognize the tRNA^{Arg} anticodon 2nd position. In Archaea, only SerRS-IIA on the right half of the code (column 4) edits, and SerRS-IIA is also found, and probably initially resided, in the left half of the code (column 2). Other than SerRS-IIA, only column 1 and 2 aaRS enzymes edit in Archaea.

Pre-life to LUCA

The pathway of the prelife to life transition on Earth is largely unknown (Cantine & Fournier, 2018; Weiss et al., 2018). Our contention has been that the key advance in evolving to cellular life was evolution of tRNA, leading to evolution of tRNAomes, the genetic code, and translation systems (Lei & Burton, 2020; Pak, Du, et al., 2018; Pak, Kim, et al., 2018). As a guesstimate, we consider the evolution of the code to be a "frozen accident" (de Pouplana et al., 2017; Kim et al., 2019; Koonin & Novozhilov, 2017) that might have taken place over about 200–300 million years. To be more accurate, the code was established systematically rather than accidently. The code was "frozen" by translational fidelity mechanisms. According to our view, evolution of the genetic code was the dominant pathway to enable life, making other metabolic, energy and motor pathways (Lane, 2014; Sojo et al., 2014) of potentially secondary importance. We have published detailed models that we consider to be highly informative and reliable for evolution of the genetic code (Kim et al., 2019; Lei & Burton, 2020).

The genetic code evolved around the tRNA anticodon (Fig. 10.8B), and tRNA evolved from a highly patterned primordial sequence that is known almost to the last nucleotide (Figs. 10.2, 10.3, 10.4, and 10.5) (Burton, 2020; Pak et al., 2017). The patterning includes sequence repeats and inverted repeats (stem-loop-stems). The prelife world, therefore, was capable of accurately producing repeating RNA sequences. At a minimum, GCG repeats (5′-acceptor stems), CGC repeats (3′-acceptor stems) and UAGCC repeats (D loop microhelix) were generated. Because the prelife world generated 31-nt minihelices (Fig. 10.3C), a capacity to "measure" the truncations of repeat units must also have existed. The prelife world must have been capable of complementary RNA replication, otherwise inverted repeats found in tRNA would not be notable. So, before evolution of the first tRNA, ribozymes must have existed to generate RNA repeats, to accomplish complementary replication and to excise functional RNAs from longer RNAs.

According to our view, the code in mRNA evolved from the code in tRNA, as we have described (Kim et al., 2019; Lei & Burton, 2020). This conclusion follows from the hypothesis that the genetic code evolved initially to synthesize polyglycine (Kim et al., 2018; Lei & Burton, 2020; Opron & Burton, 2019; Pak, Du, et al., 2018; Pak, Kim, et al., 2018). Adoption of this model yielded the following insight. The genetic code appeared to have evolved by filling in large sectors of the code, which were then invaded by incoming amino acids. Because the code initially encoded only polyglycine, tRNA^{Gly} was the first tRNA. Essentially, all anticodons must then have mutated from tRNA^{Pri}, which is a primitive tRNA^{Gly}, to all possible sequences. This is easy to imagine. The anticodon loop is exposed in tRNA, so the anticodon could mutate without affecting overall tRNA structure. Mutations in other positions of the anticodon loop (i.e., loop positions 1, 2, 6 and 7), by contrast, may disrupt the 7-nt anticodon loop conformation, which has a characteristic U-turn between loop positions 2 and 3 (Quigley & Rich, 1976). If all anticodons encoded glycine, all mRNA encoded polyglycine. Remarkably, in ancient Archaea, the tRNA^{Pri} (the primordial tRNA) is most closely related to tRNA^{Gly} (Fig. 10.4). As newly added amino acids invaded, displaced amino acids retreated, retaining the most favored anticodons and surrendering less favorable anticodons to the invader. According to this model, the entire genetic code can be populated, as the standard genetic code was populated and subsequently maintained for ~4 Ga. The rules for anticodon preference are as follows. In the 2nd and 3rd anticodon positions, the preferences are C>G>U>>A. These preferences are most apparent in the 3rd anticodon position, rather than the 2nd position, which was easier to read on the primitive ribosome. In the 1st anticodon position (the wobble position), the preference is G>(U~C)>>>>>A. In Archaea, A is strongly disfavored in the anticodon wobble position, and A is rarely or never encoded. In Bacteria and Eukarya, wobble A can be modified by deamination to inosine (Pak, Kim, et al., 2018; Saint-Léger et al., 2016). In the wobble position, only purine/pyrimidine resolution was achieved (Lei & Burton, 2020; Opron & Burton, 2019).

Evidence for this model includes the following. In Archaea, tRNA^{Gly} is closest in sequence to tRNA^{Pri} (a primordial tRNA) (Fig. 10.4). Archaeal GlyRS-IIA is the primordial aaRS from which all aaRS enzymes radiated (Fig. 10.8). Glycine, which is the first amino acid encoded, retains the best anticodons (2nd and 3rd anticodon position C). Glycine, alanine, aspartic acid, and valine appear to be the first four encoded amino acids (Chatterjee & Yadav, 2019; Kim et al., 2019; Koonin, 2017; Koonin & Novozhilov, 2009; Lei & Burton, 2020), and they occupy the most favored row 4 (3rd anticodon position C). We posit that aspartic acid entered the code before glutamic acid, and Asp retained the preferred anticodon (1st anticodon position G (Asp) appears to be preferred over 1st anticodon U/C (Glu)). Some of the last amino acids to enter the code occupy disfavored 3rd position A (Phe, Tyr, Cys, Trp). Stop codons, which are read in mRNA by protein release factors (Burroughs & Aravind, 2019), occupy disfavored row 1 (3rd position A). Other arguments for the model and its detailed description have been published elsewhere (Bernhardt, 2016; Bernhardt & Patrick, 2014; Bernhardt & Tate, 2008; Kim et al., 2019; Lei & Burton, 2020; Pak, Du, et al., 2018; Pak, Kim, et al., 2018).

So, evolution of tRNA, mRNA, and the genetic code can reasonably be understood. Much of the genetic code evolution can be inferred from the relatedness of aaRS enzymes, which is a largely solved problem (Fig. 10.8). Because our analyses are all based on existing sequences, this is a top-down analysis that penetrates deep into the prelife world. Because the standard genetic code dominates life on Earth, the top-down approach clearly identifies a winning evolutionary strategy. By contrast, bottom-up approaches may identify reasonable pathways that never became dominant or that may have gone extinct.

Polyglycine world

We posit that the genetic code initially evolved to encode polyglycine. First of all, tRNA^{Pri} is almost a tRNA^{Gly} sequence in ancient Archaea (Fig. 10.4). An archaeal typical tRNA sequence (similar to a consensus sequence) is essentially a tRNA^{Gly}, indicating that other archaeal tRNAs radiated from tRNA^{Gly} (Pak, Du, et al., 2018). Glycine is the simplest amino acid, and glycine was present early on Earth (McGeoch et al., 2020). GlyRSIIA in Archaea is the root of both the class I and class II aaRS trees (Fig. 10.7A), indicating that GlyRSIIA was the first aminoacyl-tRNA synthetase. We posit that when aaRS ribozymes were replaced by encoded protein enzymes, GlyRS-IIA was first because glycine occupied the dominant position in the code.

We further posit that the prior minihelix world that existed before tRNA world (Fig. 10.3C) also evolved to synthesize polyglycine. In tRNA world, two 31-nt minihelix sequences were preserved. We posit that numerous other 31-nt minihelices with ligated 3′-ACCA supported polyglycine synthesis. Polyglycine appears to have been of strongly selected value in the prebiotic world. Interestingly, hemolithin (a polyglycine/hydroxyglycine polymer with coordinated metals) has been identified in meteor samples. Hemolithin appears to be a prebiotic modified polyglycine transported from outer space that was not genetically encoded (McGeoch et al., 2020). We consider identification of hemolithin to be evidence of a polyglycine world before evolution of biotic systems.

- We posit Darwinian selections for prebiotic polyglycine that subsequently drove evolution of the genetic code. First, polyglycine could function as a hydrogel in forming LLPS droplets (Dunne et al., 2018; Endow et al., 2016; Harmon et al., 2017; Inoue & Keegstra, 2003). Also, polyglycine can form amyloid accretions that could act in a protocell as modified hydrogels (Lorusso et al., 2011). Furthermore, polyglycine can be a cross-linking agent to stabilize protocell internal and external architectures (Pinho et al., 2013; Scheffers & Pinho, 2005; Zapun et al., 2008). For instance, polyglycine (i.e., Gly5) is a component of bacterial peptidoglycan cell walls. Cell walls include long glycan chains (i.e., [N-acetyl-glucosamine-Nacetylmuramic acid]n) with covalently attached short peptides (i.e., L-Ala-D-Glu-L-Lys-D-Ala). In peptidoglycan, Gly5 can cross-link D-Ala and L-Lys in two nearby short glycan-linked peptides. In the ancient world, polyglycine could have functioned as a hydrogel to enhance protocell chemistry. In this regard, we note that some human transcription factors include long polyglycine tracts. Human transcription is highly dependent on hydrogel (LLPS) compartments (Boehning et al., 2018; Guo et al., 2020; Lu et al., 2018; Portz & Shorter, 2020). One example is the human androgen receptor, which includes a Gly23 tract. Forkhead Box Protein F1 has a Gly11 tract. zinc finger homeobox protein 3 has a long polyglycine tract. Alpha-fetoprotein enhancer binding protein, AT-rich interactive domain-containing protein, and SWI/SNF chromatin remodeling complex subunit OSA2 are other examples. UNC-80 (ion transport) and phosphatidylinositol 4-kinase (signaling) also have polyglycine tracts and may rely on hydrogels for functional compartmentalization. We posit that these human factors could be models for the functions of polyglycine in ancient systems. In studying ancient evolution, we posit that polyglycine included in protocell systems will improve their coacervate properties and structural stability. Silk fibroin is glycine-rich and includes polyalanine tracts (Yin et al., 2017). Fibroin forms β-sheet amyloid-like assemblies and can form hydrogels. We posit that protocell systems packed with tRNAs, polyglycine, short peptides, large RNA assemblies, and other early metabolites will be shown to have enhanced activities. Advancing to a GADV (Gly, Ala, Asp, Val) world, of course, would enhance the potential for forming hydrogels and related compartments (see below).

Evolution of the genetic code (a working model)

Fig. 10.9 shows a proposed order for the entry of amino acids into the genetic code. Fig. 10.10 gives a highly detailed model for evolution of the code, considering anticodons, codons, and aaRS enzymes (Kim et al., 2019; Lei & Burton, 2020; Pak, Kim, et al., 2018). We posit that glycine was the first encoded amino acid. Then, the code sectored on the 2nd anticodon position to encode Gly, Ala, Asp, and Val (Chatterjee & Yadav, 2019; Ikehara, 2014; Ikehara et al., 2005; Jose et al., 2017; Koonin & Novozhilov, 2009; Oba et al., 2005; Zamudio & José, 2018). The 8-aa code may have encoded Gly, Arg, Asp, Glu, Ala, Ser, Val, and Leu. The 8-aa code represents a bottle neck in evolution because the EF-Tu GTPase latch was necessary to push the code beyond 2×4 complexity (1 wobble (1st or 3rd anticodon position) + 2nd anticodon position). At the ~ 16 amino acid stage, we posit that the code may have included Gly, Arg, Asp, Glu, Asn, Gln, His, Lys, Ala, Thr, Pro, Ser, Val, Ile, and Leu. From this stage, the standard genetic code evolved, mostly by filling row 1 (disfavored 3rd anticodon position A), which was the most difficult row to fill. Our proposed order for amino acid entry is very similar to models proposed by others (Ikehara, 2014; Ikehara et al., 2005; Jose et al., 2017; Koonin & Novozhilov, 2009, 2017; Oba et al., 2005). According to our model, the code transitions from the simplest amino acids to more complex amino acids, indicating that amino acid metabolism and the genetic code

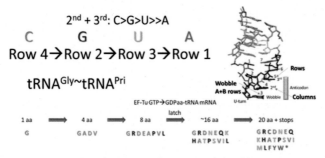

Figure 10.9 Proposed order of addition of amino acids to the genetic code. The code appears to fill by rows (anticodon 3rd position). Amino acids in 6-codon sectors that occupy more than one row (Leu, Ser and Arg) were scored on their most favored row. The *asterisk* indicates a stop codon.

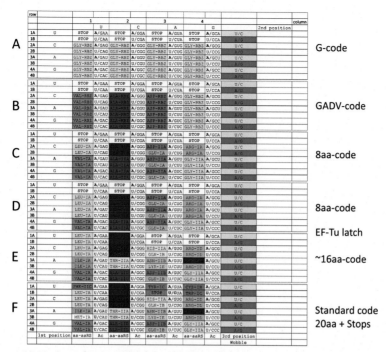

Figure 10.10 A working model for evolution of the genetic code in Archaea. Colors help to track the orders of amino acid additions into the code. We posit that the code evolved from A to F (polyglycine the standard code). Some of the last amino acids encoded on disfavored row 1 are indicated in charcoal (F). GlyRS-RBZ indicates a ribozyme GlyRS before evolution of GlyRS-IIA.

coevolved. Our model incorporates negatively charged and positively charged amino acids relatively early, in part, to evolve more complex proteins. Our model incorporates aromatic amino acids last, consistent with their late evolution, as proposed by others (Chatterjee & Yadav, 2019; Fournier & Alm, 2015).

Evolution of stop codons

The model shown in Figs. 10.9 and 10.10 was designed in part to better model the evolution of stop codons and to better describe the evolution of the 1st row (disfavored 3rd anticodon position A). Also, the model provides potential insight into evolution of 6-codon sectors for leucine, serine, and arginine. The genetic code appears to have filled from the 4th row (3rd anticodon position C) to the 2nd row (3rd anticodon position G) to the 3rd row (3rd anticodon position U) to the 1st row (3rd anticodon position A), so the 1st row was the most difficult to fill, and A was strongly disfavored in the 3rd anticodon position (Fig. 10.9). We posit the following explanation. Before evolution of the EF-Tu GTPase latch, the 1st row may have been filled with tRNAs that were utilized inefficiently, often resulting in termination of translation and relatively short peptide release. In Archaea, A is very inefficiently utilized in the 1st anticodon wobble position. We posit that A was very inefficiently utilized in the 3rd anticodon position, when the 3rd anticodon position was a wobble position, before evolution of the EF-Tu latch. We guess that row 1 tRNAs were charged with amino acids but were infrequently utilized. We posit, therefore, that stop codons recognized by protein release factors evolved after evolution of the EF-Tu latch. Because A was strongly disfavored in the 3rd anticodon position, the location of stop codons to row 1 is telling. Because row 1 (3rd anticodon position A) anticodons are disfavored and because stop codons are recognized as mRNAs, it makes sense that stop codons are located to row 1. Within the standard code, no tRNAs correspond to stop codons.

A working model

In Fig. 10.10, we show a highly detailed model for evolution of the genetic code. This is a variation of models previously published by our laboratory (Damer & Deamer, 2015; Lei & Burton, 2020). Possible advantages of this model are (1) an improved description of the evolution of stop codons; (2) an explanation for serine jumping from column 2 to column 4 of the code; (3) possible new insight into evolution of 6-codon sectors; and (4) disfavoring of A in the anticodon 3rd position. When we first tried to formulate these models, we thought that such detailed accounts of the genetic code evolution were not reasonable. Now, we are convinced that these models are highly descriptive and likely highly accurate. We were surprised at how easily these models unfolded from a very small number of reasonable initial assumptions and, also, how readily hypotheses for tRNA, aaRS, and genetic code evolution combined.

Assumptions and background

The main initial assumption that we made was that glycine was the first encoded amino acid. Furthermore, we assumed that the entire genetic code (anticodons and codons) initially encoded polyglycine (Figs. 10.9 and 10.10A). We considered that polyglycine could have multiple purposes in the ancient world. Polyglycine could function as a hydrogel, and polyglycine could act as a cross-linking agent, as it does in bacterial cell wall peptidoglycan layers (Pinho et al., 2013; Scheffers & Pinho, 2005; Zapun et al., 2008). So, polyglycine can be a structural and hydrogel (LLPS) component, enhancing protocell function (i.e., membrane function and ion transport) and rigidifying protocell membranes. One reason to believe this is a reasonable assumption is that tRNA^{Pri} (the primordial type I tRNA) is essentially tRNA^{Gly} in ancient Archaea (Fig. 10.4). As noted above, GlyRS-IIA in Archaea is the primordial aaRS enzyme (Fig. 10.9). If the entire genetic code initially encoded polyglycine, this provides a model for coevolution of mRNA and tRNA. If all mRNA codons and tRNA anticodons initially encoded glycine, it was very simple to coevolve mRNA codons and tRNA anticodons with new invasions of amino acids from outside the code.

The concept of filling in the genetic code with glycine and then adding amino acids by invasion resulted in a simple model for additions of amino acids to the code. Glycine looked like the first encoded amino acid, and glycine utilizes anticodons GCC, UCC, and CCC (2nd and 3rd anticodon position C). As we discuss, C is favored in the 2nd and 3rd anticodon positions. A is (essentially) never found in the anticodon wobble 1st position in Archaea. Furthermore, the four simplest amino acids glycine, alanine, aspartate, and valine appear to be the first four amino acids encoded, and these amino acids are all found in row 4 of the genetic code (3rd anticodon position C) (Ikehara, 2014; Ikehara et al., 2005; Oba et al., 2005). It began to look as if C was a favored base in the tRNA anticodon. Because A is so strongly disfavored in the anticodon wobble position in Archaea, we began to wonder whether A was also disfavored in the 3rd anticodon position. Phenylalanine, tyrosine, tryptophan, cysteine, and stop codons appear to be late additions to the genetic code that are found in row 1, which is 3rd anticodon position A. We began to think that A was disfavored in both the 1st (wobble) and 3rd anticodon positions. Stop codons are recognized by protein release factors as codons in mRNA, with no corresponding anticodon in tRNA, consistent with A being disfavored in the anticodon 3rd position. Protein release factors recognize stop codons to release the nascent peptide from the ribosome (Burroughs & Aravind, 2019). So, C is favored in the anticodon, in the 2nd and 3rd anticodon positions. A is disfavored in the anticodon and, probably, in all three anticodon positions, although the effect is not obvious in the 2nd anticodon position, which is the easiest to read. So, glycine occupies the most favored position in the genetic code (anticodons GCC, UCC, CCC), consistent with glycine being the first encoded amino acid.

Evolution in columns

Analysis of aaRS evolution indicates that much of the evolution of the genetic code occurred within code columns, which represent the 2nd anticodon position (Fig. 10.8). On the ancient ribosome (i.e., before LUCA), the 2nd anticodon position was the easiest to read, in part because the 1st and 3rd anticodon positions were initially wobble positions. In column 1 of the genetic code, ValRS-IA, MetRS-IA, IleRS-IA, and LeuRS-IA are all structural class IA enzymes (Fig. 10.8A). Valine, isoleucine, and leucine are hydrophobic amino acids. Valine can be converted to leucine via a 5-step pathway. Methionine is a late invader that we posit attacked an isoleucine 4-codon sector to add methionine and to evolve translation start codons, but MetRS-IA is also structural class IA, indicating additional evolution in column 1 (i.e., IleRS-IA $\Rightarrow$ MetRS-IA). In column 2 of the genetic code SerRS-IIA, ProRS-IIA and ThrRS-IIA are closely related enzymes of structural class IIA. Serine and threonine are closely related amino acids. There is substantial evidence, therefore, for coevolution of aminoacyl-tRNA synthetases and amino acids in column 2. In column 3, HisRS-IIA, AspRS-IIB, and AsnRS-IIB enzymes are reasonably closely related, and aspartate and asparagine are related amino acids. Also, in column 3, GlnRS-IB, LysRS-1E, and GluRS-IB enzymes are closely related, and glutamate and glutamine are closely related amino acids. In column 4, ArgRS-ID and CysRS-IB enzymes are closely related. In some cases, the assigned structural class for an aaRS does not represent its closest relatives. We take these data to overwhelmingly support evolution of the genetic code within columns (anticodon 2nd position) as indicated in the model in Fig. 10.10.

In the model (Figs. 10.9 and 10.10), the genetic code sectors from one that encodes only glycine (Fig. 10.10A) to one that encodes glycine, alanine, aspartate, and valine (Fig. 10.10B). Others have supported these four simple amino acids as the first four encoded amino acids (Chatterjee & Yadav, 2019; Ikehara, 2014; Ikehara et al., 2005; Koonin & Novozhilov, 2017; Oba et al., 2005). In the standard code (Fig. 10.10F), glycine retreats from occupying the entire code (Fig. 10.10A) to occupying only column 4 (Fig. 10.10B), and, finally, to occupying only row 4 and anticodons GCC, UCC and CCC (2nd and 3rd anticodon position C) (Fig. 10.10F). Aspartate retreats from initially occupying all of column 3 (Fig. 10.10B) to occupying only anticodon GUC (Fig. 10.10F). Alanine retreats from occupying all of column 2 (Fig. 10.10B) to occupying only anticodons GGC, UGC, and CGC (Fig. 10.10F). Valine retreats from occupying all of column 1 (Fig. 10.10B) to occupying only anticodons GAC, UAC and CAC (Fig. 10.10F). In this way, the first four encoded amino acids land in the 4th row after occupying the entire code. Newly added amino acids, therefore, invade previously occupied sectors of the code. Amino acids that enter the code first surrender less-favored anticodons to invaders but retain the most favorable anticodons according to clear and inviolable rules.

An evolutionary bottleneck

On a primitive ribosome, the 1st and 3rd anticodon positions were initially wobble positions. The consequence of this bottleneck in genetic code evolution is that the complexity of the code freezes at ~ 8 amino acids (Fig. 10.10C and D). Wobble positions were read with only pyrimidine-purine discrimination, and only one wobble position could be read at one time, limiting the code complexity to $2 \times 4 = 8$ assignments. Columns 1, 2, and 4, initially sectored on the 2nd and 3rd (then wobble) anticodon positions. Column 3, by contrast, initially sectored on the 1st (wobble) and 2nd anticodon positions. In columns 1, 2, and 4, it appears that incoming amino acids leucine (column 1), serine (column 2), and arginine (column 4) may have first invaded the 2nd row (3rd anticodon position G) (Fig. 10.10C). Interestingly, leucine, serine, and arginine are the three amino acids that occupy 6–codon sectors in the fully evolved genetic code. Leucine, serine, and arginine are posited to occupy row 2 first, because row 1 is difficult to occupy (3rd anticodon position A is disfavored). It appears that the tRNA anticodon bases were selected for small size (pyrimidine>purine) and stronger hydrogen bonding potential (C>G>U>>A). These rules are most apparent in the 3rd anticodon position (Figs. 10.9 and 10.10).

Evolution of 6-codon sectors

Then, we posit that leucine (column 1), serine (column 2), and arginine (column 4) may have invaded row 3 (3rd position U; Fig. 10.10D). Some reasons to think this invasion of row 3 might have occurred are as follows. First, serine jumps to row 3, column 4, from column 2 in the code. This jump is easiest to imagine if serine occupies column 2, row 3, before making its jump to column 4, row 3. According to the model, only a single base change in the tRNA anticodon (GGU $\Rightarrow$ GCU) was necessary for serine to jump. Second, arginine occupies rows 2 and 3 in the code, as if arginine made the posited invasion of row 3. In this regard, it is notable that leucine, serine, and arginine are the three amino acids to occupy 6–codon sectors in the final evolution of the standard code. Here, we suggest that 6-codon sectors may have arisen from the history of code sectoring.

Evolution of the EF-Tu latch

To proceed beyond the bottleneck of only 8 amino acids required the evolution of the EF-Tu GTPase latch. The latch sets a closed conformation of the codon-anticodon pair and the ribosome involving 16S rRNA (30S subunit) residues G530, A1492 and A1493, and 23S rRNA (50S subunit) residue A1913. Closing of the latch allows 4-base recognition at the 2nd and 3rd anticodon positions (Loveland et al., 2017, 2020; Rozov, Demeshkina, et al., 2016; Rozov et al., 2018; Rozov, Westhof, et al., 2016). The latch also improves the accuracy of the wobble position, but the wobble 1st anticodon position only has purine versus pyrimidine resolution. Evolution of the EF-Tu latch allows a genetic code with up to $2 \times 4 \times 4 = 32$ anticodon assignments.

Completion of column 1

To complete sectoring of the genetic code in column 1, we posit that isoleucine invaded row 3, displacing leucine, which retained row 2 (3rd anticodon position G (row 2) (Leu) was favored over U (row 3) (Ile)). Isoleucine formed a 4-codon sector with anticodons GAU, UAU, and CAU. When methionine invaded CAU, isoleucine retained CAU, but isoleucine codon AUA is specified by anticodon wobble C-agmatidine modification, which does not read AUG methionine codons (Mandal et al., 2020; Voorhees et al., 2013). In Archaea, generally, isoleucine UAU is not utilized. So, methionine was a late invader of an isoleucine 4-codon sector. Methionine is utilized at start codons. In column 1, row 1, phenylalanine was a late addition to the code.

Completion of column 2

To complete sectoring of column 2, serine first jumped to column 4 (Fig. 10.10D and E), then, after evolution of the EF-Tu latch, the serine sector expanded to disfavored row 1 (disfavored 3rd anticodon position A). Then proline invaded row 2, displacing serine, and threonine invaded row 3, displacing serine. Because serine jumped to column 4, row 3, serine occupied a favorable anticodon (GUC), and serine, therefore, could give up otherwise favored anticodon positions in column 2 to proline and threonine. Serine is the only amino acid to have jumped in evolution of the genetic code, indicating the overall orderly evolution of the code. Other models are possible for evolution of column 2 (Kim et al., 2019; Lei & Burton, 2020).

Evolution of column 3

Because of sectoring on the 1st anticodon (wobble) rather than the 3rd position, genetic code column 3 is the most innovated column. We posit that sectoring on the wobble position caused this innovation and the pattern of sectoring. We posit that initially, aspartate filled column 3 (Fig. 10.10B). Aspartate was displaced by the related amino acid glutamate in rows 4B, 3B and 2B. Aspartate retained rows 4A, 3A and 2A and surrendered rows 4B, 3B and 2B to glutamate, because, in the anticodon, wobble G is favored over wobble U and C, and aspartate entered the code first. Histidine displaced aspartate in row 2A. We posit that HisRS-IIA evolved from a primitive AspRS-IIA. AspRS-IIA then evolved to AspRS-IIB to suppress translation errors. In row 3A, first an amido-transferase evolved to convert aspartate bound to an emerging tRNA^{Asn} to asparagine. Subsequently, AsnRS-IIB evolved from AspRS-IIB. Note that the order of invasions and modifications are indicated by the structural classes of the aaRS enzymes (Figs. 10.8 and 10.9).

Glutamate occupied column 3, sectors 4B, 3B and 2B. GluRS-IA evolved to GluRS-IB. In sector 2B, glutamate bound to tRNA^{Gln} was modified to glutamine by an amidotransferase (Feng et al., 2005; Perona, 2005; Sheppard et al., 2007; Sheppard, Sherrer, et al., 2008; Sheppard, Yuan, et al., 2008). Subsequently, this system evolved a GlnRS-IB to substitute for the amidotransferase. We note that, metabolically, lysine can be derived from a pathway that utilizes glutamate, although the lysine carbon skeleton is derived from α-aminoadipic acid. It appears, therefore, that lysine invaded sector 3B from outside the code, displacing glutamate. Despite its different stated structural class, LysRS-IE in Archaea is very closely related to GluRS-IB and GlnRS-IB (Fig. 10.8A). Probably, lysine invasion of the code occurred before full establishment of the GlnRS-IB system, which is incompletely evolved in some Archaea. In Bacteria, LysRS-IIB is found. We posit, therefore, that LysRS-IIB was a bacterial innovation that replaced the archaeal LysRS-IE. Bacterial LysRS-IIB is closely related to AspRS-IIB and AsnRS-IIB (Fig. 10.8A). We posit that, in Bacteria, LysRS-IIB evolved within column 3 to better specify accurate tRNA^{Lys} charging. LysRS-IIB in Bacteria evolved an editing active site to remove improperly attached amino acids. It appears that LysRS-IIB mostly uses editing to discriminate against amino acids invading from outside the genetic code (Perona & Gruic-Sovulj, 2014).

Column 4

Metabolically, arginine can be derived from ornithine, which may have been a more primitive positively charged amino acid utilized in prelife (Longo et al., 2020). It is possible, therefore, that arginine replaced encoded ornithine during genetic code evolution. We posit that arginine (or ornithine) occupied column 4, rows 2 and 3, displacing glycine, which, as the first encoded amino acid, retained the most favored anticodons, GCC, UCC and CCC (2nd and 3rd anticodon C) (Fig. 10.10C and D). After evolution of the EF-Tu GTPase latch, row 1 could be occupied, and additional amino acids could be encoded. We posit that serine invasion from column 2 to column 4 occurred early in code evolution, for instance, before proline and threonine invasion of column 2. We note that serine invasion of column 4 could be initiated by a single base change in the 2nd position of the tRNA anticodon (GGU ⇒ GCU). Also, SerRS-IIA is a very different enzyme than ArgRS-ID, facilitating the invasion of column 4 by limiting tRNA charging errors. Although ArgRS-ID is classified as a structural subclass ID enzyme, ArgRS-ID is closely related to subclass IA and Cys-IB enzymes (Fig. 10.8A).

Late evolution of row 1

We posit that phenylalanine, tyrosine, cysteine, tryptophan, and stop codons, all located on disfavored row 1, were late additions to the code. Here, we suggest that these amino acids and stop codons could not be added before evolution of the EF-Tu latch. Essentially, the disfavored first row (3rd anticodon position A) could not be efficiently occupied before evolution of the latch. Prior to evolution of the latch, the stop signal is posited to have been inefficiently functioning tRNAs with 3rd position A. After evolution of the latch, row 1 tRNAs could be efficiently utilized, and leucine and serine could effectively invade row 1. Phenylalanine then invaded column 1, row 1A, displacing leucine. Leucine retained favored row 2 anticodon positions with phenylalanine invasion. Uncharacteristically, within column 2, serine surrendered more favorable anticodons to proline and threonine, but serine retained a favorable anticodon in column 4, row 3A. Also, serine utilizes a type II tRNA^{Ser}. Type II tRNA^{Ser} has an expanded variable loop that functions as a positive determinant for accurate SerRS-IIA serine addition. Because tRNA^{Ser} is a type II tRNA, in which the expanded variable loop is a SerRS-IIA-contacted determinant for accurate charging, this facilitated serine jumping in the code, from column 2 to column 4, and allowed serine to maintain a favored anticodon in column

4 (GCU) and to surrender otherwise favored anticodons in column 2 to invading proline and threonine. At about the time of the evolution of the EF-Tu latch, we posit that protein release factors evolved to take over stop codon functions, allowing proteins to become longer and more complex with accurate starts and stops (Burroughs & Aravind, 2019).

The genetic code model and perspectives

The model offered here is a variation of models published previously. The genetic code evolved around the tRNA anticodon. Taking a tRNA-centric view, therefore, simplifies the understanding of code evolution. The model presented here provides clear Darwinian selections for the locations of all amino acids in the genetic code. Amino acids enter the code by two identifiable mechanisms: (1) invasion from outside the code and (2) enzymatic modifications of amino acids bound to tRNAs (i.e., Asp$\Rightarrow$Asn, Glu$\Rightarrow$Gln, and pSer$\Rightarrow$Cys) followed by subsequent evolution of aaRS enzymes (AsnRS-IIB, GlnRS-IB, and CysRS-IB). Evolution occurred first in code columns because the 2nd anticodon position is most important and easiest to read on a primitive ribosome. Evolution also occurred by rows according to clear anticodon preference rules (Fig. 10.9). Column 3 sectored differently than columns 1, 2 and 4. Column 3 sectored early on the 1st (wobble) and 2nd anticodon positions, between aspartate and glutamate. As a result of this sectoring strategy, column 3 became the most innovated column in the code, encoding the most amino acids. Columns 1, 2, and 4, which sectored on the 2nd and 3rd anticodon positions, are characterized by larger blocks of anticodons (i.e., 4- and 6-codon sectors). Columns 1 and 2 are characterized by aaRS enzymes with editing active sites. In Archaea, ProRS-IIA is an exception. Because editing is a fidelity mechanism and because amino acids invade the code through tRNA charging errors, editing probably protects larger blocks of anticodons (i.e., 4- and 6-codon sectors). Because arginine and glycine have unique characteristics, ArgRS-ID and GlyRS-IIA were under little selection pressure to evolve editing. Arginine is a stiff and bulky, positively charged amino acid with unique hydrogen-bonding capacity. By contrast, positively charged ornithine and lysine are very flexible. Arginine, therefore, forms more structured ion pairs, particularly with aspartate (ion pairs with glutamate are more flexible). Glycine is the smallest amino acid, so a compact active site in GlyRS-IIA limits mischarging of tRNA^{Gly} with larger amino acids. We posit that aaRS editing protected 4- and 6-codon sectors in columns 1 and 2 by limiting mischarging of tRNAs. Editing was not necessary for the aaRS enzymes in column 4 because of glycine and arginine properties. Column 3 broke into 2-codon sectors because of early sectoring on the 1st anticodon (wobble) position.

Once the genetic code evolved and protein enzymes came to dominate, the potential to enrich metabolism and energy utilization exploded. Whatever systems predated current systems, therefore, were replaced by enzymatic and protein motor pathways. For these reasons, we do not favor models for genetic code evolution based primarily on metabolism. Of course, an amino acid must have been available in order to have been added to the code. On the other hand, the expanding code helped drive the evolution of metabolism to provide more amino acids because, with the advent of coding, amino acids were of enhanced selective value. With regard to energy transduction, it is clear that primitive energetic systems were sufficient to support the evolution of the standard genetic code. After code evolution, an explosion in energy transduction systems occurred, leading to modern systems. We note that multiple pathways are identified in which tRNA-bound amino acids are substrates for metabolic reactions. In the prelife world, we posit that RNA-bound peptides and amino acids were substrates for many reactions (Gospodinov & Kunnev, 2020; Kunnev & Gospodinov, 2018). One effect of covalent RNA-amino acid binding was to shield a potentially reactive group on the amino acid from unproductive side reactions.

The mutually reinforcing tRNA, tRNAome, aaRS, and genetic code evolution models presented here make many testable predictions. The model for genetic code evolution follows strongly from the aaRS evolution model (Fig. 10.8). Essentially, evolution of aaRS enzymes directs the genetic code model. Detailed hypotheses were generated for polyglycine world and GADV world, and these predictions can be tested experimentally (see below). The sequence analyses underlying the tRNA, tRNAome, and aaRS evolution models described above can be further challenged using additional sequence data and more sophisticated bioinformatics and computation. We make suggestions about RNA-linked reactions in the ancient prelife world that can be pursued. For instance, if Val-tRNA^{Val} was converted to Leu-tRNA^{Leu} through a series of tRNA-linked reactions and evolutionary steps, as we suggest, then the history of column 1 evolution becomes significantly richer and more interesting. Such a model for column 1 evolution reinforces the interaction between our views and those that support a metabolic coevolution theory. Also, such a view enriches the possibilities for RNA-linked and tRNA-linked reactions in the ancient world (Gospodinov & Kunnev, 2020; Kunnev & Gospodinov, 2018). We posit the existence of diverse ribozymes in the ancient world, some of which have not yet been generated by researchers. As an example, we imagine a telomerase-like ribozyme with a guide RNA template that accurately generated RNA repeats from an RNA 3′-end to synthesize tRNA precursor sequences (Fig. 10.3C). We also posit that diverse ribozyme aaRS enzymes with reasonable accuracy could be generated to initiate the genetic code before enough amino acids have joined the code to encode aaRS protein enzymes. We posit diverse RNA-linked reactions in the ancient world with many yet to be discovered.

The great divergence

How did Archaea and Bacteria diverge? Which domain is most similar to LUCA? Despite their many similarities, how are Archaea and Bacteria distinct? After LUCA, the great divergence occurred, which we posit resulted in the splitting of Archaea and Bacteria (Fig. 10.1). Although this point has been argued, we identify LUCA as most similar to Archaea (Long et al., 2019; Marin et al., 2017). Interestingly, a recent paper placed LUCA in the midst of the archaeal domain. We noticed that tRNAomes (all of the tRNAs of an organism) were much more compact in ancient Archaea and much more diverged in Bacteria. Radiating tRNAomes from $tRNA^{Pri}$, archaeal $tRNA^{Gly}$ is most similar to $tRNA^{Pri}$ (Fig. 10.4) (Pak, Du, et al., 2018). Archaeal tRNAomes are much more similar to $tRNA^{Pri}$ than bacterial tRNAomes. Archaeal tRNAomes are also more highly structured than bacterial tRNAomes, as if convergent and divergent evolution have scrambled bacterial tRNAomes. We posit that ancient archaeal tRNAomes are structured similarly to LUCA tRNAomes. A similar argument could be made for archaeal aaRS trees and genetic code structures. These analyses support the hypothesis that Archaea are more closely rooted to LUCA than Bacteria. Archaea and Bacteria have distinct membrane lipids and replication systems that may in some way be linked (Koonin et al., 2020). Eukaryotes made an evolutionary choice of bacterial membrane systems over archaeal systems, indicating a potential selective advantage to the adopted bacterial membrane system at least in the evolving eukaryotic system.

Models to describe genetic code evolution

We suggest that the dominant models for description of genetic code evolution be reevaluated. We found these models confusing and largely unhelpful. Formerly, views of genetic code evolution broke primarily into three main categories: (1) the stereochemical theory; (2) the coevolution (metabolism) hypothesis; and (3) the error-minimization theory (Koonin, 2017; Koonin & Novozhilov, 2009, 2017). The stereochemical theory posits that originally nucleic acids and amino acids interacted chemically, resulting in the code between tRNA anticodons and amino acids. We find the stereochemical theory to have little predictive power for evolution of the code, although the stereochemical theory appears to reasonably describe evolution of riboswitches (Bédard et al., 2020). The coevolution hypothesis has some value. Of course, amino acid metabolism and the genetic code coevolved (Preiner et al., n.d.). How could it be otherwise? If an amino acid could not be generated by primitive metabolism, it could not be added to the code. Amino acids, therefore, were added to the code from simple to complex, with glycine, alanine, aspartate, and valine the first and simplest amino acids (Ikehara, 2014; Ikehara et al., 2005; Oba et al., 2005) and phenylalanine, tyrosine, and tryptophan among the last and most complex additions (Fournier & Alm, 2015). We do not see metabolism of amino acids, however, as a strong driving force in selection of new amino acids added into the code. We also see the error minimization theory as a limiting idea. The error minimization theory indicates that the genetic code was structured to minimize translation errors. We do not think that idea is correct. Our opinion is that translational fidelity mechanisms drove the freezing of the code. Interestingly, the EF-Tu GTPase latch drove first the expansion of the code from an 8 amino acid bottleneck and later the freezing of the code at 20 amino acids plus stops (Fig. 10.10). Our view is that the standard genetic code evolved around tRNA and the tRNA anticodon (Kim et al., 2019; Lei & Burton, 2020; Pak, Du, et al., 2018; Pak, Kim, et al., 2018). The identification of 31-nt minihelices and 17-nt microhelices that can attach ACCA via a ribozyme ligase indicates a rich prebiotic chemistry involving covalent RNA-amino acid linkages and diverse ribozyme activities (Fig. 10.2) (Bédard et al., 2020; Gospodinov & Kunnev, 2020; Kim et al., 2018; Kunnev & Gospodinov, 2018; Pak et al., 2017). The capacity for doing chemistry on tRNA-amino acid linkages persists, as we describe (i.e., pSer $\Rightarrow$ Cys, Asp $\Rightarrow$ Asn, Glu $\Rightarrow$ Gln, and possibly Val $\Rightarrow$ Leu) (Feng et al., 2005; Hauenstein & Perona, 2008; Mukai et al., 2017; Rogers & Söll, 1995; Schön et al., 1988; Sheppard et al., 2007; Sheppard, Yuan, et al., 2008). We posit that ACCA bound to amino acids at its 3′-end as a substrate for chemistry before the advent of 31-nt mini-helices and tRNA. We strongly advocate for the tRNA-centric view that evolution of tRNA from ligation of three 31-nt minihelices drove the evolution of the code, mRNA, and rRNA. Therefore, tRNA was the central advance in biological intellectual property that enabled evolution of the code. We describe powerful Darwinian selections driving evolution mostly within code columns but ultimately with amino acids distributing in an ordered manner in code rows, according to clear selection rules. We strongly argue that the history of amino acid additions to the code follows these interpretable patterns. For instance, column 3 of the genetic code becomes the most innovated column because of sectoring between Asp and Glu, initially utilizing the 1st and 2nd anticodon positions rather than the 2nd and 3rd anticodon positions, as for columns 1, 2 and 4 (Fig. 10.10). Similarly, the history of evolution in columns 1, 2 and 4 appears to result in 6-codon sectors that encode leucine, serine and arginine. Of course, the failure to subdivide 6- and 4-codon sectors results in a code with fewer amino acids than could potentially be encoded. With regard to the freezing of the code, the genetic code was built by modifications of amino acids bound to tRNAs and by tRNA charging errors (invasions of amino acids from outside the code). tRNA charging errors and modifications, therefore, drove innovations of the code, and translational fidelity mechanisms froze the code. The EF-Tu latch is a major translational fidelity mechanism. The EF-Tu latch evolved to expand the code from an ~8 amino acid bottle-neck to a richer code (Fig. 10.10). Fidelity mechanisms such as the EF-Tu latch, aaRS editing, aaRS active site specificity, tRNA modifications, and tRNA specialization drove the freezing of the code at 20 amino acids + stops. In this regard, we note that aaRS editing on the left half of the genetic code appears to protect 4- and 6- codon sectors from further divisions to encode additional amino acids. Anticodons specifying specific amino acids evolved as we describe through the coevolution of tRNAomes, aaRS enzymes and translational fidelity mechanisms. We show clearly that aaRS enzymes and the genetic code were powerfully coevolved (Fig. 10.8).

Polyglycine world (a working model)

This paper reveals significant detail about the ancient prelife and protocell worlds. We describe fully the evolution of tRNAs and the genetic fragments and sequences from which tRNA was derived. We describe evolution and radiation of aaRS enzymes and the relationship between evolution of aaRS enzymes and sectoring of the genetic code. We describe how tRNA and hydrogels/LLPS could have contributed to the earliest aaRS folding. In this review, we attempt to link these ancient events to the activities of hydrogels, LLPS, membraneless compartments, and amyloids. We believe the mechanisms and descriptions will lead to advances in analyses of hydrogels in prelife reactions, protocell functions, and prokaryotic systems. We consider hydrogels and related assemblies to be a formerly largely missing consideration in analyses of ancient evolution. For some future studies, we recommend increasing the system complexity and inclusion of hydrogels to help identify reactions of interest that may lead to an understanding of earlier events. Specifically, experimental probing of the richer chemistry of a GADV world (Fig. 10.10B) would be expected to lead to insights into a prior polyglycine world (Fig. 10.10A).

Significant work has been done with coacervate systems to enhance prebiotic chemistry. Examples of coacervates include clays, polymers, and mica (Chatterjee & Yadav, 2019; Hansma, 2013; Mariscal et al., 2019). Such materials can concentrate reactants, control the access and activity of water, participate in wet-dry cycles and provide polar surfaces to help with enantiomer fractionations. Here we propose that polyglycine was a potent prebiotic hydrogel component that drove the earliest evolution of the genetic code. Above, we describe a number of human proteins that may rely on hydrogels and that have polyglycine tracts. Shorter polyglycine tracts (i.e., length 6–8) can be found in archaeal, bacterial, and phage proteins. We do not know the extent to which such short tracts can function to generate localized hydrogels. Based on our model for evolution of the genetic code, we further propose that polymers of Gly, Ala, Asp, and Val may enhance hydrogel functions in prokaryotic systems, as indicated above. In prebiotic systems, short peptide linkers can cross-link polymers to make more complex networks. Some human proteins (i.e., transcription factors) include polyalanine, polyhistidine, and polyglutamine tracts.

Hydrogels (LLPS) appear to be incompletely characterized and somewhat difficult to analyze in prokaryotic systems (Guilhas et al., 2019; Langdon & Gladfelter, 2018; Longo et al., 2020). We guess that hydrogel compartments are important for many bacterial processes including cell division, coupling of transcription and translation, nucleoid body maintenance and rearrangements, ion transport, and signaling. LLPS affects septation in Bacteria. In eukaryotic systems, LLPS has been more aggressively analyzed and is perhaps better understood. For instance, LLPS compartments tend to be larger and easier to visualize in eukaryotic systems. As a critical phase of their evolution, Eukaryotes appear to have powerfully enriched LLPS systems by increasing the use of proteins that include IDRs with the potential for covalent modification and noncovalent bonding of diverse hydrogel components (Boehning et al., 2018; Guo et al., 2020; Langdon & Gladfelter, 2018; Lu et al., 2018; McSwiggen et al., 2019; Portz & Shorter, 2020; Yoshizawa et al., 2020; Zhou et al., 2018). Examples of IDRs with these properties include histone tails and the CTD of RNA polymerase II. Covalent modifications of these disordered regions alter activities and factor binding. For instance, as RNA polymerase II traverses the transcription cycle, polymerase can move between LLPS compartments that support different phases of the cycle. Transcriptional superenhancers that direct cell-specific gene regulation programs organize hydrogel compartments, and, in some cancers, superenhancers become disorganized and sometimes fuse together (Guo et al., 2020; Yoshizawa et al., 2020). The nucleolis is a hydrogel compartment for RNA polymerase I transcription and ribosome assembly (Odeh & Shorter, 2020). . RNA polymerase III also separates into hydrogel compartments. In eukaryotic cells, hydrogel compartments appear to segregate and regulate many other activities including translational control, translational delay via microRNA, signaling and ion transport. Within cells, hydrogel compartmentalization is a mechanism to support diverse biological processes, the activity of water and the potency of acids and bases to support chemistry.

We try to imagine polyglycine world. In Fig. 10.11, we show a working model for a protocell that utilized polyglycine as a hydrogel, cross-linking agent and component, and stabilizer of protocell walls. As discussed in this paper, many of the predicted components of the protocell matrix are indicated. We posit that the interior of the protocell was packed with polyglycine hydrogels, membraneless compartments, polyglycine amyloid accretions, primitive metabolites, tRNAs, diverse ribozymes, and pre-ribosomes. Our idea of a preribosome is a pre-16S scaffold, on which to mount a mRNA (of any sequence), and a mobile and independent PTC (Zhang & Cech, 1998). We envision that tRNAs with essentially all anticodons represented are charged (essentially) only with glycine by a GlyRS-ribozyme. The idea is that the most rapid sequence to mutate successfully in tRNA is the anticodon, because most other changes cause structural defects in the L-shaped tRNA structure. We imagine that ACCA was ligated to early tRNAs, so the tRNA could be charged with glycine. ACCA is the most common 3′-end in archaeal tRNAs (Juhling et al., 2009), indicating a prelife function of ACCA ligated to RNAs. In such a system, essentially all encoded protein products were polyglycine of varying lengths. We posit that translation termination initially tended to occur at

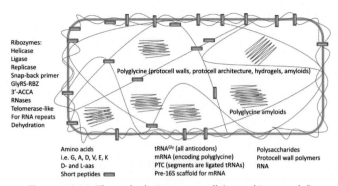

Figure 10.11 The polyglycine protocell (a working model).

NNA anticodons, because the 3rd anticodon position was a wobble position, and 3rd position A had difficulty pairing to U in mRNA. Because we imagine a minihelix world before a tRNA world (Fig. 10.3C), we posit that 31-nt minihelices with attached ACCA (i.e., via ligation) were also directed to synthesize polyglycine, and a tRNA-based polyglycine world would, therefore, have inherited many features of the prior minihelix world.

Polyglycine hydrogels and RNAs are expected to powerfully regulate the activity of water in protocells. Dehydration stimulates polymerization reactions including RNA synthesis, DNA synthesis, polypeptide synthesis, and polysaccharide (i.e., cell wall) synthesis. RNAs bind water and, therefore, include dehydrating pockets to support chemistry of nearby reactants. For instance, the PTC of the ribosome has been considered a molecular crowding and dehydration chamber to support peptide bond formation utilizing chemically diverse amino acid substrates (Bernier et al., 2018). The "trigger loop" in RNA polymerase II closes over the active site expelling water to support RNA polymerization (Mazumder et al., 2020; Wang et al., 2013). When the activity of water is decreased, acid-base reactions and polymerization reactions become more potent.

We imagine that polyglycine can be cross-linked to short polypeptide chains similar to those found in peptidoglycan bacterial cell walls and using similar chemistry. The protocell could have been encapsulated by peptidoglycan protocell walls. Capping polyglycine with other amino acids (i.e., lysine) and reduction of polyglycine C-termini to aldehyde groups would allow Schiff's base cross-linking, so a protocell matrix mostly of polyglycine with short peptide linkers could be constructed. Some reactions could be supported by the reducing environment and others by ribozymes. If protocells had cell walls, they also had long polysaccharides in addition to long

RNA polymers. We hope that the model we outline can provide some utility in developing new approaches to enrich studies of prebiotic chemistry. We see value in a top down experimental approach starting with more complex systems (i.e., GADV world), in which more diverse chemistry might be detected, and then moving to simper systems (i.e., polyglycine world).

Conclusions

Chaotic processes initiated evolution of the genetic code by providing a Darwinian selection. Specifically, polyglycine and poly-GADV (i.e., polyalanine) hydrogels, LLPS and amyloids provided compartments and coacervates to stimulate polymerization reactions and novel chemistry within protocells. By interesting contrast, tRNAs and the genetic code evolved by highly ordered and systematic processes. Based on sequences in ancient Archaea, tRNA evolution appears to be a solved and highly systematic problem. tRNA-linked amino acid reactions in Archaea demonstrate the importance of diverse RNA-linked chemistries in early evolution of life and the genetic code. Unexpectedly, divergence of aaRS enzymes strongly indicates the pathway for evolution of the genetic code. This result was somewhat unexpected because how can protein enzymes evolve before they can successfully be encoded. We posit that the answer lies in coevolution of interacting systems, as we describe here. The EF-Tu GTPase latch (closing of the 30S ribosomal subunit), which suppressed wobbling at the 3rd anticodon position, appears to have been a major driving force required for code expansion beyond the first 8 amino acids. Wobbling, therefore, was of fundamental importance in code evolution. The standard code expanded to 20 amino acids, but the maximum theoretical complexity of the code in tRNA is 32 anticodon assignments. To make additional anticodon assignments would require division of 4- and 6-codon sectors. aaRS editing and other fidelity mechanisms protected 4- and 6-codon sectors from further divisions.

Ancient evolution of the prelife $\Rightarrow$ life transition from about 4 Ga ago is becoming increasingly well understood. Evolution of the genetic code is a simpler problem in archaeal systems, which are closest to LUCA (Fig. 10.1) (Long et al., 2019; Marin et al., 2017). Coevolution of tRNA, mRNA, rRNA, aaRS enzymes, the genetic code, and ribosomes appears to be a largely outlined problem (Damer & Deamer, 2015; Lei & Burton, 2020). The pathway of evolution of translation systems was driven via the evolution of tRNA, which is a solved problem (Burton, 2020; Kim et al., 2018; Pak et al., 2017). Because of wobbling, the genetic code has a maximum complexity of 32 assignments, as in tRNA anticodons ($2 \times 4 \times 4$), not 64 assignments, as in mRNA codons ($4 \times 4 \times 4$). The standard genetic code froze at 20 amino acids + stops because of translational fidelity mechanisms (Kim et al., 2019; Lei & Burton, 2020; Pak, Kim, et al., 2018). Column 3 of the code is the most innovated column, encoding the most amino acids, because column 3 evolution was driven at an early stage by the tRNA anticodon 1st wobble position, rather than the 3rd anticodon position, as in columns 1, 2 and 4 (Fig. 10.10). aaRS enzyme structural classes and some chemically similar amino acids align with genetic code columns demonstrating the importance and centrality of the 2nd anticodon position. We posit that glycine, the first encoded amino acid, initially filled the genetic code. According to the model, after glycine, other amino acids enter the code by invading previously occupied sectors and, therefore, displacing previously encoded amino acids. Amino acids that first entered the code retained the most favored anticodons according to clear rules. Glycine lands in code column 4, row 4, indicating that C is preferred in the anticodon 2nd and 3rd positions. The genetic code, however, appears to fill via rows: row 4 ⇒ row 2 ⇒ row 3 ⇒ row 1, indicating that the 3rd anticodon position has the following preference rules: C>G>U>>A. Filling row 1 (disfavored 3rd anticodon position A) appears to have required evolution of the EF-Tu latch, a major feature of translational fidelity that allowed expansion of the code to 20 amino acids + stops from an ∼8 amino acid bottleneck. The preference rules for the anticodon wobble position are G>(U∼C)>>>>>>A. Only purine-pyrimidine discrimination was initially achieved at the anticodon wobble position. Using these simple rules, the entire genetic code was populated, as observed in the standard code (Fig. 10.10F). Other features of translation systems evolved around tRNAs. It appears that polyglycine and GADV may have constituted potent hydrogels, LLPS compartments, and amyloid accretions driving strong Darwinian selection of the standard code particularly during early stages before protein enzymes were sufficiently encoded. All features of the hypotheses arise from models for tRNA, tRNAome, aaRS and genetic code evolution, and related literature.

Acknowledgments

The authors thank Helen Hansma (UC Santa Barbara, USA) and Bruce Kowiatek (Blue Ridge Community College, WV, USA) for helpful comments on the manuscript.

References

Agris, P. F., Eruysal, E. R., Narendran, A., Väre, V. Y. P., Vangaveti, S., & Ranganathan, S. V. (2018). *RNA Biology, 15*(4—5), 537—553. https://doi.org/10.1080/15476286.2017.1356562

Agris, P. F., Narendran, A., Sarachan, K., Väre, V. Y. P., & Eruysal, E. (2017). In , *Vol 41. Enzymes* (pp. 1—50). Academic Press. https://doi.org/10.1016/bs.enz.2017.03.005

Bédard, A. S. V., Hien, E. D. M., & Lafontaine, D. A. (2020). *Biochimica et Biophysica Acta—Gene Regulatory Mechanisms, 1863*(3). https://doi.org/10.1016/j.bbagrm.2020.194501

Bernhardt, H. S. (2016). *Life, 6*(1). https://doi.org/10.3390/life6010010

Bernhardt, H. S., & Patrick, W. M. (2014). *Journal of Molecular Evolution, 78*(6), 307—309. https://doi.org/10.1007/s00239-014-9627-y

Bernhardt, H. S., & Tate, W. P. (2008). *Biology Direct, 3*. https://doi.org/10.1186/1745-6150-3-53

Bernier, C. R., Petrov, A. S., Kovacs, N. A., Penev, P. I., & Williams, L. D. (2018). *Molecular Biology and Evolution, 35*(8), 2065—2076. https://doi.org/10.1093/molbev/msy101

Boehning, M., Dugast-Darzacq, C., Rankovic, M., Hansen, A. S., Yu, T., Marie-Nelly, H., McSwiggen, D. T., Kokic, G., Dailey, G. M., Cramer, P., Darzacq, X., & Zweckstetter, M. (2018). *Nature Structural and Molecular Biology, 25*(9), 833—840. https://doi.org/10.1038/s41594-018-0112-y

Brueckner, J., & Martin, W. F. (2020). *Genome Biology and Evolution, 12*(4), 282—292. https://doi.org/10.1093/gbe/evaa047

Burroughs, A. M., & Aravind, L. (2019). *International Journal of Molecular Sciences, 20*(8), 1981. https://doi.org/10.3390/ijms20081981

Burton, Z. F. (2014). *Transcription* (Vol. 5). https://doi.org/10.4161/trns.28674

Burton, Z. F. (2020). *Journal of Molecular Evolution, 88*(3), 234—242. https://doi.org/10.1007/s00239-020-09928-2

Burton, S. P., & Burton, Z. F. (2014). *Transcription, 5*(4). https://doi.org/10.4161/21541264.2014.967599

Burton, Z. F., Opron, K., Wei, G., & Geiger, J. H. (2016). *Transcription, 7*(1), 1—13. https://doi.org/10.1080/21541264.2015.1128518

Cantine, M. D., & Fournier, G. P. (2018). *Origins of Life and Evolution of Biospheres, 48*(1), 35—54. https://doi.org/10.1007/s11084-017-9542-5

Carter, C. W., Li, L., Weinreb, V., Collier, M., Gonzalez-Rivera, K., Jimenez-Rodriguez, M., Erdogan, O., Kuhlman, B., Ambroggio, X., Williams, T., & Chandrasekharan, S. N. (2014). *Biology Direct, 9*(1). https://doi.org/10.1186/1745-6150-9-11

Chandrasekaran, S. N., Yardimci, G. G., Erdogan, O., Roach, J., & Carter, C. W. (2013). *Molecular Biology and Evolution, 30*(7), 1588—1604. https://doi.org/10.1093/molbev/mst070

Chatterjee, S., & Yadav, S. (2019). *Life, 9*(1). https://doi.org/10.3390/life9010025

Damer, B., & Deamer, D. (2015). *Life, 5*(1), 872—887. https://doi.org/10.3390/life5010872

Demongeot, J., & Seligmann, H. (2020). *Journal of Molecular Evolution, 88*(3), 243—252. https://doi.org/10.1007/s00239-020-09929-1

Di Giulio, M. (2011). *Journal of Molecular Evolution, 72*(1), 119—126. https://doi.org/10.1007/s00239-010-9407-2

Di Giulio, M. (2019). *Journal of Theoretical Biology, 480*, 99—103. https://doi.org/10.1016/j.jtbi.2019.07.020

Dunne, M., Denyes, J. M., Arndt, H., Loessner, M. J., Leiman, P. G., & Klumpp, J. (2018). *Structure, 26*(12), 1573–1582.e4. https://doi.org/10.1016/j.str.2018.07.017

Eme, L., Spang, A., Lombard, J., Stairs, C. W., & Ettema, T. J. G. (2017). *Nature Reviews Microbiology, 15*(12), 711–723. https://doi.org/10.1038/nrmicro.2017.133

Endow, J. K., Rocha, A. G., Baldwin, A. J., Roston, R. L., Yamaguchi, T., Kamikubo, H., & Inoue, K. (2016). *PLoS ONE, 11*(12). https://doi.org/10.1371/journal.pone.0167802

de Farias, S. T., Rêgo, T. G., & José, M. V. (2019). *Sci, 1*(1), 8. https://doi.org/10.3390/sci1010008.v1

Feng, L., Sheppard, K., Tumbula-Hansen, D., & Söll, D. (2005). *Journal of Biological Chemistry, 280*(9), 8150–8155. https://doi.org/10.1074/jbc.M411098200

Fournier, G. P., & Alm, E. J. (2015). *Journal of Molecular Evolution, 80*(3–4), 171–185. https://doi.org/10.1007/s00239-015-9672-1

Fournier, G. P., & Poole, A. M. (2018). *Frontiers in Microbiology, 9*. https://doi.org/10.3389/fmicb.2018.01896

Furukawa, R., Nakagawa, M., Kuroyanagi, T., Yokobori, S. I., & Yamagishi, A. (2017). *Journal of Molecular Evolution, 84*(1), 51–66. https://doi.org/10.1007/s00239-016-9768-2

Gospodinov, A., & Kunnev, D. (2020). *Life, 10*(6), 1–22. https://doi.org/10.3390/life10060081

Guilhas, B., Walter, J. C., Rech, J., David, G., Walliser, N. O., Palmeri, J., Mathieu-Demaziere, C., Parmeggiani, A., Bouet, J. Y., Gall, A. L., & Nollmann, M. (2019). *bioRxiv*. https://doi.org/10.1101/791368

Gulen, B., Petrov, A. S., Okafor, C. D., Vander Wood, D., O'Neill, E. B., Hud, N. V., & Williams, L. D. (2016). *Scientific Reports, 6*. https://doi.org/10.1038/srep20885

Guo, C., Che, Z., Yue, J., Xie, P., Hao, S., Xie, W., Luo, Z., & Lin, C. (2020). *Science Advances, 6*(14). https://doi.org/10.1126/sciadv.aay4858

Hansma, H. G. (2013). *Journal of Biomolecular Structure and Dynamics, 31*(8), 888–895. https://doi.org/10.1080/07391102.2012.718528

Harish, A. (2018). *PeerJ, 6*(10), e5770. https://doi.org/10.7717/peerj.5770

Harmon, T. S., Holehouse, A. S., Rosen, M. K., & Pappu, R. V. (2017). *eLife, 6*. https://doi.org/10.7554/eLife.30294

Hauenstein, S. I., & Perona, J. J. (2008). *Journal of Biological Chemistry, 283*(32), 22007–22017. https://doi.org/10.1074/jbc.M801839200

Ikehara, K. (2014). [GADV]-protein world hypothesis on the origin of life. *Origins of Life and Evolution of Biospheres, 44*, 299–302. https://doi.org/10.1007/s11084-014-9383-4

Ikehara, K., Pellegrino, S., Demeshkina, N., Mancera-Martinez, E., Melnikov, S., Simonetti, A., Myasnikov, A., Yusupov, M., Yusupova, G., & Hashem, Y. (2005). Possible steps to the emergence of life: the [GADV]-protein world hypothesis. *The Chemical Record, 5*, 107–118. https://doi.org/10.1002/tcr.20037

Illangasekare, M., & Yarus, M. (2012). *RNA Biology, 9*(1), 59–66. https://doi.org/10.4161/rna.9.1.18039

Inoue, K., & Keegstra, K. (2003). *Plant Journal, 34*(5), 661–669. https://doi.org/10.1046/j.1365-313X.2003.01755.x

Iyer, L. M., & Aravind, L. (2012). *Journal of Structural Biology, 179*(3), 299–319. https://doi.org/10.1016/j.jsb.2011.12.013

Jose, M. V., Zamudio, G. S., & Morgado, E. R. (2017). A unified model of the standard genetic code. *Royal Society Open Science, 4*, 160908. https://doi.org/10.1098/rsos.160908

Juhling, F., Morl, M., Hartmann, R. K., Sprinzl, M., Stadler, P. F., & Putz, J. (2009). *Nucleic Acids Research, 37*, D159–D162. https://doi.org/10.1093/nar/gkn772. Database.

Kelley, L. A., Mezulis, S., Yates, C. M., Wass, M. N., & Sternberg, M. J. E. (2015). *Nature Protocols, 10*(6), 845–858. https://doi.org/10.1038/nprot.2015.053

Kim, Y., Kowiatek, B., Opron, K., & Burton, Z. F. (2018). *International Journal of Molecular Sciences, 19*(10). https://doi.org/10.3390/ijms19103275

Kim, S., Nalaskowska, M., Germond, J. E., Pridmore, D., & Söll, D. (1996). *Nucleic Acids Research, 24*(14), 2648–2651. https://doi.org/10.1093/nar/24.14.2648

Kim, Y., Opron, K., & Burton, Z. F. (2019). *Life, 9*(2). https://doi.org/10.3390/life9020037

Köhrer, C., Mandal, D., Gaston, K. W., Grosjean, H., Limbach, P. A., & Rajbhandary, U. L. (2014). *Nucleic Acids Research, 42*(3), 1904–1915. https://doi.org/10.1093/nar/gkt1009

Koonin, E. V. (2009). *Journal of Heredity, 100*(5), 618–623. https://doi.org/10.1093/jhered/esp056

Koonin, E. V. (2017). *Life, 7*(2). https://doi.org/10.3390/life7020022

Koonin, E. V., Krupovic, M., Ishino, S., & Ishino, Y. (2020). *BMC Biology, 18*(1). https://doi.org/10.1186/s12915-020-00800-9

Koonin, E. V., & Novozhilov, A. S. (2009). *IUBMB Life, 61*(2), 99–111. https://doi.org/10.1002/iub.146

Koonin, E. V., & Novozhilov, A. S. (2017). *Annual Review of Genetics, 51*, 45–62. https://doi.org/10.1146/annurev-genet-120116-024713

Kunnev, D., & Gospodinov, A. (2018). *Life, 8*(4). https://doi.org/10.3390/life8040044

Lane, N. (2014). *Cold Spring Harbor Perspectives in Biology, 6*(5). https://doi.org/10.1101/cshperspect.a015982

Langdon, E. M., & Gladfelter, A. S. (2018). *Annual Review of Microbiology, 72*, 255–271. https://doi.org/10.1146/annurev-micro-090817-062814

Lei, L., & Burton, Z. F. (2020). *Life, 10*(3). https://doi.org/10.3390/life10030021

Longo, L. M., Despotović, D., Weil-Ktorza, O., Walker, M. J., Jabłońska, J., Fridmann-Sirkis, Y., Varani, G., Metanis, N., & Tawfik, D. S. (2020). *Proceedings of the National Academy of Sciences of the United States of America, 117*(27), 15731–15739. https://doi.org/10.1073/pnas.2001989117

Long, X., Xue, H., & Tze-Fei Wong, J. (2019). *bioRxiv.* https://doi.org/10.1101/745372

Lorusso, M., Pepe, A., Ibris, N., & Bochicchio, B. (2011). *Soft Matter, 7*(13), 6327–6336. https://doi.org/10.1039/c1sm05726j

Loveland, A. B., Demo, G., Grigorieff, N., & Korostelev, A. A. (2017). *Nature, 546*(7656), 113–117. https://doi.org/10.1038/nature22397

Loveland, A. B., Demo, G., & Korostelev, A. A. (2020). *Nature, 584*(7822), 640–645. https://doi.org/10.1038/s41586-020-2447-x

Lu, H., Yu, D., Hansen, A. S., Ganguly, S., Liu, R., Heckert, A., Darzacq, X., & Zhou, Q. (2018). *Nature, 558*(7709), 318–323. https://doi.org/10.1038/s41586-018-0174-3

Mandal, D., Köhrer, C., Su, D., Russell, S. P., Krivos, K., Castleberry, C. M., Blum, P., Limbach, P. A., Söll, D., & RajBhandary, U. L. (2010). *Proceedings of the National Academy of Sciences of the United States of America, 107*(7), 2872–2877. https://doi.org/10.1073/pnas.0914869107

Mandal, D., Kohrer, C., Su, D., Russell, S. P., Krivos, K., Castleberry, C. M., Blum, P., Limbach, P. A., Soll, D., RajBhandary, U. L., & Agmatidine. (2020). https://doi.org/10.20944/preprints202009.0162.v1

Marin, J., Battistuzzi, F. U., Brown, A. C., & Hedges, S. B. (2017). *Molecular Biology and Evolution, 34*(2), 437–446. https://doi.org/10.1093/molbev/msw245

Mariscal, C., Barahona, A., Aubert-Kato, N., Aydinoglu, A. U., Bartlett, S., Cárdenas, M. L., Chandru, K., Cleland, C., Cocanougher, B. T., Comfort, N., Cornish-Bowden, A., Deacon, T., Froese, T., Giovannelli, D., Hernlund, J., Hut, P.,

Kimura, J., Maurel, M. C., Merino, N., … James Cleaves, H. (2019). *Origins of Life and Evolution of Biospheres, 49*(3), 111–145. https://doi.org/10.1007/s11084-019-09580-x

Martinez-Rodriguez, L., Erdogan, O., Jimenez-Rodriguez, M., Gonzalez-Rivera, K., Williams, T., Li, L., Weinreb, V., Collier, M., Chandrasekaran, S. N., Ambroggio, X., Kuhlman, B., & Carter, C. W. (2015). *Journal of Biological Chemistry, 290*(32), 19710–19725. https://doi.org/10.1074/jbc.M115.642876

Mazumder, A., Lin, M., Kapanidis, A. N., & Ebright, R. H. (2020). *Proceedings of the National Academy of Sciences of the United States of America, 117*(27), 15642–15649. https://doi.org/10.1073/pnas.1920427117

McSwiggen, D. T., Hansen, A. S., Teves, S. S., Marie-Nelly, H., Hao, Y., Heckert, A. B., Umemoto, K. K., Dugast-Darzacq, C., Tjian, R., & Darzacq, X. (2019). *eLife, 8.* https://doi.org/10.7554/eLife.47098

Mukai, T., Reynolds, N. M., Crnković, A., & Söll, D. (2017). *Life, 7*(1). https://doi.org/10.3390/life7010008

Oba, T., Fukushima, J., Maruyama, M., Iwamoto, R., & Ikehara, K. (2005). *Origins of Life and Evolution of the Biosphere, 35*(5), 447–460. https://doi.org/10.1007/s11084-005-3519-5

Odeh, H. M., & Shorter, J. (2020). *Emerging Topics in Life Sciences, 4*(3), 293–305. https://doi.org/10.1042/ETLS20190167

O'Donoghue, P., & Luthey-Schulten, Z. (2003). *Microbiology and Molecular Biology Reviews, 67*(4), 550–573. https://doi.org/10.1128/mmbr.67.4.550-573.2003

O'Malley, M. A., Leger, M. M., Wideman, J. G., & Ruiz-Trillo, I. (2019). *Nature Ecology and Evolution, 3*(3), 338–344. https://doi.org/10.1038/s41559-019-0796-3

Opron, K., & Burton, Z. F. (2019). *International Journal of Molecular Sciences, 20*(1). https://doi.org/10.3390/ijms20010040

Pak, D., Du, N., Kim, Y., Sun, Y., & Burton, Z. F. (2018). *Transcription, 9*(3), 137–151. https://doi.org/10.1080/21541264.2018.1429837

Pak, D., Kim, Y., & Burton, Z. F. (2018). *Transcription, 9*(4), 205–224. https://doi.org/10.1080/21541264.2018.1467718

Pak, D., Root-Bernstein, R., & Burton, Z. F. (2017). *Transcription, 8*(4), 205–219. https://doi.org/10.1080/21541264.2017.1318811

Perona, J. J. (2005). *Structure, 13*(10), 1397–1398. https://doi.org/10.1016/j.str.2005.09.003

Perona, J. J., & Gruic-Sovulj, I. (2014). *Topics in Current Chemistry, 344*, 1–41. https://doi.org/10.1007/128_2013_456

Petrov, A. S., Bernier, C. R., Hsiao, C., Norris, A. M., Kovacs, N. A., Waterbury, C. C., Stepanov, V. G., Harvey, S. C., Fox, G. E., Wartell, R. M., Hud, N. V., & Williams, L. D. (2014). *Proceedings of the National Academy of Sciences of the United States of America, 111*(28), 10251–10256. https://doi.org/10.1073/pnas.1407205111

Petrov, A. S., Gulen, B., Norris, A. M., Kovacs, N. A., Bernier, C. R., Lanier, K. A., Fox, G. E., Harvey, S. C., Wartell, R. M., Hud, N. V., & Williams, L. D. (2015). *Proceedings of the National Academy of Sciences of the United States of America, 112*(50), 15396–15401. https://doi.org/10.1073/pnas.1509761112

Pinho, M. G., Kjos, M., & Veening, J. W. (2013). *Nature Reviews Microbiology, 11*(9), 601–614. https://doi.org/10.1038/nrmicro3088

Pittis, A. A., & Gabaldón, T. (2016). *Nature, 531*(7592), 101–104. https://doi.org/10.1038/nature16941

Portz, B., & Shorter, J. (2020). *Trends in Biochemical Sciences, 45*(1), 1–3. https://doi.org/10.1016/j.tibs.2019.10.009

de Pouplana, L. R., Torres, A. G., & Rafels-Ybern, À. (2017). *Life, 7*(2). https://doi.org/10.3390/life7020014

McGeoch, M. W., Dikler, S., & McGeoch, J. E. (2020). Hemolithin: A meteoritic protein containing iron and lithium. arXiv.org

Preiner, M., Asche, S., Becker, S., Betts, H. C., Boniface, A., Camprubi, E., Chandru, K., Erastova, V., Garg, S. G., Khawaja, N., et al. (n.d.). In Better than membranes at the origin of life? *Life (Basel), 2017*.https://doi.org/10.3390/life10030020.n127

Quigley, G. J., & Rich, A. (1976). *Science, 194*(4267), 796—806. https://doi.org/10.1126/science.790568

Rodin, A. S., Rodin, S. N., & Carter, C. W. (2009). *Journal of Molecular Evolution, 69*(5), 555—567. https://doi.org/10.1007/s00239-009-9288-4

Rogers, K. C., & Söll, D. (1995). *Journal of Molecular Evolution, 40*(5), 476—481. https://doi.org/10.1007/BF00166615

Root-Bernstein, R., Kim, Y., Sanjay, A., & Burton, Z. F. (2016). *Transcription, 7*(5), 153—163. https://doi.org/10.1080/21541264.2016.1235527

Root-Bernstein, M., & Root-Bernstein, R. (2015). *Journal of Theoretical Biology, 367*, 130—158. https://doi.org/10.1016/j.jtbi.2014.11.025

Root-Bernstein, R., & Root-Bernstein, M. (2016). *Journal of Theoretical Biology, 397*, 115—127. https://doi.org/10.1016/j.jtbi.2016.02.030

Root-Bernstein, R., & Root-Bernstein, M. (2019). *International Journal of Molecular Sciences, 20*(1). https://doi.org/10.3390/ijms20010140

Rozov, A., Demeshkina, N., Westhof, E., Yusupov, M., & Yusupova, G. (2016). *Trends in Biochemical Sciences, 41*(9), 798—814. https://doi.org/10.1016/j.tibs.2016.06.001

Rozov, A., Westhof, E., Yusupov, M., & Yusupova, G. (2016). *Nucleic Acids Research, 44*(13), 6434—6441. https://doi.org/10.1093/nar/gkw431

Rozov, A., Wolff, P., Grosjean, H., Yusupov, M., Yusupova, G., & Westhof, E. (2018). *Nucleic Acids Research, 46*(14), 7425—7435. https://doi.org/10.1093/nar/gky547

Saint-Léger, A., Bello, C., Dans, P. D., Torres, A. G., Novoa, E. M., Camacho, N., Orozco, M., Kondrashov, F. A., & De Pouplana, L. R. (2016). *Science Advances, 2*(4). https://doi.org/10.1126/sciadv.1501860

Satpati, P., Bauer, P., & Åqvist, J. (2014). *Chemistry—A European Journal, 20*(33), 10271—10275. https://doi.org/10.1002/chem.201404016

Scheffers, D. J., & Pinho, M. G. (2005). *Microbiology and Molecular Biology Reviews, 69*(4), 585—607. https://doi.org/10.1128/MMBR.69.4.585-607.2005

Schön, A., Hottinger, H., & Söll, D. (1988). *Biochimie, 70*(3), 391—394. https://doi.org/10.1016/0300-9084(88)90212-X

Sheppard, K., Akochy, P. M., Salazar, J. C., & Söll, D. (2007). *Journal of Biological Chemistry, 282*(16), 11866—11873. https://doi.org/10.1074/jbc.M700398200

Sheppard, K., Sherrer, R. L., & Söll, D. (2008). *Journal of Molecular Biology, 377*(3), 845—853. https://doi.org/10.1016/j.jmb.2008.01.064

Sheppard, K., Yuan, J., Hohn, M. J., Jester, B., Devine, K. M., & Soll, D. (2008). *Nucleic Acids Research, 36*(6), 1813—1825. https://doi.org/10.1093/nar/gkn015

Sojo, V., Pomiankowski, A., Lane, N., & Penny, D. (2014). *PLoS Biology, 12*(8), e1001926. https://doi.org/10.1371/journal.pbio.1001926

Turk, R. M., Chumachenko, N. V., & Yarus, M. (2010). *Proceedings of the National Academy of Sciences of the United States of America, 107*(10), 4585—4589. https://doi.org/10.1073/pnas.0912895107

Voorhees, R. M., Mandal, D., Neubauer, C., Köhrer, C., Rajbhandary, U. L., & Ramakrishnan, V. (2013). *Nature Structural and Molecular Biology, 20*(5), 641—643. https://doi.org/10.1038/nsmb.2545

Wang, B., Predeus, A. V., Burton, Z. F., & Feig, M. (2013). *Biophysical Journal, 105*(3), 767—775. https://doi.org/10.1016/j.bpj.2013.05.060

Weiss, M. C., Preiner, M., Xavier, J. C., Zimorski, V., & Martin, W. F. (2018). *PLoS Genetics, 14*(8). https://doi.org/10.1371/journal.pgen.1007518

Yarus, M. (2011). *Philosophical Transactions of the Royal Society B: Biological Sciences,* *366*(1580), 2902–2909. https://doi.org/10.1098/rstb.2011.0139

Yin, Z., Wu, F., Xing, T., Yadavalli, V. K., Kundu, S. C., & Lu, S. (2017). *RSC Advances,* *7*(39), 24085–24096. https://doi.org/10.1039/c7ra02682j

Yoshizawa, T., Nozawa, R. S., Jia, T. Z., Saio, T., & Mori, E. (2020). *Biophysical Reviews,* *12*(2), 519–539. https://doi.org/10.1007/s12551-020-00680-x

Zachar, I., & Boza, G. (2020). *Cellular and Molecular Life Sciences,* 77(18), 3503–3523. https://doi.org/10.1007/s00018-020-03462-6

Zamudio, G. S., & José, M. V. (2018). *Origins of Life and Evolution of Biospheres, 48*(1), 83–91. https://doi.org/10.1007/s11084-017-9552-3

Zapun, A., Vernet, T., & Pinho, M. G. (2008). *FEMS Microbiology Reviews, 32*(2), 345–360. https://doi.org/10.1111/j.1574-6976.2007.00098.x

Zhang, B., & Cech, T. R. (1998). *Chemistry and Biology, 5*(10), 539–553. https://doi.org/10.1016/S1074-5521(98)90113-2

Zhou, H. X., Nguemaha, V., Mazarakos, K., & Qin, S. (2018). *Trends in Biochemical Sciences, 43*(7), 499–516. https://doi.org/10.1016/j.tibs.2018.03.007

CHAPTER 11

Evolution of the genetic code*

Lei Lei[1] and Zachary F. Burton[2]
[1]School of Biological Sciences, University of New England, Biddeford, ME, United States; [2]Department of Biochemistry and Molecular Biology, Michigan State University, East Lansing, MI, United States

Introduction

A model is presented for evolution of the genetic code based on analyses of tRNA and aminoacyl-tRNA synthetase (aaRS) evolution. The model is highly detailed and provides simple rules for filling code sectors. Strong selection rules are also apparent for the earliest evolution of the code.

A primordial tRNA^{Pri} was comprised of ordered sequences, GCG, CGC, and UAGCC repeats and inverted repeats with 7-nt U-turn loops (homologous 17-nt anticodon and T stem-loop-stems) (Burton, 2020; Kim et al., 2018, 2019; Lei & Burton, 2020; Pak et al., 2017; Root-Bernstein et al., 2016). With the exception of a few anticodon loop and T loop bases, the tRNA^{Pri} sequence is completely known. Three tRNA evolution models are considered here (Burton, 2020; Demongeot & Seligmann, 2020a, 2020c; Di Giulio, 2019, 2020; Jacques Demongeot & Seligmann, 2021; Lei & Burton, 2020). Only the 3−31-nt minihelix model can be correct. The 3−31-nt minihelix model has been referred to as a theorem (Burton, 2020). There are no theorems in evolutionary biology, but the 3−31-nt minihelix model for tRNA evolution is very close to being one.

AaRS enzymes attach amino acids to the 3′-end of tRNAs (Perona & Gruic-Sovulj, 2014). Much has been published on evolution of aaRS (i.e., GlyRS-IIA; IIA indicates the class (I or II) and structural subclass (i.e., A-E)) (Aravind et al., 2002; Carter & Wills, 2019; Kim et al., 2019; Koonin & Novozhilov, 2017; Lei & Burton, 2020; O'Donoghue & Luthey-Schulten, 2003; Perona & Hadd, 2012; Wolf et al., 1999). We have simplified the understanding of aaRS evolution and brought it in line with the evolution of the genetic code (Kim et al., 2019; Lei & Burton, 2020).

Evolution of tRNA and the genetic code provides new models for evolution of life on Earth and the prelife to life transition. In agreement

* This work was previously published: Transcription. 2021; 12(1): 28−53. Published online 2021 May 18. doi: 10.1080/21541264.2021.1927652. PMCID: PMC8172153. PMID: 34000965.

The Makings of a Clinical Protocol
ISBN 978-0-323-95749-6
https://doi.org/10.1016/B978-0-323-95749-6.00015-6

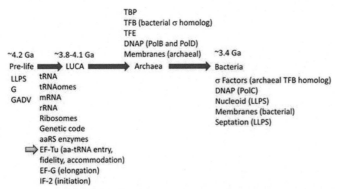

Figure 11.1 A working model for the prelife to life transition and for divergence of Archaea and Bacteria. A major driving force for the divergence of Bacteria and Archaea is posited to have been divergence of transcription systems. EF-Tu is highlighted (*small green arrow*) because EF-Tu evolution allowed expansion of the genetic code from an 8-aa bottleneck to the standard code. Abbreviations: *LLPS*, liquid—liquid phase separation; *G*, glycine; *GADV*, glycine, alanine, aspartic acid, valine; *EF*, translational elongation factor; *IF*, translational initiation factor; *TBP*, TATA-box binding protein; *TF*, transcription factor; *DNAP (Pol)*, DNA polymerase; *1 Ga*, 1 billion years ago. (For interpretation of the references to colour in this figure legend, the reader is referred to the web version of this article.)

with some others, we posit that Archaea are the oldest organisms, and Archaea are the most similar to the last universal common (cellular) ancestor (LUCA) (Fig. 11.1) (Battistuzzi et al., 2004; Kim et al., 2019; Lei & Burton, 2020; Long et al., 2020; Marin et al., 2017). Our interest in this issue comes from studies of earliest evolution of transcription (Burton & Burton, 2014; Burton et al., 2016) and translation systems (Kim et al., 2019; Lei & Burton, 2020; Opron & Burton, 2019). We find that tRNAomes (all the tRNAs of an organism) are simpler in archaeal systems relative to bacterial systems (Burton, 2020; Opron & Burton, 2019; Pak, Du, et al., 2018; Kim et al., 2018). AaRS enzymes are closer to root sequences in archaeal systems, and aaRS evolution falls more in line with the simpler archaeal genetic code (Kim et al., 2019; Lei & Burton, 2020; Opron & Burton, 2019; Pak, Kim, & Burton, 2018). Archaeal TFB is a homolog of bacterial σ factors (Burton & Burton, 2014; Iyer & Aravind, 2012), indicating that evolution of bacterial transcription systems may have largely drove divergence of Archaea and Bacteria (Lei & Burton, 2021). In this work, we concentrate on archaeal systems for these reasons.

Methods

Sequences of tRNAs were obtained from the tRNA database (Juhling et al., 2009) and the genomic tRNA database (Chan & Lowe, 2009, 2016).

Typical tRNA diagrams were generated using tRNAdb tools and modified as necessary. Longer V loops in type II tRNA^{Leu} and tRNA^{Ser} were analyzed using WebLogo 3.7.4 (Schneider & Stephens, 1990). Molecular graphics was done using UCSF ChimeraX (Goddard et al., 2018; Pettersen et al., 2021). Evolution of aaRS enzymes was analyzed using the Phyre2 protein fold recognition server. Phyre2 identifies nearest and distant matches to a seed sequence, in the RCSB protein data bank. Phyre2 is a very useful tool for identifying close and distant protein family members that are related by both structure and sequence. Using only sequence-based tools, it is difficult to relate aaRS enzymes, which were driven to differentiate in order to establish and maintain translational accuracy of tRNA charging. Phyre2 was used to build a lineage of class I aaRS enzymes in which all class I aaRS were connected by both close and distant metrics (Phyre2 homology scores). Phyre2 could be used to build a model for a class II aaRS lineage. Some distantly related class II aaRS, however, could not be scored to one another. For instance, class IIA and class IID enzymes could not be scored as homologs without connecting intermediate class II aaRS. Because class I and class II aaRS have different folds, Phyre2 could not be used to identify homology of class I and class II aaRS enzymes. We used standard NCBI (National Center for Biotechnology Information) tools, such as Blast, to relate GlyRS-IIA to IleRS-IA and ValRS-IA (Kim et al., 2019; Lei & Burton, 2020).

Prelife evolution of transcription, metabolism, and translation

There are some shared concepts comparing the earliest evolution of transcription and translation systems. On Earth, complexity was often generated from repetition of a motif. We imagine a prelife mechanism to duplicate and multimerize RNAs (i.e., by ribozyme-mediated ligation and replication), often resulting in duplication of a common or related sequence (Burton, 2014; Burton & Burton, 2014; Burton et al., 2016; Lei & Burton, 2021). If the RNA was protein-encoding, dimerization would generate a protein motif duplication. In this way, a β-β-α-β unit was duplicated to create a β-β-α-β — β-β-α-β motif refolded into a 6-β-sheet barrel (Alva et al., 2008, 2009; Coles et al., 2005, 2006). Double-Ψ-β-barrels were generated in this way. Cellular RNAPs are 2-double-Ψ-β-barrel type enzymes (Iyer & Aravind, 2012; Iyer et al., 2004; Koonin et al., 2020; Lei & Burton, 2021). PolD is a 2-double-Ψ-β-barrel type replicative DNAP from Archaea and may be the most ancient replicative DNA polymerase (Koonin et al., 2020; Madru et al., 2020; Raia et al., 2019; "Two-Barrel" Polymerases Superfamily: Structure, Function and Evolution," 2019). RNA template-dependent RNAPs can also be of the 2-double-Ψ-β-barrel type, and these appear to be the oldest form of the enzyme class (Salgado et al., 2006). In Archaea, TFB includes a duplication of a helix-turn-helix motif ((HTH)$_2$), also referred to as a cyclin-like repeat. In Bacteria, σA is a homolog of TFB that appears to be derived from a (HTH)$_4$ repeat, which probably arose as a TFB (HTH)$_2$ duplication (Burton, 2014; Burton & Burton, 2014; Burton et al., 2016). BP was generated by duplication of a motif encoding multiple β-sheets and may be coevolved with DNA (Brindefalk et al., 2013). We consider TFB and TBP to be founding general transcription factors. It appears that Bacteria evolved sigma factors from TFB and lost TBP and that these were defining events in the divergence of Archaea and Bacteria. In many ways, Bacteria seem to be more successful prokaryotes than Archaea. For instance, many Archaea appear to be pushed into extremophile environments on Earth. We posit that Bacteria evolved from Archaea, and that Archaea are most similar to LUCA (Fig. 11.1) (Lei & Burton, 2021). For many factors and functions, Bacteria are simplified relative to Archaea, presumably due to genetic loss.

Much of core metabolism, including the glycolysis pathway and the citric acid cycle, evolved around (β-α)$_8$ barrels (i.e., glycolysis; TIM (triosephosphate isomerase) barrels) and (β-α)$_8$ sheets (i.e., citric acid cycle; Rossmann folds). We posit that Rossmann fold (β-α)$_8$ sheets were refolded from (β-α)$_8$ barrels. We posit that (β-α)$_8$ barrels and sheets were initially generated from two serial duplications of β-α-β-α motifs (Burton et al., 2016; Lei & Burton, 2021).

The evolution of tRNA from ligation of 3–31-nt minihelices, two of which were identical, is described below. Because of our experience with evolution of transcription systems, we searched for repeating motifs in tRNAs and found them easily. In prelife, mostly, ribozyme-dependent mechanisms must have existed to replicate 31-nt minihelices and tRNAs. We posit that genetic code evolution was mostly driven by replication of tRNAs, mutation of the tRNA anticodon, and coevolution of tRNAs with aaRS. AaRS enzymes evolved by a chaotic pathway are described below. Class I aaRS have their active site mounted on a platform of parallel β-sheets (Perona & Gruic-Sovulj, 2014). For this reason, class I aaRS is often referred to as "Rossmann folds," but this is improper, as described below. Remarkably, the lineages of aaRS enzymes in Archaea give the pattern of genetic code evolution. The somewhat more complex pattern of genetic code evolution in Bacteria can be derived from the older archaeal pattern (Kim et al., 2019; Lei & Burton, 2020).

Evolution of tRNA

Concepts

We posit that evolution of tRNA, from an RNA—and ribozyme-dominated world, laid the foundation for evolution of the genetic code. We posit that the genetic code sectored according to the tRNA anticodon. Initially, code columns were selected because the central anticodon base (2nd position) was easiest to read on a primitive ribosome. Initially, both the 1st and 3rd anticodon positions were read as wobble positions with pyrimidine-purine discrimination. Evolution of EF-Tu suppressed wobbling at the 3rd anticodon position allowing expansion of the code. Because of the pathway to evolution of tRNA, the anticodon is read in a register of 3-nt, so 2-nt code registers could not be supported using tRNAs. Wobbling and code degeneracy are described by the evolution of tRNA reading on a primitive ribosome and coevolution of EF-Tu. Sequence analyses of type II tRNAs with longer V loops in ancient Archaea provides reasonable models for serine jumping during evolution of the code. Sequences of tRNAs in Archaea show that tRNA evolved from repeat and inverted repeat sequences. Therefore, before evolution of tRNA and protein encoding, there must have been ribozyme-based mechanisms to generate RNA repeats and inverted repeats. In this way, tRNA is a central key to understand the prelife to life transition and evolution of the genetic code. TRNA is uniquely suited as a genetic adapter to support evolution of a genetic code. To generate a genetic code with a different adapter than tRNA presents many problems that we would not know how to solve.

The 3—31-nt minihelix model

The pathway of evolution of tRNA has been controversial, but tRNA evolution is essential to grasp, in order to understand the prelife to life transition and evolution of the genetic code. Here, three models are considered, but we focus on one model, the 3—31-nt minihelix model (Burton, 2020; Lei & Burton, 2020). So far as we can discern, the 3—31-nt minihelix model is the only viable model, and alternate models are falsified. The alternate models can be described as the Uroboros model (Demongeot & Seligmann, 2019, 2020a, 2020b) and the 2-minihelix model (Di Giulio,

2009, 2019, 2020). Both of these models are accretion models, in which tRNA evolves in expanding and/or contracting segments. Because proposed expansion and contraction segments in tRNA would derive from highly ordered sequences (i.e., repeats and inverted repeats), no accretion model can reasonably describe early tRNA evolution. In an accretion model, expansions and contractions would need to result in ordered sequences (Burton, 2020; Kim et al., 2019; Lei & Burton, 2020). Furthermore, analyses of archaeal tRNA sequences provide an irrefutable record of the 3—31-nt minihelix model.

The 3—31-nt minihelix model is summarized in Fig. 11.2. 3—31-nt minihelices of mostly known sequence were ligated to form a 93-nt tRNA precursor, which was then processed by internal 9-nt deletion(s) into type I and type II tRNAs (Burton, 2020). 31-nt minihelices were comprised of a 5′-7-nt acceptor stem, a 17-nt core, and a 3′-7-nt acceptor stem. The sequence of the 5′-7-nt acceptor stem was originally GCGGCGG, which is a truncated GCG repeat. The sequence of

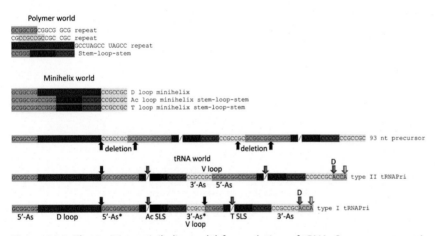

Figure 11.2 The 3—31-nt minihelix model for evolution of tRNA. Sequences are primordial but remain represented in archaeal tRNAs. A 93-nt tRNA precursor formed from ligation of 3—31-nt minihelices, as shown. Type I and type II tRNAs were processed by 9-nt internal deletion(s) from the 93-nt precursor. Minihelix world was preceded by polymer world. Colors: green) 5′-acceptor stems (5′-As) and derived 5′-As*; magenta) D loop; cyan) 5′-stem for the anticodon (Ac) and T stem-loop-stems (SLS); red) anticodon and T loops; purple) anticodon; cornflower blue) 3′-stems for the anticodon and T stem-loop-stem; yellow) 3′-As* and 3′-As sequences. *Arrow colors: red*) internal deletion endpoints; *blue*) U-turns; *cyan*) discriminator (D); and *gold*) amino acid placements. (For interpretation of the references to colour in this figure legend, the reader is referred to the web version of this article.)

the 3′-7-nt acceptor stem was originally CCGCCGC, which is a complementary, truncated CGC repeat. For the D loop, the 17-nt core sequence was originally UAGCCUAGCCUAGCCUA, which is a truncated UAGCC repeat. Remarkably, the anticodon and T loop 17-nt core sequences were both originally close to CCGGGUU/AAAAACCCGG (/indicates a U-turn in a 7-nt loop). These 17-nt sequences form a stem-loop-stem with 5-nt 5′-stems (CCGGG), a 7-nt U-turn loop (∼UU/AAAAA) and 5-nt 3′-stems (CCCGG). There is only slight sequence ambiguity in the primordial 7-nt U-turn loop, not in the stems.

Type II tRNAs have a longer V loop (V for variable) relative to type I tRNAs (Kim et al., 2018). The model shown in Fig. 11.2 describes the evolution of both type I and type II tRNAs. To generate a type II tRNA, the 93-nt tRNA precursor was processed by a single 9-nt internal deletion, as shown. The 9-nt deletion occurred within ligated 3′- and 5′-7-nt acceptor stems. The 5-nt segment that remains after deletion (originally GGCGG) became the last 5-nt of the D loop region just before the anticodon stem-loop-stem. The original type II tRNA, therefore, was 84-nt before addition of 3′-ACCA (A is the primordial discriminator base), presumably via ligation. The type II tRNA V loop, therefore, was initially 14-nt (7-nt + 7-nt). 14-nt remains a dominant length of tRNA^{Leu} V loops in Archaea.

To generate type I tRNA from the 93-nt precursor, required two internal 9-nt deletions, as shown (Burton, 2020). The 5′-9-nt deletion was identical to the processing event in type II tRNAs. The 3′-9-nt internal deletion in type I tRNA was also within ligated 3′- and 5′-7-nt acceptor stem segments. The 9-nt deletion generated the type I tRNA V loop, which was initially 5-nt (originally CCGCC). The primordial type I tRNA was 75-nt before addition of 3′-ACCA. Folding into the tRNA L-shaped structure brought the D loop into contact with the V loop and the D stem in contact with the V region, causing a small number of systematic changes in tRNA sequences (Burton, 2020). The 5′-acceptor stem remnant (5′-As*) initially changed from GGCGG to GGGCG to form the D stem and to break base pairing contacts to the V loop (numbered V1−V5). The V loop (3′-As*) changed with time from CCGCC to ∼UGGUC. The V1 base is often U to form a wobble pair with G26. The V5 base tends to remain C to form the Levitt reverse Watson-Crick base pair to G15. Statistical tests support the homology of bases 3−7 of the 5′-acceptor stem to the 5′-As* sequence, with a P-value of .001 (highest indication of homology). Statistical tests support the homology of bases 1−5 of the 3′-acceptor stem to the 3′-As* sequence (the V loop), with a *P*-value of .001 (highest indication of homology) (Pak et al., 2017).

Type I tRNA

Fig. 11.3A shows a type I tRNA colored according to the 3—31-nt minihelix model. Notice that the anticodon stem-loop-stem and the T stem-loop-stem are homologs (cyan-red-cornflower blue segments) with 7-nt U-turn loops, as indicated in the model (Fig. 11.2). As in the model, the 5′-acceptor stem is a homolog of the 5′-As* sequence, and the 3′-As* is a homolog of the 3′-acceptor stem sequence (compare green and yellow segments).

Fig. 11.3B shows a typical tRNA diagram (as DNA sequence) from *Pyrococcus furiosis* (Juhling et al., 2009), which is an ancient Archaeon. The typical tRNA sequence is almost identical to the tRNA^{Pri} sequence shown in the model (Fig. 11.2), indicating that the model is correct. The D loop sequence is UAGCNUAGCC, indicating conservation of the UAGC-CUAGCC repeat sequence of tRNA^{Pri}. The homology of the anticodon stem-loop-stem (CCGGNCU/NNNGANCCGG) and the T stem-loop-stem (CCGGGUU/CAAAUCCCGG) is obvious by inspection. Statistical tests support this homology, with a *P*-value of .001 (highest indication of homology) (Pak et al., 2017). The typical 5′-acceptor stem sequence is

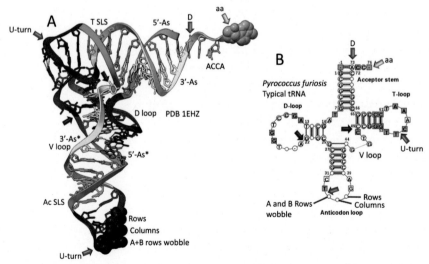

Figure 11.3 Type I tRNA. (A) structure of a type I tRNA colored according to the model. (B) A typical type I tRNA diagram from *Pyrococcus furiosis*. Typical tRNA diagrams from the tRNAdb website are exported as DNA sequences rather than RNA.

GCGGCGG, identical to the tRNA^{Pri} sequence. The 3′-acceptor stem sequence is CCGCNNC, consistent with the tRNA^{Pri} sequence (CCGCCGC). Acceptor stem sequences vary among tRNAs because acceptor stems are determinants for amino acid placements by aaRS enzymes at 3′-ACCA. The anticodon is highlighted (purple) because the genetic code evolved around the tRNA anticodon and its reading on the evolving ribosome.

TRNA^{Gly} was the first tRNA

The closest tRNA in Archaea to tRNA^{Pri} is tRNA^{Gly} (Fig. 11.4) (Burton, 2020; Kim et al., 2019; Lei & Burton, 2020; Opron & Burton, 2019; Pak, Du, et al., 2018). Fig. 11.4A shows a typical tRNA^{Gly} from three Pyrococcus species. The typical sequence of tRNA^{Gly} is almost identical to tRNA^{Pri}. The sequence alignment is shown in Fig. 11.4B. The tRNA^{Pri}, typical tRNA^{Gly} (Pyrococcus), and typical tRNA (*Pyrococcus furiosis*) are nearly identical sequences. This result indicates that tRNA^{Gly} was the first

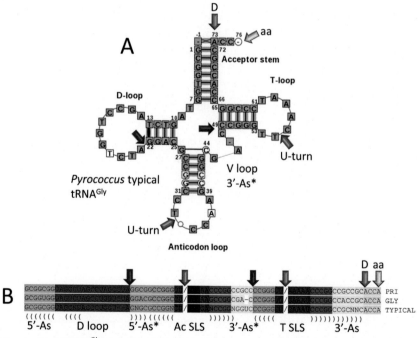

Figure 11.4 TRNA^{Gly} was the first tRNA. (A) A typical tRNA^{Gly} from three *Pyrococcus* species (*P. furiosis*, *P. abyssi* and *P. horikoshii*; nine sequences). (B) TRNA^{Pri}, tRNA^{Gly}, and tRNA^{Typical} (*P. furiosis*) are close homologs. TRNA secondary structure is indicated.

tRNA and that all archaeal tRNAs radiated from tRNA^{Gly}. This observation is relevant to the evolution of aaRS enzymes and the genetic code, as described below.

The typical D loop sequence of Pyrococcus tRNA^{Gly} is UAGU-CUAGCCUGGUCUA (D1 to D17) versus UAGCCUAGCCUAGC-CUA in tRNA^{Pri}. These sequences are nearly identical. The D12 A→G shift from the primordial sequence appears to be universal in Archaea. D12G (19 G in standard tRNA numbering) intercalates into the T loop between 57A and 58A ((54-UU/CAAAU-60); standard numbering) and forms a specific H-bond contact, explaining the A→G sequence change. Apparently, 19 G is a preferred intercalating base to the primordial 19A. Note that the lengths of the D loop are identical in tRNA^{Gly} and tRNA^{Pri}. Standard tRNA numbering can be somewhat confusing compared to tRNA^{Pri} because standard numbering is based on a 2-nt deletion in the D loop in eukaryotic tRNAs.

Type II tRNAs

In Archaea, tRNA^{Leu} and tRNA^{Ser} are type II tRNAs, with longer V loops (Kim et al., 2018). As we have shown previously, tRNA^{Leu} is most similar in overall sequence to type II tRNA^{Pri} (Fig. 11.5). Fig. 11.5A shows a tRNA^{Leu} from *Pyrococcus horikoshii* (PDB 1WZ2). The typical V loop has the 14-nt sequence UCCCGUAGGGGUUC (V1−V14). The V loop sequence varies from the primordial 14-nt CCGCCGCGCGGCGG because the primordial sequence can pair along its entire length, which would be awkward for tRNA folding and for functional contacts (i.e., with aaRS enzymes). Instead, the tRNA^{Leu} V loop evolved to form a short stem (i.e., 3-nt) and loop (i.e., 4-nt). Also, V1C typically became V1U to form a wobble pair with 26 G, and V14G became V14C to form a Levitt reverse Watson-Crick base pair with 15 G. Statistical tests support the model that the V loop is derived from a 3′-acceptor stem ligated to a 5′-acceptor stem (Fig. 11.2), with a *P*-value of .001 (the highest indication of homology) (Kim et al., 2018). TRNA^{Leu} has a 5′-acceptor stem with the typical sequence GCGGGGG versus GCGGCGG in tRNA^{Pri}. As in tRNA^{Gly} (Fig. 11.4), the typical Pyrococcus tRNA^{Leu} D loop (Fig. 11.5B) includes no deleted bases, with the 17-nt typical sequence UUGCCGAGCCUG-GUCAA vs. UAGCCUAGCCUAGCCUA in tRNA^{Pri}. These sequences are very similar.

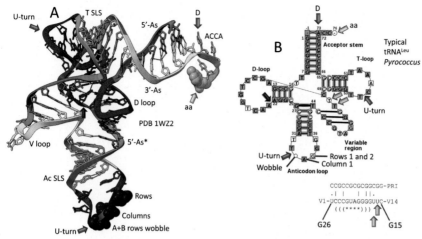

Figure 11.5 Type II tRNA. (A) Structure of tRNA^Leu from *Pyrococcus horikoshii*. The larger V loop of type II tRNA is the 7-nt yellow segment linked to the 7-nt green segment. (B) Typical tRNA^Leu from three *Pyrococcus* species (*P. furiosis*, *P. abyssi*, and *P. horikoshii*; 15 sequences). *Light green arrows* indicate unpaired bases (V12U and V13U) just before the Levitt base pair (V14C = 15 G) (*thin red line*). *Parentheses* indicate paired bases; * indicates loop. (For interpretation of the references to colour in this figure legend, the reader is referred to the web version of this article.)

The tRNA^Ser V loop evolved from the tRNA^Leu and tRNA^Pri sequences, in order to be distinguished from the tRNA^Leu V loop. Notably, SerRS-IIA interacts with the tRNA^Ser V loop directly as a determinant in order to recognize and accurately charge tRNA^Ser (Perona & Gruic-Sovulj, 2014). By contrast, the tRNA^Leu V loop is an antideterminant to reduce inaccurate aminoacylation of tRNA^Leu by SerRS-IIA. LeuRS-IA recognizes the opposite face of tRNA^Leu from the expanded V loop, so LeuRS-IA recognizes other determinants such as the acceptor stem and discriminator base. A comparison of tRNA^Ser and tRNA^Leu V loop sequences is shown in Fig. 11.6. Fig. 11.6A is a typical tRNA^Ser diagram from 3-Pyrococcus species. Fig. 11.6B shows an alignment of typical tRNA^Ser and tRNA^Leu V loop sequences versus tRNA^Pri. Fig. 11.6C shows the comparison of tRNA^Ser and tRNA^Leu V loop sequence logos. We could not find a suitable tRNA^Ser structure (i.e., from an ancient Archaeon) to compare to the tRNA^Leu structure in Fig. 11.5A, for instance, to compare the expected different trajectories of the V loops.

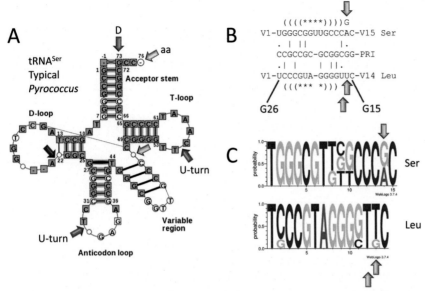

Figure 11.6 Comparison of tRNA^{Leu} and tRNA^{Ser} V loops. (A) A typical tRNA^{Ser} from three *Pyrococcus* species (*P. furiosis, P. abyssi,* and *P. horikoshii*; 12 sequences) (as DNA sequence). (B) The alignment compares V loops of tRNA^{Pri} to tRNA^{Leu} and tRNA^{Ser} typical sequences. *Parentheses* indicate stems; * indicates loops. (C) Sequence logos comparing tRNA^{Leu} and tRNA^{Ser} V loops (as DNA sequences). *Light green arrows* indicate unpaired bases after the V loop stem. (For interpretation of the references to colour in this figure legend, the reader is referred to the web version of this article.)

Accurate recognition of tRNA^{Ser} by SerRS-IIA is important in understanding the evolution of the genetic code. Serine is the only amino acid that sectors within two code columns, as described below. We posit that differences in sequence, stem positions, and unpaired bases were important to discriminate the tRNA^{Ser} and tRNA^{Leu} V loops. Specifically, in Pyrococcus, tRNA^{Ser} has a single unpaired base just 3′ of its short stem, while tRNA^{Leu} has two unpaired bases (light green arrows in Figs. 11.5 and 11.6). This difference should change the trajectory of the type II V loop stems. Also, the tRNA^{Ser} V loop is G-rich at its 5′ stem, while the tRNA^{Leu} V loop is C-rich at its 5′ stem. These and possibly other differences in type II V loops are expected to contribute to discrimination. The sequence logos in Fig. 11.6C show that, in their major features, V loops are highly conserved in Pyrococcus. Solution of a structure of tRNA^{Ser} from a Pyrococcus species would contribute to this discussion.

Evolution of aaRS enzymes

Concepts

Aminoacyl–tRNA synthetases (aaRS) place amino acids at the 3′-end of tRNAs (Perona & Gruic-Sovulj, 2014). The idea behind this paper is that insight can be gained into the evolution of the genetic code based on coevolution of aaRS enzymes, tRNAomes, and EF-Tu (Kim et al., 2019; Lei & Burton, 2020). Using the Phyre2 protein fold recognition server (Kelley et al., 2015), we were able to establish simplified pathways of aaRS evolution that appear to describe routes for genetic code evolution often within code columns (2nd anticodon position). Also, we noted sequence homology between class I and class II aaRS enzymes with different folds (Pak, Kim, & Burton, 2018), which initially was unexpected. Class I aaRS enzymes have been termed "Rossmann folds," but this is a mischaracterization, as we describe.

Class II and class I aaRS enzymes

There are two structural classes of aaRS enzymes, termed class II and class I, and multiple structural subclasses (i.e., A–E) (Perona & Gruic-Sovulj, 2014). Some aaRS enzymes have a separate active site from the aminoacylating active site that removes noncognate amino acids from their cognate tRNA. This process is referred to as "proofreading" or "editing." Remarkably, in Archaea, aaRS enzymes that edit noncognate amino acids are found in the left half of the genetic code, in columns 1 and 2. Column 1 encodes hydrophobic amino acids Val, Met, Ile, Leu, and Phe. Column 2 encodes neutral amino acids Ala, Thr, Pro, and Ser. Ser is found also in column 4. We posit that Ser jumped from column 2 to column 4 in establishment of the code. In Archaea, ProRS-IIA does not include an editing active site, but ProRS-IIA does edit in Bacteria. Because neutral and hydrophobic amino acids are limited in forming specific hydrogen bonds and ion pairs, editing may be necessary to reduce tRNA charging errors.

Evolution of aaRS enzymes is described in Fig. 11.7. In Fig. 11.7A, a detail of an alignment of IleRS-IA from Methanobacterium bryantii to GlyRS-IIA from Methanobacterium congolense is shown. The e-value for the alignment is 4×10^{-12}, so, qualitatively, about a 1 chance in 2.5×10^{11} of the alignment being due to random chance rather than homology. Class I and class II aaRS enzymes, however, have incompatible folds, so these enzymes were not thought to be homologous. Class II aaRS appears to be the older fold, indicating that class I aaRS may be derived from class II aaRS (Perona & Gruic-Sovulj, 2014). In class II aaRS, the active site is mounted on a scaffold of antiparallel β-sheets. In class I aaRS, by contrast, the active site is mounted on a scaffold of parallel β-sheets. Commonly, class I aaRS is referred to as "Rossmann folds," although there is no genetic relation of class I aaRS and Rossmann fold enzymes. Rossmann fold enzymes derive from $(β–α)_8$ sheets that appear to derive from refolding $(β–α)_8$ barrels (i.e., TIM barrels; TIM for triosephosphate isomerase) (Burton et al., 2016).

Falsification of the Carter-Rodin-Ohno hypothesis

Fig. 11.7B is a schematic of a multiple sequence alignment comparing GlyRS-IIA, IleRS-IA, and ValRS-IA in ancient Archaea (Kim et al., 2019; Lei & Burton, 2020; Opron & Burton, 2019; Pak, Du, et al., 2018).

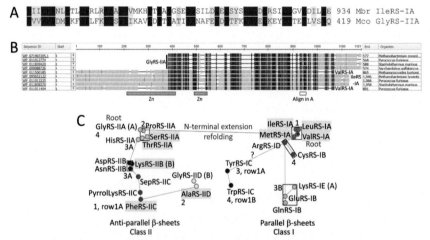

Figure 11.7 Evolution of aaRS enzymes. (A) Local alignment of a class I IleRS-IA and a class II GlyRS-IIA aaRS. Identities are *shaded red* and similarities are *shaded yellow*. Mbr) Methanobacterium bryantii; Mco) Methanobacterium congolense. (B) Class II and class I aaRS with incompatible folds are simple sequence homologs. *Red boxes* indicate sequence homologies in the multiple alignment. Two Zn fingers are indicated. The position of the alignment in panel A is shown. (C) Evolution of aaRS enzymes based on Phyre2 homology scores [1,3]. Distances in the map represent structural and genetic relatedness. In Archaea, related aaRS enzymes mostly cluster according to genetic code columns (boxes; numbers indicate genetic code columns; some rows are specified). *Yellow* highlighting indicates aaRS with editing active sites. *Red arrow* indicates homology of GlyRS-IIA and ValRS-IA. (A) Archaea-specific; (B) Bacteria-specific. (?) indicates that TyrRS-IC and TrpRS-IC may be derived from a primitive ArgRS-ID. (For interpretation of the references to colour in this figure legend, the reader is referred to the web version of this article.)

Fig. 11.7B also shows how the aligned genes encoding class II and class I aaRS compare. Archaeal GlyRS-IIA is a simple sequence homolog of IleRS-IA and ValRS-IA (i.e., Fig. 11.7A) showing that class IA aaRS were derived from GlyRS-IIA by N-terminal extension (i.e., upstream transcription and in-frame translation start) and refolding. Probably, this sequence comparison can only be done successfully with aaRS enzymes from ancient Archaea. One reason this homology is relevant is that a model (referred to as the Carter-Rodin-Ohno model) has been proposed that class I and class II aaRS were derived from "molten globule" smaller "Urzymes" encoded on complementary DNA strands (Carter, 2014, 2017; Carter & Wills, 2019; Carter et al., 2014; Kelley et al., 2015; Rodin et al., 2009) [13, 51—54]. These primitive Urzymes were posited to have expanded to full-length aaRS enzymes. Molten globule Urzymes must expand to a more complex version to take on their eventual specificity and refined functions. The Carter-Rodin-Ohno model is certainly incorrect. GlyRS-IIA is the root of all aaRS evolution, including all class II enzymes and all class I enzymes.

ValRS-IA and IleRS-IA enzymes include an N-terminal extension relative to GlyRS-IIA (Fig. 11.7B). The N-terminal extension includes part of the class I aaRS active site scaffold and, in ancient Archaea, also, a Zn-finger lacking in GlyRS-IIA (Pak, Kim, & Burton, 2018). Also, GlyRS-IIA, IleRS-IA, and ValRS-IA can share a Zn finger, as indicated. It is not possible that class II and class I aaRS are simultaneously simple homologs, as we show here, and also that class II and class I aaRS were generated as molten globule Urzymes from an ancestral bidirectional gene. Rather, GlyRS-IIA was a large and complex protein, not a molten globule, at the time of its refolding to (probably) a primitive ValRS-IA. Furthermore, ValRS-IA was a large and complex protein and not a molten globule Urzyme, from its first formation. We speculate that GlyRS-IIA and ValRS-IA assumed their initial and incompatible folds based, in part, on Zn and tRNA binding. Class II and class I aaRS bind opposite faces of their cognate tRNAs (Perona & Gruic-Sovulj, 2014).

Lineages of aaRS enzymes

Fig. 11.7C summarizes the following: (1) lineage information of aaRS enzymes; (2) aaRS enzymes with editing active sites; and (3) relationships of the aaRS lineages to the pattern of the genetic code (Kim et al., 2019; Lei & Burton, 2020; Opron & Burton, 2019; Pak, Kim, & Burton, 2018). We used the Phyre2 protein fold recognition server [35] in order to determine close and distant structural and sequence homologs among class II and class I aaRS (Lei & Burton, 2020; Opron & Burton, 2019; Pak, Kim, & Burton, 2018). What Phyre2 does is to score nearest and more distant homologs using both structure and sequence. As seed sequences, we used aaRS enzymes mostly from *Pyrococcus furiosis*, an ancient Archaeon. The Phyre2 server searches all sequences in the Protein Data Bank to find matches. Based on the homology scores, the lineage in Fig. 11.7C was drawn. Distances in the map represent evolutionary distances, so clustered nodes indicate closely related enzymes. The map represents both close and distant relationships in the placements of the nodes (Kim et al., 2019; Lei & Burton, 2020). AaRS enzymes with editing active sites are highlighted in yellow. Remarkably, the map closely matches the evolution and structure of the genetic code, indicating that the analysis of aaRS enzymes is reliable. Other approaches have not been as informative for the structure and evolution of the code (Aravind et al., 2002; Hartman & Smith, 2019; O'Donoghue & Luthey-Schulten, 2003; Smith & Hartman, 2015; Wetzel, 1995; Wolf et al., 1999).

The genetic code evolved primarily in columns, which represent the 2nd anticodon position. The anticodon central position is most important for translational accuracy. Closely related aaRS enzymes, therefore, tend to group within columns. This observation is explained in detail below. Strikingly, the pattern of aaRS evolution in Fig. 11.7C describes a history of genetic code evolution.

EF-Tu and coding degeneracy

We posit that the translation functions of EF-Tu describe the evolution of coding degeneracy (Opron & Burton, 2019). EF-Tu is a GTPase RNA chaperone that binds aminoacylated tRNA (aa-tRNA) and docks the aa-tRNA-EF-Tu complex on the ribosome. EF-Tu (translational elongation factor Tu) is a homolog of GTPases IF-2 (translational initiation factor 2) and EF-G (translational elongation factor G) (Maracci & Rodnina, 2016). These homologous GTPases occupy the same site on the ribosome during different phases of protein synthesis: initiation (IF-2), tRNA loading, clamping, accommodation (EF-Tu), and elongation (EF-G). EF-Tu is the major factor regulating translational fidelity on the ribosome (Loveland et al., 2017, 2020; Rozov, Demeshkina, et al., 2016; Rozov, Westhof, et al., 2016; Rozov et al., 2015). With the incoming aa-tRNA-EF-Tu in the hybrid A/T ribosome docking site, EF-Tu hydrolyzes GTP and sets the aa-tRNA-mRNA "latch" or clamp. The latch tightens the tRNA anticodon-mRNA codon attachment. Specifically, the latch checks for

Watson-Crick geometry at the 2nd and 3rd anticodon positions. The latch allows wobbling at the 1st anticodon position, the wobble position. At a wobble position, without modification of the wobble tRNA base, only pyrimidine-purine discrimination is achieved. The aa–tRNA–mRNA connection is monitored (latched) by 50S subunit 23S rRNA A1913 and by 30S subunit S12 and 16S rRNA G530, A1492 and A1493 (Thermus thermophilus ribosome numbering). After setting the latch, EF–Tu dissociates, and the released aminoacylated 3′-end of the aa–tRNA makes a long rotation into the peptidyl-transferase center (the A/A site), where peptide bond formation occurs. If a latched aa–tRNA–mRNA connection cannot be formed because of a base mismatch or inappropriate wobble pair, the inaccurately loaded aa–tRNA dissociates.

Before evolution of EF–Tu, therefore, the tRNA 3rd anticodon position could not have been read with 4-base accuracy. So, the 3rd anticodon position must have been a wobble position, limiting the complexity of the genetic code to $2 \times 4 = 8$-aa complexity. We posit that, before EF–Tu evolved, only one wobble position (the 1st or 3rd anticodon position) could be efficiently read at a time. Also, in Archaea, A is not read in the anticodon wobble position, so A (i.e., row 1 of the genetic code is 3rd position A), in a wobble position, formed an inefficiently utilized tRNA that functioned as a primitive translation stop signal. After evolution of EF–Tu, the 3rd anticodon position was locked down by the latch, and the maximum complexity of the genetic code became $2 \times 4 \times 4 = 32$-assignments. Because of translational fidelity mechanisms, the standard genetic code froze at a complexity of 21-assignments: 20-aa + stops. EF–Tu allowed the genetic code, therefore, to escape an 8-aa bottleneck and expand to the standard code, as described below. Significantly, we posit that coding degeneracy evolved as a natural consequence of how tRNA was read on the primitive ribosome. EF–Tu evolved to expand the genetic code beyond 8-aa. Of course, it is possible that protein EF–Tu evolved to replace a ribozyme with some shared properties in locking down the 3rd anticodon position. At this time, it is difficult to know how complex the genetic code needed to become to encode functional enzymes. Here, we indicate that a code of 8-aa may have been sufficient to encode a primitive EF–Tu enzyme. Note that the 8-aa code we describe includes bending (G), bulky hydrophobics (A, V, L), hydrogen bonding (S), positive (R), and negative (D, E) amino acids. In Archaea, aa–tRNAs can be modified. Amination of D and E and addition of C for metal chelation could enrich an evolving code. C entered the code by a circuitous path described below.

Evolution of the genetic code

Overview

The genetic code evolved around the tRNA anticodon. In the wobble position of tRNA, only purine-pyrimidine resolution was typically achieved. Because of this limitation in reading tRNA on a ribosome, the genetic code evolved to have a maximum potential complexity of 32-assignments rather than 64-assignments, as in DNA and mRNA. In Fig. 11.8 we show an annotated standard genetic code for Archaea with a maximum complexity of 32-assignments. The code is shown as a codon-anticodon table because the tRNA anticodon limits the complexity of the code. The amino acids are colored according to closely related aaRS enzymes (Fig. 11.7C) to emphasize that most evolution occurred in genetic code columns, which represent the central position of the anticodon. AaRS enzymes that have editing active sites are highlighted in gray. Note that, in Archaea, aaRS that edit are limited to hydrophobic and neutral amino acids found in the leftmost two columns of the code. SerRS-IIA, which edits, is found in both column 2 and column 4. We posit that serine first invaded column 2 and then jumped to column 4 (see below).

We posit an approximate order of addition for amino acids entering the genetic code (Fig. 11.9). Glycine appears to be the first amino acid to enter the code (Bernhardt & Patrick, 2014; Bernhardt & Tate, 2008; Kim et al., 2019; Lei & Burton, 2020). Two reasons to consider glycine as the first encoded amino acid are: (1) tRNA^{Gly} appears to be the first tRNA

row	1st									3rd
		1		2		3		4		Col
		U		C		A		G		2nd
1A	U	PHE-IIC	A/GAA	SER-IIA	A/GGA	TYR-IC	A/GUA	CYS-IB	A/GCA	U/C
1B		LEU-IA	U/CAA	SER-IIA	U/CGA	STOP	U/CUA	TRP-IC	U/CCA	A/G
2A	C	LEU-IA	A/GAG	PRO-IIA	A/GGG	HIS-IIA	A/GUG	ARG-ID	A/GCG	U/C
2B		LEU-IA	U/CAG	PRO-IIA	U/CGG	GLN-IB	U/CUG	ARG-ID	U/CCG	A/G
3A	A	ILE-IA	A/GAU	THR-IIA	A/GGU	ASN-IIB	A/GUU	SER-IIA	A/GCU	U/C
3B		MET-IA	U/CAU	THR-IIA	U/CGU	LYS-IE	U/CUU	ARG-ID	U/CCU	A/G
4A	G	VAL-IA	A/GAC	ALA-IID	A/GGC	ASP-IIB	A/GUC	GLY-IIA	A/GCC	U/C
4B		VAL-IA	U/CAC	ALA-IID	U/CGC	GLU-IB	U/CUC	GLY-IIA	U/CCC	A/G
		aa-aaRS	Ac	aa-aaRS	Ac	aa-aaRS	Ac	aa-aaRS	Ac	

Figure 11.8 The standard genetic code in Archaea as a codon-anticodon table with a complexity of 32-assignments. Amino acids and aaRS (aa-aaRS) are colored according to closely related aaRS enzymes to emphasize evolution within code columns (Col). Anticodon (Ac) bases in red are rarely or never used in Archaea. Boxes highlighted in gray have aaRS enzymes with separate editing active sites. Codon 5′→3′ positions are labeled 1st, 2nd and 3rd.

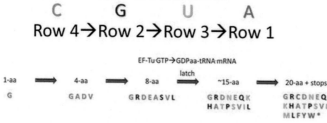

Figure 11.9 Proposed order of addition of amino acids into the genetic code. Amino acids appear to fill the code mostly by rows. 6-codon sectors for Leu, Ser, and Arg were scored for their most favored anticodon. Ser jumps from column 2 to favored column 4, changing the Ser most-favored row assignment. Row 1 F, Y, C, and W and row 3 M are posited to be the last additions to the code.

(Figs. 11.2 and 11.4) GlyRS-IIA appears to be the first aaRS enzyme (Fig. 11.7C). Glycine appears to occupy the most favored position in the code (anticodons GCC, UCC, and CCC; 2nd and 3rd anticodon position C). Glycine, alanine, aspartic acid, and valine (GADV) appear to be the first four amino acids to enter the code (Ikehara, 2005, 2009, 2014, 2016, 2019; Ikehara & Niihara, 2007; Ikehara et al., 2002; Oba et al., 2005). GADV are the four simplest amino acids chemically. These amino acids occupy row 4 of the code that appears to be the most favored row (3rd anticodon position C). We posit that Arg, Glu, Ser, and Leu enter the code next. Arg, Ser, and Leu end up occupying 6-codon sectors in the code. This is described in more detail below.

As described above, to progress beyond an 8-aa code required evolution of EF-Tu to suppress wobbling at the 3rd anticodon position (Opron & Burton, 2019). After filling rows 4 and 2, row 3 and finally row 1 could be filled. Row 1 was disfavored because, initially, the 3rd anticodon position was a wobble position, and A is not allowed in a wobble anticodon position in Archaea. Because row 1 was disfavored, stop codons located to row 1. Stop codons are read by protein release factors that bind to mRNA codons (Burroughs & Aravind, 1981), so no tRNA is associated with stop codons, except in suppressor tRNA strains. We judge the order of amino acid additions that we propose to be consistent with the following rules for tRNA anticodon position preferences: (1) for the 2nd and 3rd anticodon positions, C>G>U$\gg$A; preferences are more extreme for the 3rd anticodon position because the 2nd anticodon position is central and easier to read; and (2) for the 1st anticodon (wobble) position, G>(U$\sim$C).

The genetic code evolved mostly within columns (a working model is summarized in Fig. 11.10) Please refer back to Figs. 11.7, 11.8, 11.9, and 11.10 as reference figures for the details of amino acid placements in the code. Genetic code columns represent the 2nd position in the tRNA anticodon, which is the most important position for translational accuracy. In the final steps, the genetic code filled row 1, which was initially disfavored. Row 1 was difficult to fill, because of wobbling of the 3rd anticodon position. Wobbling at the 3rd anticodon position was suppressed by evolution of EF-Tu (Kim et al., 2019; Lei & Burton, 2020; Loveland et al., 2017, 2020; Opron & Burton, 2019). We posit systematic rules for population of the code with amino acids (see above). These rules reflect preferences for the sequence of the tRNA anticodon. We posit that the genetic code evolved around the tRNA anticodon and the evolution of its reading on the ribosome. This mode of thinking describes the following features of the genetic code: (1) evolution in columns (Figs. 11.7C and 11.8); (2) evolution in rows (Fig. 11.9); (3) the order of additions of amino

Column 1:

Val (A, ValRS-IA)→Leu (A, LeuRS-IA)→Ile (A, IleRS-IA)→Met (A, MetRS-IA); Leu (A, LeuRS-IA)→Phe (A, PheRS-IC)

Column 2:

Ala (A, AlaRS-IIA)→Ser (A, SerRS-IIA)→Pro (A, ProRS-IIA); Ser (A→G, SerRS-IIA)→Thr (G→U, ThrRS-IIA); AlaRS-IIA→AlaRS-IID (before LUCA)

Column 3:

Asp (A, AspRS-IIA)→Glu (A, GluRS-IA); Asp (A→G, AspRS-IIA)→His (C, HisRS-IIA); Asp (G, AspRS-IIA→Asp (G, AspRS-■); Asp (G, AspRS-■)→A-t'ase→Asn (G, AsnRS-■); Glu (A, GluRS-IB)→Lys (G, LysRS-IE); Glu (A, GluRS-IB)→A-t'ase→Gln (A, GlnRS-IB)

Column 4:

Gly (A, GlyRS-IIA)→Arg (G, ArgRS-■)→Cys S-thase→Cys (U, CysRS-■)

Filling Row 1:

Phe (A, PheRS-IC)→Tyr (A, TyrRS-IC)→Trp (A, TrpRS-IC)

Figure 11.10 Summary of the proposed order of events in evolution of the genetic code, mostly by columns. In parentheses (discriminator base (3'-XCCA; X = the discriminator) from Pyrococcus, aaRS). Colors are used as a guide for closely related aaRS enzymes and tRNA discriminator sequences. Classic identifications of aaRS subclass (i.e., ArgRS-ID) are not necessarily reliable. *Yellow* highlighting indicates tRNA-mediated enzymatic reactions: A-t'ase (Asp-tRNA^{Asn} and Glu-tRNA^{Gln} amidotransferase); Cys S-thase) Sep-tRNA^{Cys} Cys synthase (Sep for o-phosphoserine). See the text for details. *Arrows* indicate the order of amino acid entries into the code, mostly within columns, and not necessarily the lineage of aaRS evolution. TRNAs and aaRS can be reassigned in evolution. (For interpretation of the references to colour in this figure legend, the reader is referred to the web version of this article.)

acids (Fig. 11.9); (4) late occupancy of row 1 (Figs. 11.7C and 11.9); (5) the complexity of the code; (6) evolution of 6-, 4-, 3-, 2- and 1-codon sectors; (7) evolution of stop codons; (8) coevolution of translation factors such as EF-Tu; (9) coevolution of aaRS enzymes with the code (Figs. 11.7C and 11.8); (5) the complexity of the code; (6) evolution of 6-, 4-, 3-, 2- and 1-codon sectors; (7) evolution of stop codons; (8) coevolution of translation factors such as EF-Tu; (9) coevolution of aaRS enzymes with the code (Figs. 11.7C and 11.8); (10) complexity and structure in the 3rd genetic code column; (11) selections of code structures and amino acid placements; (12) serine jumping during code evolution; (13) evolution of translational fidelity; and (14) freezing of the code. The model, therefore, is highly predictive and descriptive for the final structure and sectoring of the code. Fig. 11.11 describes a proposed order of addition of amino acids into the code. We know of no comparable model for genetic code evolution.

Polyglycine world

We posit that glycine was the first encoded amino acid (Bernhardt & Patrick, 2014; Bernhardt & Tate, 2008), and that the genetic code first evolved to synthesize polyglycine (Fig. 11.11) (Kim et al., 2019; Lei & Burton, 2020; Opron & Burton, 2019; Pak, Du, et al., 2018). Initially, this was an assumption, but it turned out to be such a useful assumption, it should not be rejected easily. Also, if one were to choose another amino

		1		2		3		4		
		U		C		A		G		
1A	U	STOP	A/GAA	STOP	A/GGA	STOP	A/GUA	STOP	A/GCA	U/C
1B		STOP	U/CAA	STOP	U/CGA	STOP	U/CUA	STOP	U/CCA	A/G
2A	C	GLY-RBZ	A/GAG	GLY-RBZ	A/GGG	GLY-RBZ	A/GUG	GLY-RBZ	A/GCG	U/C
2B		GLY-RBZ	U/CAG	GLY-RBZ	U/CGG	GLY-RBZ	U/CUG	GLY-RBZ	U/CCG	A/G
3A	A	GLY-RBZ	A/GAU	GLY-RBZ	A/GGU	GLY-RBZ	A/GUU	GLY-RBZ	A/GCU	U/C
3B		GLY-RBZ	U/CAU	GLY-RBZ	U/CGU	GLY-RBZ	U/CUU	GLY-RBZ	U/CCU	A/G
4A	G	GLY-RBZ	A/GAC	GLY-RBZ	A/GGC	GLY-RBZ	A/GUC	GLY-RBZ	A/GCC	U/C
4B		GLY-RBZ	U/CAC	GLY-RBZ	U/CGC	GLY-RBZ	U/CUC	GLY-RBZ	U/CCC	A/G
		aa-aaRS	Ac	aa-aaRS	Ac	aa-aaRS	Ac	aa-aaRS	Ac	

Figure 11.11 Polyglycine world. A is inefficiently read in a wobble position (at this stage, both 1st and 3rd anticodon positions were wobble positions). Aa-aaRS) amino acid-aminoacyl-tRNA synthetase; RBZ) ribozyme; Ac) anticodon. Letters around the periphery indicate codon (mRNA) sequence. Colored shading for amino acids is maintained, so that placements of amino acids can be tracked. *Red* letters indicate anticodons that are not used or are used inefficiently. (For interpretation of the references to colour in this figure legend, the reader is referred to the web version of this article.)

acid (i.e., Ala or Pro) as the first encoded amino acid, we do not believe reasonable rules can be as easily established for filling the code. Selecting Gly, as the initial encoded amino acid, however, a reasonable mechanism and simple rules for populating the code became apparent. We posit that the entire primitive code, including all anticodons and all codons, evolved to encode glycine. Row 1 tRNAs (1st anticodon position wobble A and 3rd anticodon position (initially) wobble A) were utilized inefficiently, so these tended to function as stops. We posit that wobbling at the 3rd anticodon position was suppressed by evolution of EF-Tu. In Archaea, A is not allowed in a wobble position (Pak, Kim, & Burton, 2018). Basically, we posit that a ribozyme mechanism existed to replicate templated tRNA (and other RNA) sequences. The anticodon is the easiest feature of tRNA to mutate without consequence for folding, so proliferation of tRNAs and mutations rapidly resulted in all possible anticodon sequences. We posit that a GlyRS ribozyme (GlyRS-RBZ) charged diverse $tRNA^{Gly}$ with glycine because the code was not yet sufficiently evolved to generate protein catalysts. Hemolithin, recovered from meteorite samples, is a polyglycine peptide from outer space, indicating that a polyglycine world existed, even beyond an Earth environment (McGeoch et al., 2020).

One advantage of this mode of thinking is that it gives insight into the sectoring of the genetic code. If the entire primitive code encoded glycine, then invasion by other amino acids caused the glycine sector to contract. Invasion of the code by newly encoded amino acids, therefore, resulted in shrinking of previously occupied sectors. We then realized that clear rules could be stated for how incoming amino acids displaced previously added amino acids. Currently, glycine occupies the most favored anticodon positions in the code, which are GCC, UCC, and CCC. If amino acids that entered the code at an early time protected the most advantageous sectors, then C was favored in the 2nd and 3rd anticodon positions. The rules for the 2nd and 3rd anticodon positions began to emerge as C>G>U≫A (Lei & Burton, 2020; Kim et al., 2019). Preferences were strongest for the 3rd anticodon position, because the 2nd anticodon position was easier to read.

GADV world

We posit that polyglycine world gave way to GADV world (GADV for glycine, alanine, aspartic acid, and valine) (Fig. 11.12) (Ikehara, 2005, 2009, 2014, 2016, 2019; Ikehara & Niihara, 2007; Ikehara et al., 2002). Positing GADV world explains why the genetic code sectored in columns (Figs. 11.7C and 11.8). Glycine, alanine, aspartic acid, and valine are the

			U		C		A		G	
			1		2		3		4	
1A	U	STOP	A/GAA	STOP	A/GGA	STOP	A/GUA	STOP	A/GCA	U/C
1B		STOP	U/CAA	STOP	U/CGA	STOP	U/CUA	STOP	U/CCA	A/G
2A	C	VAL-RBZ	A/GAG	ALA-RBZ	A/GGG	ASP-RBZ	A/GUG	GLY-RBZ	A/GCG	U/C
2B		VAL-RBZ	U/CAG	ALA-RBZ	U/CGG	ASP-RBZ	U/CUG	GLY-RBZ	U/CCG	A/G
3A	A	VAL-RBZ	A/GAU	ALA-RBZ	A/GGU	ASP-RBZ	A/GUU	GLY-RBZ	A/GCU	U/C
3B		VAL-RBZ	U/CAU	ALA-RBZ	U/CGU	ASP-RBZ	U/CUU	GLY-RBZ	U/CCU	A/G
4A	G	VAL-RBZ	A/GAC	ALA-RBZ	A/GGC	ASP-RBZ	A/GUC	GLY-RBZ	A/GCC	U/C
4B		VAL-RBZ	U/CAC	ALA-RBZ	U/CGC	ASP-RBZ	U/CUC	GLY-RBZ	U/CCC	A/G
		aa-aaRS	Ac	aa-aaRS	Ac	aa-aaRS	Ac	aa-aaRS	Ac	

Figure 11.12 GADV-world. Explaining evolution in code columns.

simplest amino acids chemically, so it is reasonable that these might be the first four amino acids to enter the code (Ikehara, 2005, 2009, 2014, 2016, 2019; Ikehara & Niihara, 2007; Ikehara et al., 2002; Oba et al., 2005). Also, GADV are amino acids that end up on the 4th row of the code, which corresponds to 3rd anticodon position C, in keeping with the C>G>U≫A rule for the 2nd and 3rd anticodon positions. At the GADV stage, we posit that tRNAs were charged by ribozymes.

The 8-aa bottleneck

Before evolution of EF–Tu, the genetic code froze at 8-aa (Fig. 11.13). EF–Tu suppresses wobbling at the 3rd anticodon position. Before EF–Tu, therefore, 3rd anticodon position A could not easily be read on the primitive ribosome. We note here that 1st anticodon position wobble A is not utilized in Archaea (Kim et al., 2019; Lei & Burton, 2020; Pak, Kim, & Burton, 2018). In Bacteria and Eukaryotes, wobble A is encoded only

			U		C		A		G	
			1		2		3		4	
1A	U	STOP	A/GAA	STOP	A/GGA	STOP	A/GUA	STOP	A/GCA	U/C
1B		STOP	U/CAA	STOP	U/CGA	STOP	U/CUA	STOP	U/CCA	A/G
2A	C	LEU-IA	A/GAG	SER-IIA	A/GGG	ASP-IIA	A/GUG	ARG-IA	A/GCG	U/C
2B		LEU-IA	U/CAG	SER-IIA	U/CGG	GLU-IA	U/CUG	ARG-IA	U/CCG	A/G
3A	A	VAL-IA	A/GAU	ALA-IIA	A/GGU	ASP-IIA	A/GUU	GLY-IIA	A/GCU	U/C
3B		VAL-IA	U/CAU	ALA-IIA	U/CGU	GLU-IA	U/CUU	GLY-IIA	U/CCU	A/G
4A	G	VAL-IA	A/GAC	ALA-IIA	A/GGC	ASP-IIA	A/GUC	GLY-IIA	A/GCC	U/C
4B		VAL-IA	U/CAC	ALA-IIA	U/CGC	GLU-IA	U/CUC	GLY-IIA	U/CCC	A/G
		aa-aaRS	Ac	aa-aaRS	Ac	aa-aaRS	Ac	aa-aaRS	Ac	

Figure 11.13 The 8-aa bottleneck. Columns 1, 2 and 4 sectored differently than column 3 because of the selection of the wobble position (3rd anticodon position for columns 1, 2 and 4; 1st anticodon position for column 3).

when it is modified to inosine, which broadens the set of recognized co-dons (Demongeot & Norris, 2019; Rafels-Ybern et al., 2018, 2019; Saint-Léger et al., 2016). Interestingly, columns 1, 2 and 4 in the code sectored by one mechanism, and column 3, which became the most innovated column, encoding the most amino acids, sectored by a slightly different mechanism. We posit that the genetic code froze at 8-aa because both anticodon 1st and 3rd positions were wobble positions, and wobble positions are read with pyrimidine-purine (2-assignment) resolution. The 2nd anticodon position was much easier to read because it is the middle position. We further posit that only a single wobble position could be read at a time. Because, at a wobble position, only purine versus pyrimidine discrimination was initially achieved, the maximum complexity of the code was $2 \times 4 = 8$-aa. In columns 1, 2, and 4, the 2nd and 3rd anticodon positions were primarily read. Interestingly, leucine, serine, and arginine are the amino acids that maintained 6-codon sectors (3 boxes in the genetic code tables shown). By contrast, in column 3, the 1st and 2nd anticodon positions were primarily read, giving the striped pattern of the related amino acids Asp and Glu. These differences in wobble selection gave rise to 6-codon sectors, for leucine, serine and arginine, and high innovation in column 3 (encoding many amino acids). Because of the geometry of the 7-nt U-turn anticodon loop in tRNA, the reading register for the primitive ribosome was always 3-nt.

In Fig. 11.14, we posit that leucine, serine, and arginine invaded row 3 of the code, displacing valine, alanine, and glycine into favored row 4 (3rd anticodon position C). Positing this invasion of row 3 helps to describe the evolution of 6-codon sectors, the placement of threonine in the code and the jumping of serine from column 2 to column 4. Because valine, alanine,

		1		2		3		4		
		U		C		A		G		
1A	U	STOP	A/GAA	STOP	A/GGA	STOP	A/GUA	STOP	A/GCA	U/C
1B		STOP	U/CAA	STOP	U/CGA	STOP	U/CUA	STOP	U/CCA	A/G
2A	C	LEU-IA	A/GAG	SER-IIA	A/GGG	ASP-IIA	A/GUG	ARG-IA	A/GCG	U/C
2B		LEU-IA	U/CAG	SER-IIA	U/CGG	GLU-IA	U/CUG	ARG-IA	U/CCG	A/G
3A	A	LEU-IA	A/GAU	SER-IIA	A/GGU	ASP-IIA	A/GUU	ARG-IA	A/GCU	U/C
3B		LEU-IA	U/CAU	SER-IIA	U/CGU	GLU-IA	U/CUU	ARG-IA	U/CCU	A/G
4A	G	VAL-IA	A/GAC	ALA-IIA	A/GGC	ASP-IIA	A/GUC	GLY-IIA	A/GCC	U/C
4B		VAL-IA	U/CAC	ALA-IIA	U/CGC	GLU-IA	U/CUC	GLY-IIA	U/CCC	A/G
		aa-aaRS	Ac	aa-aaRS	Ac	aa-aaRS	Ac	aa-aaRS	Ac	

Figure 11.14 The 8-aa bottleneck: Leu, Ser, and Arg invaded row 3, helping to describe evolution of 6-codon sectors.

		1		2		3		4		
		U		C		A		G		
1A	U	LEU-IA	A/GAA	SER-IIA	A/GGA	STOP	A/GUA	ARG-IA	A/GCA	U/C
1B		LEU-IA	U/CAA	SER-IIA	U/CGA	STOP	U/CUA	ARG-IA	U/CCA	A/G
2A	C	LEU-IA	A/GAG	SER-IIA	A/GGG	ASP-IIA	A/GUG	ARG-IA	A/GCG	U/C
2B		LEU-IA	U/CAG	SER-IIA	U/CGG	GLU-IA	U/CUG	ARG-IA	U/CCG	A/G
3A	A	LEU-IA	A/GAU	SER-IIA	A/GGU	ASP-IIA	A/GUU	SER-IIA	A/GCU	U/C
3B		LEU-IA	U/CAU	SER-IIA	U/CGU	GLU-IA	U/CUU	ARG-IA	U/CCU	A/G
4A	G	VAL-IA	A/GAC	ALA-IIA	A/GGC	ASP-IIA	A/GUC	GLY-IIA	A/GCC	U/C
4B		VAL-IA	U/CAC	ALA-IIA	U/CGC	GLU-IA	U/CUC	GLY-IIA	U/CCC	A/G
		aa-aaRS	Ac	aa-aaRS	Ac	aa-aaRS	Ac	aa-aaRS	Ac	

Figure 11.15 The 8-aa bottleneck. Ser jumped from column 2 to column 4, invading the Arg sector. We posit evolution of a primitive EF-Tu or a ribozyme with EF-Tu-like activity suppressing wobbling at the 3rd anticodon position.

and glycine retained the favored 4th row (3rd anticodon position C), these invasions were tolerated.

In Fig. 11.15, we posit that serine jumped from column 2 into column 4. This event is described in more detail below. Leucine, serine, and arginine began to invade disfavored row 1. This is considered in the model because leucine and serine end up in row 1 of the code. Also, a primitive ArgRS-ID may have evolved to TyrRS-IC, TrpRS-IC, and CysRS-IB (Fig. 11.7C) that end up in row 1. Evolution of a primitive EF-Tu at the 8-aa stage, or else evolution of a ribozyme to partly lock down the 3rd anticodon wobble position, might have this effect. We do not propose a specific order of events. We note that, at this stage, row 1 probably included tRNAs that were charged but not efficiently utilized, until EF-Tu or a corresponding ribozyme evolved. Ser jumping from column 2 to column 4 required only a single base change in the tRNA anticodon at the 2nd position (GGU → GCU). The jump was favorable for placement of serine in the code because column 4 (2nd anticodon position C) was favored over column 2 (2nd anticodon position G).

Evolution of EF-Tu suppressed wobbling at the 3rd anticodon position and broke the 8-aa bottleneck

We posit that evolution of EF-Tu converted the 3rd anticodon position from a wobble position with pyrimidine–purine (2-assignment) resolution to a position with 4-base (A, G, C, U) resolution. This advance expanded the genetic code from a maximum complexity of $2 \times 4 = 8$-aa to a maximum complexity of $2 \times 4 \times 4 = 32$-assignments. It should be noted that, in mRNA, the maximum complexity of the code is $4 \times 4 \times 4 = 64$-assignments. The complexity of the code, which froze at

20-aa + stops = 21-assignments, was limited to a large extent by reading of tRNA on the ribosome.

An intermediate state of the code with ∼17-aa is posited in Fig. 11.16. In column 1, Ile was added as a 4-codon sector. Because Leu entered the code before Ile, and because Ile displaced Leu, Ile occupied row 3 (3rd anticodon position U), which was disfavored compared to row 2 (3rd anticodon position G) (C>G>U≫A). In column 2, Thr replaced Ser in row 3, and Pro replaced Ser in row 2. Because Thr and Ser are related amino acids, it is easy to see how SerRS-IIA could have duplicated and a copy could have diverged to evolve ThrRS-IIA (Fig. 11.7C). In column 1, Ile was added as a 4-codon sector. Because Leu entered the code before Ile, and because Ile displaced Leu, Ile occupied row 3 (3rd anticodon position U), which was disfavored compared to row 2 (3rd anticodon position G) (C>G>U≫A). In column 2, Thr replaced Ser in row 3, and Pro replaced Ser in row 2. Because Thr and Ser are related amino acids, it is easy to see how SerRS-IIA could have duplicated and a copy could have diverged to evolve ThrRS-IIA (Fig. 11.7C). Ser now occupied row 1. Column 3 filled to become the most innovated column encoding the most amino acids. The two founding amino acids in column 3, Asp, and Glu remained on favored row 4 (3rd anticodon position C). We posit that Asp, which entered the code first, occupied row 4A and Glu occupied row 4B because tRNA wobble G was favored over wobble U/C. We posit that AspRS was originally AspRS-IIA and evolved to AspRS-IIB after evolution of HisRS-IIA. AsnRS-IIB was later derived from AspRS-IIB. This is a continuing evolution in some archaeal species. GluRS-IB gave rise to LysRS-IE and GlnRS-IB (Fig. 11.7C). GluRS-IB → GlnRS-IB is ongoing in some archaeal species.

		1		2		3		4		
		U		C		A		G		
1A	U	PHE-IIC	A/GAA	SER-IIA	A/GGA	TYR-IC	A/GUA	TYR-IC	A/GCA	U/C
1B		LEU-IA	U/CAA	SER-IIA	U/CGA	STOP	U/CUA	TYR-IC	U/CCA	A/G
2A	C	LEU-IA	A/GAG		A/GGG	HIS-IIA	A/GUG	ARG-ID	A/GCG	U/C
2B		LEU-IA	U/CAG		U/CGG	GLN-IB	U/CUG	ARG-ID	U/CCG	A/G
3A	A	ILE-IA	A/GAU		A/GGU	ASN-IIB	A/GUU	SER-IIA	A/GCU	U/C
3B		ILE-IA	U/CAU		U/CGU	LYS-IE	U/CUU	ARG-ID	U/CCU	A/G
4A	G	VAL-IA	A/GAC	ALA-IID	A/GGC	ASP-IIB	A/GUC	GLY-IIA	A/GCC	U/C
4B		VAL-IA	U/CAC	ALA-IID	U/CGC	GLU-IB	U/CUC	GLY-IIA	U/CCC	A/G
		aa-aaRS	Ac	aa-aaRS	Ac	aa-aaRS	Ac	aa-aaRS	Ac	

Figure 11.16 An intermediate ∼17-aa stage and evolution of high innovation in column 3. We speculate that TyrRS-IC may be derived from a primitive ArgRS-ID and that tyrosine may have jumped from column 4 to column 3.

We begin to model the filling of disfavored row 1. We posit that phenylalanine and tyrosine may have entered the code at about this stage. PheRS-IIC may be derived from a primitive AspRS-IIB in several steps (Fig. 11.7C). TRNA^{Phe}, in Pyrococcus, appears to be derived from tRNA^{Lys} (Pak, Du, et al., 2018). These identifications do not provide a simple model for placement of phenylalanine in the code. TyrRS-IC may be derived from a primitive ArgRS-ID, filling column 4, row 1, and jumping to column 3 by a single base change in the anticodon. We posit that pressure is building to evolve modern stop codons at this stage of evolution.

Filling disfavored row 1, evolution of stop codons, and Met invasion of an Ile sector

To evolve the standard genetic code (Fig. 11.17), then required filling in disfavored row 1 (3rd anticodon position A). We posit that Phe invaded column 1, row 1A, displacing Leu. Tyr invaded column 3, row 1A, perhaps as described above. Stop codons located to columns 3 and 4, row 1B. Cys invaded column 4, row 1A. We posit that CysRS-IB was derived from duplication and repurposing of a primitive ArgRS-ID. The classic naming of these aaRS enzymes is deceptive. CysRS-IB and ArgRS-ID are closely related enzymes, despite their IB and ID structural subclass designations (Fig. 11.7C). Trp invaded column 4, row 1B. There are few 1-codon sectors in the genetic code, but Trp shares a sector with a stop codon, which is read in mRNA by a protein release factor (Burroughs & Aravind, 1981), so there is no competing tRNA occupying the Trp sector (column 4, row 1B).

		1				3		4		
			U		C		A		G	
1A	U	PHE-IIC	A/GAA	SER-IIA	A/GGA	TYR-IC	A/GUA	CYS-IB	A/GCA	U/C
1B		LEU-IA	U/CAA	SER-IIA	U/CGA	STOP	U/CUA	TRP-IC	U/CCA	A/G
2A	C	LEU-IA	A/GAG	PRO-IIA	A/GGG	HIS-IIA	A/GUG	ARG-ID	A/GCG	U/C
2B		LEU-IA	U/CAG	PRO-IIA	U/CGG	GLN-IB	U/CUG	ARG-ID	U/CCG	A/G
3A	A	ILE-IA	A/GAU		A/GGU	ASN-IIB	A/GUU	SER-IIA	A/GCU	U/C
3B		MET-IA	U/CAU		U/CGU	LYS-IE	U/CUU	ARG-ID	U/CCU	A/G
4A	G	VAL-IA	A/GAC	ALA-IID	A/GGC	ASP-IIB	A/GUC	GLY-IIA	A/GCC	U/C
4B		VAL-IA	U/CAC	ALA-IID	U/CGC	GLU-IB	U/CUC	GLY-IIA	U/CCC	A/G
		aa-aaRS	Ac	aa-aaRS	Ac	aa-aaRS	Ac	aa-aaRS	Ac	

Figure 11.17 The standard genetic code. Amino acids in charcoal and gray were the last to enter the code. Disfavored row 1 was filled. Stop codons evolved on disfavored row 1. Met invaded the Ile sector. The code now has stops and starts.

Evolution of row 1 appears somewhat chaotic. Some chaos might be expected in filling the last available positions in the code. It appears from Fig. 11.7C that TyrRS-IC and TrpRS-IC might have been derived from a primitive ArgRS-ID. Tyrosine and tryptophan may have been two of the last amino acids added to the code (Fournier & Alm, 2015). CysRS-IB appears to be derived from a primitive ArgRS-ID. PheRS-IIC appears to have evolved in steps from a primitive AspRS-IIB. In Pyrococcus, tRNA^{Phe} (discriminator A) appears to be derived from tRNA^{Lys} (discriminator G). TRNA^{Tyr} (discriminator A) appears to be derived from tRNA^{Asn} (discriminator G). TRNA^{Trp} (discriminator A) appears to be derived from tRNA^{Pro} (discriminator A). TRNA^{Cys} (discriminator U) appears to be derived from tRNA^{Thr} (discriminator U) (Pak, Du, et al., 2018). We guess that all of these tRNAs were assigned and then reassigned, and, therefore, their apparent derivations do not indicate the precise steps in adding these amino acids to the code. Reassignments of aaRS enzymes and tRNAs in evolution enhance translational accuracy by suppressing mischarging of tRNAs.

Cys appears to have entered the genetic code via a circuitous path (Hauenstein & Perona, 2008; Mukai et al., 2017; Turanov et al., 2011). Notably, some Archaea generate Cys from Sep-tRNA^{Cys} charged by SepRS-IIC (Sep for o-phosphoserine). Modification of amino acids bound to tRNAs is a repeated theme in ancient Archaea that may reflect chemistry from a prelife world (Gospodinov & Kunnev, 2020; Kunnev & Gospodinov, 2018). This is also how Asn and Gln entered the code. Asp-tRNA^{Asn} and Glu-tRNA^{Gln} were aminated by Asp-tRNA^{Asn} and Glu-tRNA^{Gln} amidotransferase (Feng et al., 2004, 2005; Min et al., 2002; Raczniak et al., 2001; Rampias et al., 2010; Salazar et al., 2001; Tumbula-Hansen et al., 2002; Wu et al., 2009). TRNA-linked and RNA-linked chemistry must have been common in the prelife world before evolution to cellular life (Gospodinov & Kunnev, 2020; Kunnev & Gospodinov, 2018).

In column 1, row 3B, Met invaded a 4-codon Ile sector. MetRS-IA and IleRS-IA are closely related enzymes (Fig. 11.7C). Furthermore, tRNA^{Met} and tRNA^{Ile} are closely related tRNAs, in ancient Archaea such as Pyrococcus (Pak, Du, et al., 2018). In keeping with our contention that 1-codon sectors were difficult to form and maintain, tRNA^{Ile}(CAU) and tRNA^{Met}(CAU) are both utilized in Archaea. The UAU anticodon, however, is rarely used. In tRNA^{Ile}(CAU), wobble C is converted to agmatidine to recognize Ile codon AUA but not Met codon AUG (Mandal et al., 2010; Numata, 2015; Osawa et al., 2011; Phillips & de Crécy-Lagard,

2011; Satpati et al., 2014; Voorhees et al., 2013). In tRNA^{Met}(CAU), wobble C is lightly modified and recognizes Met codon AUG but not Ile codon AUA. Wobble anticodon modification, therefore, resolves the ambiguity of tRNA^{Ile}(CAU) and tRNA^{Met}(CAU) but anticodon UAU was generally lost in the process. Met provides translation starts, in addition to bringing another amino acid into the code.

Code punctuation

We posit that genetic code starts and stops were late additions to the code. Translation initiation in Archaea has recently been reviewed (Schmitt et al., 2020). Archaea utilize Met-tRNA^{iMet} (iMet for initiator methionine) and a set of translation initiation factors. Bacteria utilize fMet-tRNA^{iMet} (fMet for N-formyl-methionine) and a simplified initiation mechanism. In Pyrococcus, tRNA^{iMet}(CAU), tRNA^{Met}(CAU), and tRNA^{Ile}(CAU) are discriminated mostly based on acceptor stems. For tRNA^{iMet}, the 5′-acceptor stem sequence is AGCGGG(G), with the 3′-G uncharacteristically unpaired (opposite 3′-acceptor stem (G)CCCGCU). For tRNA^{Met}, the 5′-acceptor stem sequence is GCCGGGG, with all bases paired. For tRNA^{Ile}(CAU), the 5′-acceptor stem sequence is GGGCCCG, with all bases paired. It appears that selection of methionine as the initiating amino acid in Archaea was a complex coevolution of Met-tRNA^{iMet} and initiation factors. We guess that the need for a translation initiation start signal was a powerful driving force to evolve this system. For instance, to initiate translation at internal mRNA sites for gene expression regulation may have required evolution of the Met-tRNA^{iMet} translation initiation system. We guess that Bacteria simplified the archaeal system in evolution, adopting fMet-tRNA^{iMet} and shedding initiation factors during divergence from Archaea.

Evolution of translation stops appears to have been a complex process with multiple stages (Burroughs & Aravind, 1981). We posit that initially the problem was generating longer peptides to form more complex proteins. The system appears to have started with inefficient tRNAs, that is, 1st or 3rd anticodon A, as primitive stop signals. The system suppressed 3rd anticodon position wobbling by evolving EF-Tu. Finally, protein translation release factors evolved to recognize stop codons in mRNA. Suppression of wobbling at the 3rd anticodon position by EF-Tu expanded the genetic code and may have driven evolution of protein release factors and stop codons.

Serine jumping from column 2 to column 4

We posit that jumping across columns was rare in establishment of the code. The advantage for Ser to jump from column 2 to column 4 was that serine obtained a favored anticodon. GCU was favored over GGU, because 2nd anticodon position C was favored over G. Serine could invade the arginine sector because ArgRS-ID reads type I tRNA^{Arg} and cannot read type II tRNA^{Ser}. Also, SerRS-IIA must bind the type II tRNA^{Ser} V loop to add Ser. There is no advantage to serine invading the leucine sector, and such an invasion would be problematic, partly because both tRNA^{Leu} and tRNA^{Ser} are type II tRNAs. Invasion of row 3 would not be advantageous for serine and would also eliminate an amino acid from the code.

Summary

In summary, a highly detailed working model is possible for evolution of the genetic code. The model is mostly based on tRNA and aaRS sequence analyses. The genetic code evolved around the tRNA anticodon. The model tracks evolution of aaRS enzymes, indicating that both the genetic code model and the model for aaRS evolution are substantially correct. When we started this work, we did not think such a detailed and predictive model was possible or reasonable. Now, we consider this a very strong model for further analysis of the code.

Discussion

tRNA evolution

Remarkably, tRNA evolution was determined to the last nucleotide (Fig. 11.2) (Burton, 2020). The solution to tRNA evolution was possible because tRNA^{Pri} sequences were highly ordered and these repeats and inverted repeats can still be detected in ancient Archaea. No accretion model can describe tRNA evolution, because of conservation of highly regular tRNA^{Pri} sequences. For an accretion model to have credence, tRNAs would need to expand, inserting ordered and preordained sequences, which seems unlikely if not impossible. Only the 3–31-nt minihelix model describes tRNA evolution. Accretion models, such as the Uroboros (Demongeot & Seligmann, 2020a) and 2-minihelix (Di Giulio, 2019, 2020) models are, therefore, falsified. Also, the genetic code evolved around the tRNA anticodon (Burton, 2020; Kim et al., 2019; Lei & Burton, 2020; Opron & Burton, 2019). The genetic code complexity is determined by the way the tRNA was read on a ribosome, explaining why the genetic code went from being frozen at 8-aa complexity (Figs. 11.13, 11.14, and 11.15), before attaining the standard code (Fig. 11.17) that has 21-assignments (20-aa + stops).

aaRS evolution

Analysis of aaRS evolution tracks evolution of the genetic code (Fig. 11.7C) (Kim et al., 2019; Lei & Burton, 2020; Opron & Burton, 2019). This analysis is easiest to do in archaeal systems because Archaea are the oldest organisms, and Bacteria are more derived (Fig. 11.1). Because GlyRS-IIA, ValRS-IA, and IleRS-IA are simple sequence homologs (Fig. 11.7A and B), the Carter-Ohno-Rodin hypothesis for aaRS evolution (Carter, 2014, 2017; Carter & Wills, 2019; Carter et al., 2014) is falsified. Class I and class II aaRS were not generated from opposite DNA strands of a primordial, bidirectional gene encoding molten globule Urzymes.

Genetic code evolution

A highly detailed model has been generated that describes standard genetic code evolution in Archaea (Figs. 11.10, 11.11, 11.12, 11.13, 11.14, 11.15, 11.16, and 11.17). Every aspect of code evolution is described by this model. Simple rules were developed to describe sectoring of the code. The model can be modified to generate the more derived genetic codes of Bacteria and Eukarya.

Freezing the code

We posit that new amino acids were introduced through tRNA charging errors and through aa-tRNA linked chemistry, and that translational fidelity mechanisms froze the code (Kim et al., 2019; Lei & Burton, 2020). Based on archaeal systems, Asn, Gln, and Cys appear to have initially entered the code through enzymatic mechanisms in which aa-tRNAs were modified. Subsequently, the tRNA-linked reactions were replaced by evolution of AsnRS-IIB, GlnRS-IB, and CysRS-IB. Other amino acids may also have entered the code via tRNA-linked reactions. For instance, Arg may have replaced ornithine early in code evolution. Ornithine can be converted enzymatically to Arg in two steps (Longo et al., 2020). Similarly, Leu may have been synthesized from tRNA-linked Val in 5 enzymatic steps. Because of initial wobbling in the 1st and 3rd anticodon positions, EF-Tu evolution was necessary to expand the code beyond 8-aa. Subsequently, EF-Tu contributed to freezing of the code by enforcing translational accuracy. Some aaRS have proofreading (editing) to remove inappropriately added amino acids from their cognate tRNAs (Perona & Gruic-Sovulj, 2014). Remarkably, the aaRS that edit are limited in Archaea to amino acids located in the left half of the code (columns 1 and 2; Figs. 11.7C and 11.8). Hydrophobic and neutral amino acids locate to columns 1 and 2 of the code, so editing helped with translational accuracy for amino acids with limited chemical character, such that these amino acids could not be as easily specified in the aaRS active site. Editing helped to freeze the code by protecting 6- and 4-codon sectors in the left half of the code. To add additional amino acids required splitting larger sectors of the code. 6-codon sectors encoding Leu, Ser, and Arg resulted from the history of evolution, as described. Splitting a 2-codon sector into two 1-codon sectors was problematic because of tRNA wobble ambiguity (generally, in Archaea, tRNA wobble U∼C and only G is allowed, not A). High innovation in column 3 of the code resulted from the history of column 3 sectoring and the initial selection of the wobble base (1st anticodon position). In the case of Ile and Met, co-occupancy of the CAU anticodon through wobble modifications resulted in suppression of the Ile UAU anticodon, as described.

A model for evolution of protocells

Fig. 11.18 shows a model for evolution of the first cells. We posit that a number of ribozymes must have been present in order for tRNA to evolve prior to evolution of complex proteins (Kim et al., 2019; Lei & Burton, 2020). TRNA^{Pri} was comprised completely of ordered sequences, notably, repeats and inverted repeats, so ribozymes must have existed to generate repeats, inverted repeats, 31-nt minihelices and tRNA (Burton, 2020) (Fig. 11.2). Furthermore, polyglycine and GADV polymers appear to have been the first products of the evolving genetic code. Therefore, a selection for polyglycine and GADV polymers would describe the selective pressure for evolution of the code before complex proteins became possible. We posit that polyglycine and GADV polymers formed essential structures in protocells. Structures included cell walls, internal structures, amyloid plaques and LLPS (liquid—liquid phase separation) compartments. We note that (Gly)$_5$ is a component of bacterial cell walls (Vollmer, 2008; Vollmer et al., 2008) and may be a relic of a prelife world.

Amyloid plaques form from assemblies of long, mis-associated β-sheets. In eukaryotic cells, amyloid plaques are a symptom of neurological diseases (Elbaum-Garfinkle, 2019; Puzzo et al., 2020; Singh et al., 2020). We posit

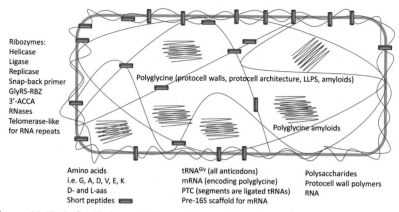

Ribozymes:
Helicase
Ligase
Replicase
Snap-back primer
GlyRS-RBZ
3′-ACCA
RNases
Telomerase-like
for RNA repeats

Polyglycine (protocell walls, protocell architecture, LLPS, amyloids)

Polyglycine amyloids

Amino acids
i.e. G, A, D, V, E, K
D- and L-aas
Short peptides

tRNA^{Gly} (all anticodons)
mRNA (encoding polyglycine)
PTC (segments are ligated tRNAs)
Pre-16S scaffold for mRNA

Polysaccharides
Protocell wall polymers
RNA

Figure 11.18 A detailed model for evolution of protocells, polyglycine- and GADV-world. *PTC*, peptidyl transferase center.

that amyloid plaques may be generated from misregulation of LLPS membraneless compartments, which are a normal feature of eukaryotic cells. In the ancient world, amyloid plaques would have regulated hydration in protocells to enhance diverse chemistries such as polymerization reactions. LLPS compartments stimulate diverse chemistries in eukaryotic cells and may function similarly in prokaryotic cells (Boehning et al., 2018; Guo et al., 2020; Portz & Shorter, 2020). We posit that polyglycine and GADV LLPS compartments were essential features of early protocells. Amyloids and LLPS are posited to have been selected because they supported novel and essential protocell chemistries, partly by regulating hydration and dehydration. We posit that amyloids and LLPS provided the selective driving force for the early evolution of the genetic code before complex proteins could be encoded.

Other models

Here, we briefly contrast our genetic code evolution model with some other models that have been advanced. Koonin and Novozhilov review a number of genetic code models (Koonin & Novozhilov, 2017). Much emphasis has been given to three models: (1) stereochemical; (2) error-minimization; and (3) coevolution. We do not consider these models to be highly predictive (Lei & Burton, 2020). A new approach has been proposed based on codon energetics (Klump et al., 2020). Computational approaches may be of interest (Yarus, 2021a, 2021b, 2021c), although the concept of late evolution of degeneracy seems unlikely compared to our model that degeneracy is a natural result of evolution of anticodon reading on the ribosome and evolution of EF-Tu. Kunnev and Gospodinov have proposed models that include RNA-aa linked reactions in prelife, as we also support (Gospodinov & Kunnev, 2020; Kunnev & Gospodinov, 2018). Rogers has put forth a model in some ways similar to ours (Rogers, 2019). Another somewhat similar model to ours has been advanced by Chatterjee and Yadav (Chatterjee & Yadav, 2019). A different view of serine sectoring than that we propose was recently advanced (Inouye et al., 2020). Simply stated, our models are more detailed than others and make many more specific predictions. We provide a clear selection strategy and a set of rules for the placements of all amino acids in the standard code. We enrich the discussion of Kunnev and Gospodinov on tRNA- and RNA-linked reactions in the ancient world before the first true cells. Our approach is centered on the tRNA anticodon, and others should adopt the anticodon-centered view. For instance, evolving directly to codons is a mistake, because the genetic code was limited in complexity by the tRNA anticodon (Kim et al., 2019; Lei & Burton, 2020). Emphasis on the tRNA anticodon and reading of the anticodon on the primitive and modern ribosomes (i.e., +/− EF-Tu) also describes degeneracy of the code. Furthermore, we consider filling the code piecemeal to be a mistake. We take a more orderly approach to code-filling that is based on clear rules for anticodon sequence preference. We provide strong selections for the initial steps of code evolution before complex proteins can be encoded. Our model for tRNA evolution reaches far back into the prelife world with many predictions for prelife ribozymes and, surprisingly, ordered prelife RNA chemistry. Remarkably, existing tRNA sequence provides a record of chemistry in the prelife world.

Author contributions

The authors wrote the paper, made the figures, and did the research.

References

Alva, V., Dunin-Horkawicz, S., Habeck, M., Coles, M., & Lupas, A. N. (2009). The GD box: A widespread noncontiguous supersecondary structural element. *Protein Science, 18*(9), 1961−1966. https://doi.org/10.1002/pro.207

Alva, V., Koretke, K. K., Coles, M., & Lupas, A. N. (2008). Cradle-loop barrels and the concept of metafolds in protein classification by natural descent. *Current Opinion in Structural Biology, 18*(3), 358−365. https://doi.org/10.1016/j.sbi.2008.02.006

Aravind, L., Anantharaman, V., & Koonin, E. V. (2002). Monophyly of class I aminoacyl tRNA synthetase, USPA, ETFP, photolyase, and PP-ATPase nucleotide-binding domains: Implications for protein evolution in the RNA world. *Proteins: Structure, Function and Genetics, 48*(1), 1–14. https://doi.org/10.1002/prot.10064

Battistuzzi, F. U., Feijao, A., & Hedges, S. B. (2004). A genomic timescale of prokaryote evolution: Insights into the origin of methanogenesis, phototrophy, and the colonization of land. *BMC Evolutionary Biology, 4.* https://doi.org/10.1186/1471-2148-4-44

Bernhardt, H. S., & Patrick, W. M. (2014). Genetic code evolution started with the incorporation of glycine, followed by other small hydrophilic amino acids. *Journal of Molecular Evolution, 78*(6), 307–309. https://doi.org/10.1007/s00239-014-9627-y

Bernhardt, H. S., & Tate, W. P. (2008). Evidence from glycine transfer RNA of a frozen accident at the dawn of the genetic code. *Biology Direct, 3.* https://doi.org/10.1186/1745-6150-3-53

Boehning, M., Dugast-Darzacq, C., Rankovic, M., Hansen, A. S., Yu, T., Marie-Nelly, H., McSwiggen, D. T., Kokic, G., Dailey, G. M., Cramer, P., Darzacq, X., & Zweckstetter, M. (2018). RNA polymerase II clustering through carboxy-terminal domain phase separation. *Nature Structural and Molecular Biology, 25*(9), 833–840. https://doi.org/10.1038/s41594-018-0112-y

Brindefalk, B., Dessailly, B. H., Yeats, C., Orengo, C., Werner, F., & Poole, A. M. (2013). Evolutionary history of the TBP-domain superfamily. *Nucleic Acids Research, 41*(5), 2832–2845. https://doi.org/10.1093/nar/gkt045

Burroughs, A., & Aravind, L. (1981). The origin and evolution of release factors: Implications for translation termination, ribosome rescue, and quality control pathways. *International Journal of Molecular Science, 20*(8).

Burton, Z. F. (2014). The old and new testaments of gene regulation: Evolution of multi-subunit RNA polymerases and co-evolution of eukaryote complexity with the RNAP II CTD. *Transcription, 5.* https://doi.org/10.4161/trns.28674

Burton, Z. F. (2020). The 3-minihelix tRNA evolution theorem. *Journal of Molecular Evolution, 88*(3), 234–242. https://doi.org/10.1007/s00239-020-09928-2

Burton, & Burton, Z. F. (2014). The sigma enigma: Bacterial sigma factors, archaeal TFB and eukaryotic TFIIB are homologs. *Transcription, 5*(4).

Burton, Z. F., Opron, K., Wei, G., & Geiger, J. H. (2016). A model for genesis of transcription systems. *Transcription,* 7(1), 1–13. https://doi.org/10.1080/21541264.2015.1128518

Carter, C. W. Jr. (2014). Urzymology: Experimental access to a key transition in the appearance of enzymes. *Journal of Biological Chemistry, 289*(44), 30213–30220. https://doi.org/10.1074/jbc.R114.567495

Carter, C. W. Jr. (2017). Coding of class I and II aminoacyl-tRNA synthetases. In , *Vol 966. Advances in experimental medicine and biology* (pp. 103–148). Springer New York LLC. https://doi.org/10.1007/5584_2017_93

Carter, C. W. Jr., Li, L., Weinreb, V., Collier, M., Gonzalez-Rivera, K., Jimenez-Rodriguez, M., Erdogan, O., Kuhlman, B., Ambroggio, X., Williams, T., & Chandrasekharan, S. N. (2014). The Rodin-Ohno hypothesis that two enzyme superfamilies descended from one ancestral gene: An unlikely scenario for the origins of translation that will not be dismissed. *Biology Direct, 9*(1), 11. https://doi.org/10.1186/1745-6150-9-11

Carter, C. W. Jr., & Wills, P. R. (2019). Class I and II aminoacyl-tRNA synthetase tRNA groove discrimination created the first synthetase–tRNA cognate pairs and was therefore essential to the origin of genetic coding. *IUBMB Life, 71*(8), 1088–1098. https://doi.org/10.1002/iub.2094

Chan, P. P., & Lowe, T. M. (2009). GtRNAdb: A database of transfer RNA genes detected in genomic sequence. *Nucleic Acids Research, 37,* D93–D97. https://doi.org/10.1093/nar/gkn787. Database.

Chan, P. P., & Lowe, T. M. (2016). GtRNAdb 2.0: An expanded database of transfer RNA genes identified in complete and draft genomes. *Nucleic Acids Research, 44*(1), D184—D189. https://doi.org/10.1093/nar/gkv1309

Chatterjee, S., & Yadav, S. (2019). The origin of prebiotic information system in the peptide/RNA world: A simulation model of the evolution of translation and the genetic code. *Life, 9*(1). https://doi.org/10.3390/life9010025

Coles, M., Djuranovic, S., Söding, J., Frickey, T., Koretke, K., Truffault, V., Martin, J., & Lupas, A. N. (2005). AbrB-like transcription factors assume a swapped hairpin fold that is evolutionarily related to double-Psi β barrels. *Structure, 13*(6), 919—928. https://doi.org/10.1016/j.str.2005.03.017

Coles, M., Hulko, M., Djuranovic, S., Truffault, V., Koretke, K., Martin, J., & Lupas, A. N. (2006). Common evolutionary origin of swapped-hairpin and double-psi β barrels. *Structure, 14*(10), 1489—1498. https://doi.org/10.1016/j.str.2006.08.005

Demongeot, J., & Norris, V. (2019). Emergence of a "CYCLOSOME" in a primitive network capable of building "infinite" proteins. *Life, 9*(2), 51. https://doi.org/10.3390/life9020051

Demongeot, J., & Seligmann, H. (2019). The uroboros theory of life's origin: 22-nucleotide theoretical minimal RNA rings reflect evolution of genetic code and tRNA-rRNA translation machineries. *Acta Biotheoretica, 67*(4), 273—297. https://doi.org/10.1007/s10441-019-09356-w

Demongeot, J., & Seligmann, H. (2020a). RNA rings strengthen hairpin accretion hypotheses for tRNA evolution: A reply to Commentaries by Z.F. Burton and M. Di Giulio. *Journal of Molecular Evolution, 88*(3), 243—252. https://doi.org/10.1007/s00239-020-09929-1

Demongeot, J., & Seligmann, H. (2020b). The primordial tRNA acceptor stem code from theoretical minimal RNA ring clusters. *BMC Genetics, 21*(1). https://doi.org/10.1186/s12863-020-0812-2

Demongeot, J., & Seligmann, H. (2020c). Theoretical minimal RNA rings mimick molecular evolution before tRNA-mediated translation: Codon-amino acid affinities increase from early to late RNA rings. *Comptes Rendus—Biologies, 343*(1), 111—122. https://doi.org/10.5802/crbiol.1

Demongeot, J., & Seligmann, H. (2021). Codon assignment evolvability in theoretical minimal RNA rings. *Gene, 769*, 145208. https://doi.org/10.1016/j.gene.2020.145208

Di Giulio, M. (2009). A comparison among the models proposed to explain the origin of the tRNA molecule: A synthesis. *Journal of Molecular Evolution, 69*(1), 1—9. https://doi.org/10.1007/s00239-009-9248-z

Di Giulio, M. (2019). A comparison between two models for understanding the origin of the tRNA molecule. *Journal of Theoretical Biology, 480*, 99—103. https://doi.org/10.1016/j.jtbi.2019.07.020

Di Giulio, M. (2020). An RNA ring was not the progenitor of the tRNA molecule. *Journal of Molecular Evolution, 88*(3), 228—233. https://doi.org/10.1007/s00239-020-09927-3

Elbaum-Garfinkle, S. (2019). Matter over mind: Liquid phase separation and neurodegeneration. *Journal of Biological Chemistry, 294*(18), 7160—7168. https://doi.org/10.1074/jbc.REV118.001188

Feng, L., Sheppard, K., Namgoong, S., Ambrogelly, A., Polycarpo, C., Randau, L., Tumbula-Hansen, D., & Söll, D. (2004). Aminoacyl-tRNA synthesis by pre-translational amino acid modification. *RNA Biology, 1*(1), 15—19. https://doi.org/10.4161/rna.1.1.953

Feng, L., Sheppard, K., Tumbula-Hansen, D., & Söll, D. (2005). Gln-tRNAGln formation from Glu-tRNAGln requires cooperation of an asparaginase and a Glu-tRNAGln kinase. *Journal of Biological Chemistry, 280*(9), 8150—8155. https://doi.org/10.1074/jbc.M411098200

Fournier, G. P., & Alm, E. J. (2015). Ancestral reconstruction of a pre-LUCA aminoacyl-tRNA synthetase ancestor supports the late addition of Trp to the genetic code. *Journal of Molecular Evolution, 80*(3–4), 171–185. https://doi.org/10.1007/s00239-015-9672-1

Goddard, T. D., Huang, C. C., Meng, E. C., Pettersen, E. F., Couch, G. S., Morris, J. H., & Ferrin, T. E. (2018). UCSF ChimeraX: Meeting modern challenges in visualization and analysis. *Protein Science, 27*(1), 14–25. https://doi.org/10.1002/pro.3235

Gospodinov, A., & Kunnev, D. (2020). Universal codons with enrichment from GC to AU nucleotide composition reveal a chronological assignment from early to late along with LUCA formation. *Life, 10*(6), 1–22. https://doi.org/10.3390/life10060081

Guo, C., Che, Z., Yue, J., et al. (2020). ENL initiates multivalent phase separation of the super elongation complex (SEC) in controlling rapid transcriptional activation. *Science Advances, 6*(14), eaay4858.

Hartman, H., & Smith, T. F. (2019). Origin of the genetic code is found at the transition between a thioester world of peptides and the phosphoester world of polynucleotides. *Life, 9*(3). https://doi.org/10.3390/life9030069

Hauenstein, S. I., & Perona, J. J. (2008). Redundant synthesis of cysteinyl-tRNACys in Methanosarcina mazei. *Journal of Biological Chemistry, 283*(32), 22007–22017. https://doi.org/10.1074/jbc.M801839200

Ikehara, K. (2005). Possible steps to the emergence of life: The [GADV]-protein world hypothesis. *Chemical Record, 5*(2), 107–118. https://doi.org/10.1002/tcr.20037

Ikehara, K. (2009). Pseudo-replication of [GADV]-proteins and origin of life. *International Journal of Molecular Sciences, 10*(4), 1525–1537. https://doi.org/10.3390/ijms10041525

Ikehara, K. (2014). [GADV]-Protein world hypothesis on the origin of life. *Origins of Life and Evolution of Biospheres, 44*(4), 299–302. https://doi.org/10.1007/s11084-014-9383-4

Ikehara, K. (2016). Evolutionary steps in the emergence of life deduced from the bottom-up approach and GADV hypothesis (top-down approach). *Life, 6*(1). https://doi.org/10.3390/life6010006

Ikehara, K. (2019). The origin of tRNA deduced from *Pseudomonas aeruginosa* 5′ anticodon-stem sequence: Anticodon-stem loop hypothesis. *Origins of Life and Evolution of Biospheres, 49*(1–2), 61–75. https://doi.org/10.1007/s11084-019-09573-w

Ikehara, K., & Niihara, Y. (2007). Origin and evolutionary process of the genetic code. *Current Medicinal Chemistry, 14*(30), 3221–3231. https://doi.org/10.2174/092986707782793853

Ikehara, K., Omori, Y., Arai, R., & Hirose, A. (2002). A novel theory on the origin of the genetic code: A GNC-SNS hypothesis. *Journal of Molecular Evolution, 54*(4), 530–538. https://doi.org/10.1007/s00239-001-0053-6

Inouye, M., Takino, R., Ishida, Y., & Inouye, K. (2020). Evolution of the genetic code; evidence from serine codon use disparity in *Escherichia coli*. *Proceedings of the National Academy of Sciences, 117*(46), 28572–28575. https://doi.org/10.1073/pnas.2014567117

Iyer, Lakshminarayan M., & Aravind, L. (2012). Insights from the architecture of the bacterial transcription apparatus. *Journal of Structural Biology, 179*(3), 299–319. https://doi.org/10.1016/j.jsb.2011.12.013

Iyer, L. M., Koonin, E. V., & Aravind, L. (2004). Evolution of bacterial RNA polymerase: Implications for large-scale bacterial phylogeny, domain accretion, and horizontal gene transfer. *Gene, 335*(1–2), 73–88. https://doi.org/10.1016/j.gene.2004.03.017

Juhling, M., & Hartmann, R. (2009). Compilation of tRNA sequences and tRNA genes. *Nucleic Acids Research, 37*, 159–162.

Kelley, L. A., Mezulis, S., Yates, C. M., Wass, M. N., & Sternberg, M. J. E. (2015). The Phyre2 web portal for protein modeling, prediction and analysis. *Nature Protocols, 10*(6), 845–858. https://doi.org/10.1038/nprot.2015.053

Kim, Y., Kowiatek, B., Opron, K., & Burton, Z. (2018). Type-II tRNAs and evolution of translation systems and the genetic code. *International Journal of Molecular Sciences, 19*(10), 3275. https://doi.org/10.3390/ijms19103275

Kim, Y., Opron, K., & Burton, Z. F. (2019). A trna- and anticodon-centric view of the evolution of aminoacyl-trna synthetases, trnaomes, and the genetic code. *Life, 9*(2). https://doi.org/10.3390/life9020037

Klump, H. H., Völker, J., & Breslauer, K. J. (2020). Energy mapping of the genetic code and genomic domains: Implications for code evolution and molecular Darwinism. *Quarterly Reviews of Biophysics.* https://doi.org/10.1017/S0033583520000098

Koonin, E. V., Krupovic, M., Ishino, S., & Ishino, Y. (2020). The replication machinery of LUCA: Common origin of DNA replication and transcription. *BMC Biology, 18*(1). https://doi.org/10.1186/s12915-020-00800-9

Koonin, E. V., & Novozhilov, A. S. (2017). Origin and evolution of the universal genetic code. *Annual Review of Genetics, 51,* 45—62. https://doi.org/10.1146/annurev-genet-120116-024713

Kunnev, D., & Gospodinov, A. (2018). Possible emergence of sequence specific RNA aminoacylation via peptide intermediary to initiate darwinian evolution and code through origin of life. *Life, 8*(4). https://doi.org/10.3390/life8040044

Lei, L., & Burton, Z. F. (2020). Evolution of life on earth: TRNA, aminoacyl-tRNA synthetases and the genetic code. *Life, 10*(3). https://doi.org/10.3390/life10030021

Lei, L., & Burton, Z. F. (2021). Early evolution of transcription systems and divergence of Archaea and Bacteria. *Frontiers in Molecular Biosciences, 8.* https://doi.org/10.3389/fmolb.2021.651134

Longo, L. M., Despotović, D., Weil-Ktorza, O., Walker, M. J., Jabłońska, J., Fridmann-Sirkis, Y., Varani, G., Metanis, N., & Tawfik, D. S. (2020). Primordial emergence of a nucleic acid-binding protein via phase separation and statistical ornithine-to-arginine conversion. *Proceedings of the National Academy of Sciences of the United States of America, 117*(27), 15731—15739. https://doi.org/10.1073/pnas.2001989117

Long, X., Xue, H., & Wong, J. T.-F. (2020). Descent of Bacteria and Eukarya from an archaeal root of life. *Evolutionary Bioinformatics, 16.* https://doi.org/10.1177/1176934320908267, 117693432090826.

Loveland, A. B., Demo, G., Grigorieff, N., & Korostelev, A. A. (2017). Ensemble cryo-EM elucidates the mechanism of translation fidelity. *Nature, 546*(7656), 113—117. https://doi.org/10.1038/nature22397

Loveland, A. B., Demo, G., & Korostelev, A. A. (2020). Cryo-EM of elongating ribosome with EF-Tu•GTP elucidates tRNA proofreading. *Nature, 584*(7822), 640—645. https://doi.org/10.1038/s41586-020-2447-x

Madru, C., Henneke, G., Raia, P., Hugonneau-Beaufet, I., Pehau-Arnaudet, G., England, P., Lindahl, E., Delarue, M., Carroni, M., & Sauguet, L. (2020). Structural basis for the increased processivity of D-family DNA polymerases in complex with PCNA. *Nature Communications, 11*(1). https://doi.org/10.1038/s41467-020-15392-9

Mandal, D., Köhrer, C., Su, D., Russell, S. P., Krivos, K., Castleberry, C. M., Blum, P., Limbach, P. A., Söll, D., & RajBhandary, U. L. (2010). Agmatidine, a modified cytidine in the anticodon of archaeal tRNA Ile , base pairs with adenosine but not with guanosine. *Proceedings of the National Academy of Sciences, 107*(7), 2872—2877. https://doi.org/10.1073/pnas.0914869107

Maracci, C., & Rodnina, M. V. (2016). Review: Translational GTPases. *Biopolymers, 105*(8), 463—475. https://doi.org/10.1002/bip.22832

Marin, J., Battistuzzi, F. U., Brown, A. C., & Hedges, S. B. (2017). The timetree of prokaryotes: New insights into their evolution and speciation. *Molecular Biology and Evolution, 34*(2), 437—446. https://doi.org/10.1093/molbev/msw245

McGeoch, M.W., Dikler, S., McGeoch, J.E.M. (2020). Hemolithin: a meteoritic protein containing iron and lithium. arXiv:2002.11688v1.

Min, B., Pelaschier, J. T., Graham, D. E., Tumbula-Hansen, D., & Söll, D. (2002). Transfer RNA-dependent amino acid biosynthesis: An essential route to asparagine formation.

Proceedings of the National Academy of Sciences of the United States of America, 99(5), 2678—2683. https://doi.org/10.1073/pnas.012027399

Mukai, T., Crnković, A., Umehara, T., Ivanova, N. N., Kyrpides, N. C., Söll, D., Harwood, C. S., Ibba, M., & Whitman, W. (2017). RNA-dependent cysteine biosynthesis in Bacteria and Archaea. *mBio, 8*(3). https://doi.org/10.1128/mbio.00561-17

Numata, T. (2015). Mechanisms of the tRNA wobble cytidine modification essential for AUA codon decoding in prokaryotes. *Bioscience, Biotechnology and Biochemistry, 79*(3), 347—353. https://doi.org/10.1080/09168451.2014.975185

Oba, T., Fukushima, J., Maruyama, M., Iwamoto, R., & Ikehara, K. (2005). Catalytic activities of [GADV]-peptides: Formation and establishment of [GADV]-protein world for the emergence of life. *Origins of Life and Evolution of the Biosphere, 35*(5), 447—460. https://doi.org/10.1007/s11084-005-3519-5

O'Donoghue, P., & Luthey-Schulten, Z. (2003). On the evolution of structure in aminoacyl-tRNA synthetases. *Microbiology and Molecular Biology Reviews, 67*(4), 550—573. https://doi.org/10.1128/mmbr.67.4.550-573.2003

Opron, K., & Burton, Z. F. (2019). Ribosome structure, function, and early evolution. *International Journal of Molecular Sciences, 20*(1). https://doi.org/10.3390/ijms20010040

Osawa, T., Kimura, S., Terasaka, N., Inanaga, H., Suzuki, T., & Numata, T. (2011). Structural basis of tRNA agmatinylation essential for AUA codon decoding. *Nature Structural and Molecular Biology, 18*(11), 1275—1280. https://doi.org/10.1038/nsmb.2144

Pak, D., Du, N., Kim, Y., Sun, Y., & Burton, Z. F. (2018). Rooted tRNAomes and evolution of the genetic code. *Transcription, 9*(3), 137—151. https://doi.org/10.1080/21541264.2018.1429837

Pak, D., Kim, Y., & Burton, Z. F. (2018). Aminoacyl-tRNA synthetase evolution and sectoring of the genetic code. *Transcription, 9*(4), 205—224. https://doi.org/10.1080/21541264.2018.1467718

Pak, D., Root-Bernstein, R., & Burton, Z. F. (2017). tRNA structure and evolution and standardization to the three nucleotide genetic code. *Transcription, 8*(4), 205—219. https://doi.org/10.1080/21541264.2017.1318811

Perona, J. J., & Gruic-Sovulj, I. (2014). Synthetic and editing mechanisms of aminoacyl-tRNA synthetases. *Topics in Current Chemistry, 344*, 1—41. https://doi.org/10.1007/128_2013_456

Perona, J. J., & Hadd, A. (2012). Structural diversity and protein engineering of the aminoacyl-tRNA synthetases. *Biochemistry, 51*(44), 8705—8729. https://doi.org/10.1021/bi301180x

Pettersen, E. F., Goddard, T. D., Huang, C. C., Meng, E. C., Couch, G. S., Croll, T. I., Morris, J. H., & Ferrin, T. E. (2021). UCSF ChimeraX: structure visualization for researchers, educators, and developers. *Protein Science, 30*(1), 70—82. https://doi.org/10.1002/pro.3943

Phillips, G., & de Crécy-Lagard, V. (2011). Biosynthesis and function of tRNA modifications in Archaea. *Current Opinion in Microbiology, 14*(3), 335—341. https://doi.org/10.1016/j.mib.2011.03.001

Portz, B., & Shorter, J. (2020). Switching condensates: The CTD code goes liquid. *Trends in Biochemical Sciences, 45*(1), 1—3. https://doi.org/10.1016/j.tibs.2019.10.009

Puzzo, D., Argyrousi, E. K., Staniszewski, A., Zhang, H., Calcagno, E., Zuccarello, E., Acquarone, E., Fa, M., Puma, D. D. L., Grassi, C., D'Adamio, L., Kanaan, N. M., Fraser, P. E., & Arancio, O. (2020). Tau is not necessary for amyloid-ß-induced synaptic and memory impairments. *Journal of Clinical Investigation, 130*(9), 4831—4844. https://doi.org/10.1172/JCI137040

Raczniak, G., Becker, H. D., Min, B., & Söll, D. (2001). A single amidotransferase forms asparaginyl-tRNA and glutaminyl-tRNA in *Chlamydia trachomatis*. *Journal of Biological Chemistry, 276*(49), 45862–45867. https://doi.org/10.1074/jbc.M109494200

Rafels-Ybern, À., Torres, A. G., Camacho, N., Herencia-Ropero, A., Frigolé, H. R., Wulff, T. F., Raboteg, M., Bordons, A., Grau-Bove, X., Ruiz-Trillo, I., & De Pouplana, L. R. (2019). The expansion of inosine at the wobble position of tRNAs, and its role in the evolution of proteomes. *Molecular Biology and Evolution, 36*(4), 650–662. https://doi.org/10.1093/molbev/msy245

Rafels-Ybern, À., Torres, A. G., Grau-Bove, X., Ruiz-Trillo, I., & Ribas de Pouplana, L. (2018). Codon adaptation to tRNAs with Inosine modification at position 34 is widespread among Eukaryotes and present in two Bacterial phyla. *RNA Biology, 15*(4–5), 500–507. https://doi.org/10.1080/15476286.2017.1358348

Raia, P., Carroni, M., Henry, E., Pehau-Arnaudet, G., Brûlé, S., Béguin, P., Henneke, G., Lindahl, E., Delarue, M., & Sauguet, L. (2019). Structure of the DP1-DP2 PolD complex bound with DNA and its implications for the evolutionary history of DNA and RNA polymerases. *PLoS Biology, 17*(1). https://doi.org/10.1371/journal.pbio.3000122

Rampias, T., Sheppard, K., & Söll, D. (2010). The archaeal transamidosome for RNA-dependent glutamine biosynthesis. *Nucleic Acids Research, 38*(17), 5774–5783. https://doi.org/10.1093/nar/gkq336

Rodin, A. S., Rodin, S. N., & Carter, C. W., Jr. (2009). On primordial sense-antisense coding. *Journal of Molecular Evolution, 69*(5), 555–567.

Rogers, S. O. (2019). Evolution of the genetic code based on conservative changes of codons, amino acids, and aminoacyl tRNA synthetases. *Journal of Theoretical Biology, 466*, 1–10. https://doi.org/10.1016/j.jtbi.2019.01.022

Root-Bernstein, R., Kim, Y., Sanjay, A., et al. (2016). tRNA evolution from the proto-tRNA minihelix world. *Transcription, 7*(5), 153–163.

Rozov, A., Demeshkina, N., Westhof, E., Yusupov, M., & Yusupova, G. (2015). Structural insights into the translational infidelity mechanism. *Nature Communications, 6*(1). https://doi.org/10.1038/ncomms8251

Rozov, A., Demeshkina, N., Westhof, E., Yusupov, M., & Yusupova, G. (2016). New structural insights into translational miscoding. *Trends in Biochemical Sciences, 41*(9), 798–814. https://doi.org/10.1016/j.tibs.2016.06.001

Rozov, A., Westhof, E., Yusupov, M., & Yusupova, G. (2016). The ribosome prohibits the G•U wobble geometry at the first position of the codon-anticodon helix. *Nucleic Acids Research, 44*(13), 6434–6441. https://doi.org/10.1093/nar/gkw431

Saint-Léger, A., Bello, C., Dans, P. D., Torres, A. G., Novoa, E. M., Camacho, N., Orozco, M., Kondrashov, F. A., & De Pouplana, L. R. (2016). Saturation of recognition elements blocks evolution of new tRNA identities. *Science Advances, 2*(4). https://doi.org/10.1126/sciadv.1501860

Salazar, J. C., Zúiga, R., Raczniak, G., Becker, H., Söll, D., & Orellana, O. (2001). A dual-specific Glu-tRNAGln and Asp-tRNAAsn amidotransferase is involved in decoding glutamine and asparagine codons in *Acidithiobacillus ferrooxidans*. *FEBS Letters, 500*(3), 129–131. https://doi.org/10.1016/S0014-5793(01)02600-X

Salgado, P. S., Koivunen, M. R. L., Makeyev, E. V., Bamford, D. H., Stuart, D. I., & Grimes, J. M. (2006). The structure of an RNAi polymerase links RNA silencing and transcription. *PLoS Biology, 4*(12), 2274–2281. https://doi.org/10.1371/journal.pbio.0040434

Satpati, P., Bauer, P., & Åqvist, J. (2014). Energetic tuning by tRNA modifications ensures correct decoding of isoleucine and methionine on the ribosome. *Chemistry—A European Journal, 20*(33), 10271–10275. https://doi.org/10.1002/chem.201404016

Sauguet, L. (2019). The extended "Two-Barrel" polymerases superfamily: Structure, function and evolution. *Journal of Molecular Biology, 431*(20), 4167–4183.

Schmitt, E., Coureux, P. D., Kazan, R., Bourgeois, G., Lazennec-Schurdevin, C., & Mechulam, Y. (2020). Recent advances in archaeal translation initiation. *Frontiers in Microbiology, 11.* https://doi.org/10.3389/fmicb.2020.584152

Schneider, T. D., & Stephens, R. M. (1990). Sequence logos: A new way to display consensus sequences. *Nucleic Acids Research, 18*(20), 6097–6100. https://doi.org/10.1093/nar/18.20.6097

Singh, V., Xu, L., Boyko, S., Surewicz, K., & Surewicz, W. K. (2020). Zinc promotes liquid–liquid phase separation of tau protein. *Journal of Biological Chemistry, 295*(18), 5850–5856. https://doi.org/10.1074/jbc.AC120.013166

Smith, T. F., & Hartman, H. (2015). The evolution of Class II Aminoacyl-tRNA synthetases and the first code. *FEBS Letters, 589*(23), 3499–3507. https://doi.org/10.1016/j.febslet.2015.10.006

Tumbula-Hansen, D., Feng, L., Toogood, H., Stetter, K. O., & Söll, D. (2002). Evolutionary divergence of the archaeal aspartyl-tRNA synthetases into discriminating and nondiscriminating forms. *Journal of Biological Chemistry, 277*(40), 37184–37190. https://doi.org/10.1074/jbc.M204767200

Turanov, A. A., Xu, X. M., Carlson, B. A., Yoo, M. H., Gladyshev, V. N., & Hatfield, D. L. (2011). Biosynthesis of selenocysteine, the 21st amino acid in the genetic code, and a novel pathway for cysteine biosynthesis. *Advances in Nutrition, 2*(2), 122–128. https://doi.org/10.3945/an.110.000265

Vollmer, W. (2008). Structural variation in the glycan strands of bacterial peptidoglycan. *FEMS Microbiology Reviews, 32*(2), 287–306. https://doi.org/10.1111/j.1574-6976.2007.00088.x

Vollmer, W., Blanot, D., & De Pedro, M. A. (2008). Peptidoglycan structure and architecture. *FEMS Microbiology Reviews, 32*(2), 149–167. https://doi.org/10.1111/j.1574-6976.2007.00094.x

Voorhees, R. M., Mandal, D., Neubauer, C., Köhrer, C., Rajbhandary, U. L., & Ramakrishnan, V. (2013). The structural basis for specific decoding of AUA by isoleucine tRNA on the ribosome. *Nature Structural and Molecular Biology, 20*(5), 641–643. https://doi.org/10.1038/nsmb.2545

Wetzel, R. (1995). Evolution of the aminoacyl-tRNA synthetases and the origin of the genetic code. *Journal of Molecular Evolution, 40*(5), 545–550. https://doi.org/10.1007/BF00166624

Wolf, Y. I., Aravind, L., Grishin, N. V., & Koonin, E. V. (1999). Evolution of Aminoacyl-tRNA synthetases-analysis of unique domain architectures and phylogenetic trees reveals a complex history of horizontal gene transfer events. *Genome Research, 9*(8), 689–710.

Wu, J., Bu, W., Sheppard, K., Kitabatake, M., Kwon, S. T., Söll, D., & Smith, J. L. (2009). Insights into tRNA-dependent amidotransferase evolution and catalysis from the structure of the Aquifex aeolicus enzyme. *Journal of Molecular Biology, 391*(4), 703–716. https://doi.org/10.1016/j.jmb.2009.06.014

Yarus, M. (2021a). Crick wobble and superwobble in standard genetic code evolution. *Journal of Molecular Evolution, 89*(1–2), 50–61. https://doi.org/10.1007/s00239-020-09985-7

Yarus, M. (2021b). Evolution of the standard genetic code. *Journal of Molecular Evolution, 89*(1–2), 19–44. https://doi.org/10.1007/s00239-020-09983-9

Yarus, M. (2021c). Optimal evolution of the standard genetic code. *Journal of Molecular Evolution, 89*(1–2), 45–49. https://doi.org/10.1007/s00239-020-09984-8

Index

'*Note:* Page numbers followed by "f" indicate figures and "t" indicate tables.'

A

8-aa bottleneck, 269—273, 269f—272f
Acceptor stems, 95—96
Accretion models, 172—173
Alternate class I and class II folding,
    45—46
Amino acid, 247
  guidelines for placements of, 63
  placements, 213—214, 213f
Aminoacyl-tRNA synthetase (aaRS), 69,
    82, 210—218, 247
  accuracy, 216—218
  alternate class I and class II folding,
    45—46
  anticodon-codon interaction, ribosome
    proofreading of, 51
  anticodon wobble preference, 53
  Arg coding, 40
  class I and class II, 259—260
    evolution, 30—33, 31f
    homology, 30—33, 31f
    incompatibility, 33—35, 35f
  1-codon sectors, resistance to,
    48—49
  coevolution of, 42—48
  concepts, 259
  correlation of, 51—52
  editing hypothesis, 40—45, 42f
  evolution, 30—33, 277
  genetic code
    maximal size of, 46
    universality, 46—48
  homology, 30—33
    modeling, 53—54
  Ile-Met sector, 49
  lineages of, 261—262
  mitochondria, synonymous anticodon
    wobble preference in, 39
  NCBI blast, 53
  proofread, 36, 37f
  ribozyme-catalyzed tRNA
    aminoacylation, 30
  standard genetic code, 35—36, 49—51
  Staphylothermus marinus, 32—33
  statistical methods, 54
  synonymous anticodon preferences,
    36—39
    I>>G>>A, 36—38
    U>C, 38—39
  Thermus thermophilus, 33
  tRNA, 42—45, 53
    anticodon wobble ambiguity, positive
    selection of, 48
Animate transition, inanimate to,
    110—112, 110f
Anticodon-codon interaction, ribosome
    proofreading of, 51
Anticodon loop, 2
Anticodon position, 271—273
Anticodon preference rules, 62
Anticodon wobble position, adenine in,
    10—11, 11f
Anticodon wobble preference, 53
Archaea and bacteria, 150—158,
    150f—151f
  divergence of, 158
  DPBB loops, 155—158, 155f
  evolution of, 153—155
  promoter-specific regulatory HTH
    factors, 152—153
Archaeal species, 4
Archaeal tRNAs, 99—100, 99f
  radiated from tRNAGly, 197—198,
    199f
Arg coding, 40
Artificial intelligence, 194—195

B

Bacterial tRNA, 100—103, 102f
  anticodon modification, 14—15

Breast cancer stage 1, Nationwide implementation
U.S. FDA, protocol first-line as per, 186

C
Carter-Rodin-Ohno hypothesis, falsification of, 260—261, 260f
Catalytic subunits, 144—148, 144f, 146f—148f
Chronic lymphocytic leukemia (CLL), 179—181
Class I and class II
evolution, 30—33, 31f
homology, 30—33, 31f
incompatibility, 33—35, 35f
Cloverleaf tRNA, 19—21, 95—96
Code
freezing, 277—278
punctuation, 275
Coding degeneracy, 262—263
6-codon sectors, 227—228
1-codon sectors, resistance to, 48—49
Column 2 to column 4, serine jumping from, 276
Cradle-loop barrels, 140
Cyclin-like repeats (CLR), 150—151

D
Deoxyribonucleic acid (DNA), 117—118
Dihydrofolate reductase (DHFR), 117—118
Divergence, 233
D-loop microhelix, 96
DNA polymerases (DNAPs)
catalytic subunits, 144—148, 144f, 146f—148f
2-double-Ψ-β-barrel type, 138—143
Saccharolobus shibatae, 138—140
transcription by multisubunit, 2-Mg mechanism of, 148—149, 149f
2-double-Ψ-β-barrel type, 138—143

E
Editing hypothesis, 40—45, 42f
EF-Tu, 228, 262—263
Escherichia coli, 70—71

Eukaryota, 193—194
tRNA anticodon modification, 14—15
Expanded V loops
archaeal tRNAs, 99—100, 99f
bacterial tRNAs, 100—103, 102f
kinship of, 105, 106f
last universal common ancestor (LUCA), 97—98
3-minihelix model, 97—98
statistical analyses, 104—105, 104t
type II tRNAs with, 198—200, 199f
Explosive evolution, 20

F
Filling disfavored row 1, 273—275
Folic acid, 117f, 118—119
Frozen accident, 69—71, 88

G
GADV world, 268—269, 269f
General transcription factors (GTFs)
archaeal and bacterial, 150—158, 150f—151f
divergence of, 158
DPBB loops, 155—158, 155f
evolution of, 153—155
promoter-specific regulatory HTH factors, 152—153
Genetic code
8-aa bottleneck, 269—273, 269f—272f
alternate genetic code models, 83—85
alternate representations of, 85
amino acids, 63, 247
aminoacyl-tRNA synthetase (aaRS), 210—218, 247
accuracy, 216—218
amino acid placements, 213—214, 213f
class I, 259—260
class II, 259—260
code sectoring based on, 214—216
concepts, 259
evolution, 277
lineages of, 261—262
structural classes, 211—212, 212f
aminoacyl-tRNA synthetases (aaRS), 69, 82

another planet, life on, 85—86
anticodon
    position, 271—273
    preference rules, 62
archaeal domain, 64
archaeal tRNAs radiated from
        tRNAGly, 197—198, 199f
artificial intelligence, 194—195
assumptions, 224—225
Carter-Rodin-Ohno hypothesis,
        falsification of, 260—261, 260f
code
    freezing, 277—278
    punctuation, 275
coding degeneracy, 262—263
6-codon sectors, 227—228
column, 226—227
column 1, 77—79
column 2, 76—77
    to column 4, serine jumping from,
        276
    sectoring, 81
column 3, 79—80, 229
column 4, 77—79, 230
within columns, 74—76, 75f
completion
    column 1, 228
    column 2, 228—229
EF-Tu, 228, 262—263
Eukaryota, 193—194
evolution, 277
evolutionary bottleneck, 227
evolution of, 71—74, 74f, 202—204,
        222—233, 222f
evolution within, 80—81
expanded V loop, type II tRNAs with,
        198—200, 199f
filling disfavored row 1, 273—275
frozen accident, 69—71, 88
GADV world, 268—269, 269f
great divergence, 233
helix-turn-helix (HTH) factors, 193
Ile sector, Met invasion of, 273—275
last universal common (cellular) ancestor
        (LUCA), 218—220, 247—248
late additions to, 82—83
life transition, prelife to, 195—201

tRNA, evolution of, 196
    type I tRNA, evolution of, 196—197,
        197f—198f
maximal size of, 46
metabolism, 249—251
methods, 248—249
minihelix and polymer worlds, 65—67
models
    demonstration of, 200—201
    to describe, 233—235
    perspectives, 231—233
order, 67—69, 68f
other models, 279—280
overview, 264—267, 264f—266f
polyglycine world, 71, 220—222,
        235—239, 238f, 267—268
predictions, 86—88
protocells, 278—279, 279f
ribosomes, 204—210
    amalgamated tRNAs, 205—207,
        206f
    prokaryotic ribosome, 207—210
    rRNA, 205—207, 206f
row 1, late evolution of, 230—231
stop codons, 82, 223—224,
        273—275
transcription, 249—251
translation, 249—251
tRNA evolution, 64—65, 251—258,
        276—277
    concepts, 251
    3—31-nt minihelix model, 251—254,
        252f
    TRNAGly, 255—256, 255f
    type II tRNAs, 256—258,
        257f—258f
    type I tRNA, 254—255, 254f
universality, 46—48
working model, 224—233
Glycine, 71

H
Helix-turn-helix (HTH) factors, 193
Homology, 30—33
    modeling, 53—54
Hydrochloric acid (HCl), 120
Hydrophobic amino acids, 78

I

Ile–Met sector, 49
Ile sector, Met invasion of, 273–275
Initial protocol prospectus, 183

K

Kowiatek protocol, 181
  case studies, 181
    CBC highlights immediately
      post-chemo, 181–182

L

Last universal common ancestor
    (LUCA), 2, 97–98, 218–220,
    247–248
Life transition, prelife to, 195–201
  tRNA, evolution of, 196
  type I tRNA, evolution of, 196–197,
    197f–198f

M

Megaloblastic anemia
  deoxyribonucleic acid (DNA),
    117–118
  dihydrofolate reductase (DHFR),
    117–118
  experimental, 119–120
  folic acid, 117f, 118–119
  hydrochloric acid (HCl), 120
  nonenzymatic mechanism, 123
  phosphatidylcholine (PC), 117–118,
    118f
  phosphatidylethanolamine (PE),
    118–119, 119f
  precipitation, 122
  S-adenosylmethionine (SAM-e),
    118–119
  sodium hydroxide (NaOH), 120
Methylating agents, 180f, 181
  CBC highlights
    3/14/2022, 183
    4/13/2021, 183
    8/23/2021, 183
    COVID-19 restrictions, 182
    discussion of results, 182
    post-chemo, 181–182
    statistical analysis of, 182

change in, 182
cobalamin, 180f
folic acid, 180f
on the horizon, 188
Micelle self-division, 188, 189f
research, 187
Minihelix and polymer worlds, 65–67
2-minihelix model, 172–175
3-minihelix model, 97–98
3 minihelix tRNA evolution theorem
    models, 163–164
  numbering of, 164–175, 164f,
    166f–168f
  predictions of, 173–174
Mitochondria, synonymous anticodon
    wobble preference in, 39
Mutagenesis, 9–10

N

NCBI blast, 53
  searches, 22
Nonenzymatic mechanism, 123
Nonenzymatic methylation, 125–127
  cytosine
    experimental, 129, 129f–130f
    nonenzymatic methylation, 125–127
    prospectus, 134
    results, 131–134, 131f–133f
    ribonucleic acid (RNA), 125–127
    S-adenosylmethionine (SAM), 126f
    translation initiation, 127–128

P

Peptidyl transferase center (PTC),
    65–66
Phosphatidylcholine (PC), 117–118,
    118f
Phosphatidylethanolamine (PE),
    118–119, 119f
Polyglycine world, 65, 71, 220–222,
    235–239, 238f, 267–268
  code sectoring based on, 214–216
  hypothesis, 7–8, 7f
  structural classes, 211–212, 212f
Precipitation, 122
Promoter-Proximal Element (PPE),
    154–155

Prospectus, 134
Protocells, 278–279, 279f
P-value statistical analysis, 171
Pyrobaculum aerophilum, 165
Pyrococcus, 100
Pyrococcus furiosis, 70–71, 165–167
Pyrococcus species, 6–7

R
Radar graphs, 8–9, 8f
Ribonucleic acid (RNA),
        125–127
    scaffolds, 66
Ribonucleic acid polymerases (RNAPs)
    catalytic subunits, 144–148, 144f,
        146f–148f
    2-double-Ψ-β-barrel type,
        138–143
    Saccharolobus shibatae, 138–140
    transcription by multisubunit, 2-Mg
        mechanism of, 148–149, 149f
Ribosomes, 204–210
    amalgamated tRNAs, 205–207,
        206f
    prokaryotic ribosome, 207–210
    rRNA, 205–207, 206f
Ribozyme-catalyzed tRNA
        aminoacylation, 30
Row 1, late evolution of, 230–231
Rugged evolution, 16

S
Saccharolobus shibatae, 138–140, 144
S-adenosylmethionine (SAM), 118–119,
        126f, 179–181, 179f
Sandwich-barrel hybrid motif (SBHM),
        142
Sectoring, 17
Sequence logos, 22
Sodium hydroxide (NaOH), 120
Standard genetic code, 35–36, 49–51
Staphylothermus marinus, 32–33,
        165–167
Statistical methods, 54
Stop codons, 223–224, 273–275
Synonymous anticodon preferences,
        36–39

T
Thermus thermophilus, 33
Transcription, 249–251
    elongation, 157–158
    multisubunit, 2-Mg mechanism of,
        148–149, 149f
Translation, 249–251
    initiation, 127–128
tRNA, 42–45, 53
    anticodon wobble ambiguity, positive
        selection of, 48
    evolution, 251–258, 276–277
        concepts, 251
        3–31-nt minihelix model, 251–254,
            252f
        TRNAGly, 255–256, 255f
        type II tRNAs, 256–258, 257f–258f
        type I tRNA, 254–255, 254f
    evolution of, 64–65
TRNAGly, 255–256, 255f
tRNAomes rooting
    anticodon loop, 2
    anticodon wobble position, adenine in,
        10–11, 11f
    archaeal species, 4
    bacterial tRNA anticodon modification,
        14–15
    Cloverleaf tRNA, 19–21
    computational approaches, 20–21
    degeneracy, 17
    divergence, 9–10
    eukaryotic tRNA anticodon modifica-
        tion, 14–15
    evolutionary archetype, 19–20
    evolutionary trees, 16–17, 22–23
    explosive evolution, 20
    lineages in, 4–7, 5f
    LUCA, 2
    methods, 21–23
    model for, 15
    mutagenesis, 9–10
    NCBI blast searches, 22
    polyglycine hypothesis, 7–8, 7f
    Pyrococcus species, 6–7
    radar graphs, 8–9, 8f
    rugged evolution, 16
    sectoring, 17

tRNAomes rooting (*Continued*)
  sequence logos, 22
  two minihelix tRNA evolution models,
    15
  wobble pairing, 17–18
Two minihelix tRNA evolution models,
  15
Type II tRNAs, 256–258, 257f–258f
  acceptor stems, 95–96
  animate transition, inanimate to,
    110–112, 110f
  Cloverleaf tRNA, 95–96
  D-loop microhelix, 96
  expanded V loops
    archaeal tRNAs, 99–100, 99f
    bacterial tRNAs, 100–103, 102f
    kinship of, 105, 106f
    last universal common ancestor
      (LUCA), 97–98
    3-minihelix model, 97–98
    statistical analyses, 104–105, 104t
  genetic code, 108–109
  last universal common ancestor
    (LUCA), 97–98

  materials, 112
  methods, 112
  3-minihelix model, 97–98
  model for, 98
  tRNA
    evolution models, 107–108
    sequence, 109
Type I tRNA, 254–255, 254f

U
U.S. nationwide implementation
  U.S. FDA, protocol first-line as per,
    183, 186f
  U.S. FDA, protocol second-line as per,
    184, 186f
  U.S. FDA, protocol third-line as per,
    184–186, 186f

V
V loop, 169–170

W
Wobble pairing, 17–18

Printed in the United States
by Baker & Taylor Publisher Services